The Origin of Plants

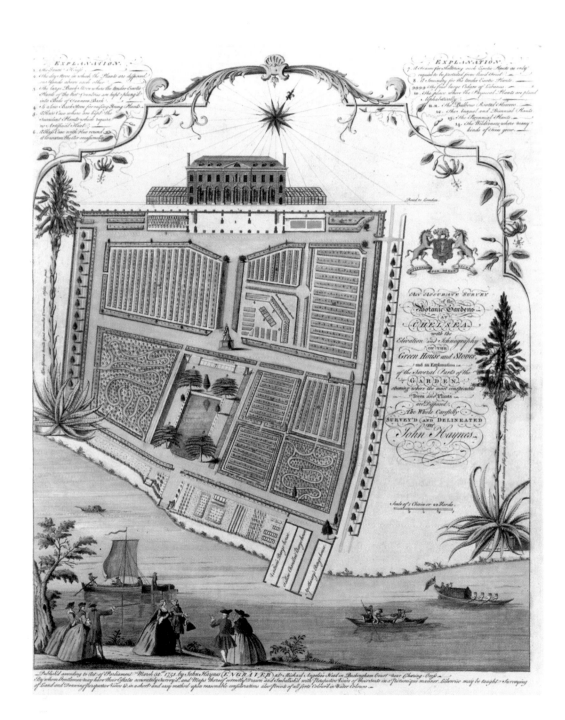

The Victorian idea of creating a floral clock from
living plants is given a new twist at Knoll Gardens in
Dorset, where the owners make a daily calendar bed
to welcome their visitors.

The Origin of Plants

*The People and Plants that
have shaped Britain's Garden History
since the Year 1000*

Maggie Campbell-Culver

HEADLINE

First published in 2001
by HEADLINE BOOK PUBLISHING

10 9 8 7 6 5 4 3 2 1

British Library Cataloguing in Publication Data
Campbell-Culver, Maggie
Origin of plants: the people and the plants that shaped Britain
1. Botany – Great Britain – History 2. Botanists – Great 3. Plants – Origin
1. Title
581.9'41
ISBN 0 7472 7214 X

Typeset by Letterpart Limited, Reigate, Surrey
Designed by Anthony Cohen
Picture research by Melanie Watson
Colour reproduction by Spectrum Colour
Printed and bound by Rotolitho Lombarda, Italy

HEADLINE BOOK PUBLISHING
A division of Hodder Headline
338 Euston Road
London NW1 3BH

www.headline.co.uk
www.hodderheadline.com

Contents

Picture Credits

Acknowledgements

If you have two pennies, spend one on a loaf and one on a flower.
The bread will give you life, the flower a reason for living.

Chinese proverb

THE IDEA OF LOOKING at the history of plant introductions into this country began with my involvement with the Plant Heritage Display for the National Council for the Conservation and Plants and Gardens at the Hampton Court International Flower Show where, for about ten years, a group of us would meet each year to help set up and run the marquee for the week-long show. Knowing relatively little about plants and their history (when one thinks how much there is to know), I quickly came to realise the truth of John Ruskin's aphorism that the best way to learn about a subject is to write a book about it. This is how the book was born and I would like especially to thank those who were there at its conception: Jane Green, Neil Lucas, Graham Pattison, Ann Rawlings, and Edna and Mike Squires.

I wish also to acknowledge particularly the support of Charmian and John Hussey and the kind help that I have received from Mavis Batey, past President of the Garden History Society, and my colleagues in the Society, as well as all my gardening friends up and down the country including Dr Julian Ball, Dominic Cole, Merrial Connell, Sylvia Fitzgerald (the former Archivist and Chief Librarian at the Royal Botanic Gardens, Kew), Hazel Fryer, the late Daphne Lawry (Secretary of the Cornwall Gardens Trust), Denise Manning, Nigel Mathews, Simon Milne of Hillier's Arboretum, Douglas and Mary Pett, Paddy Powell, Tim Smit of The Eden Project, Rosemary and Thomas Stokoe, Philip White of Hestercombe Gardens, and all the members of the Cornwall Garden Society and the Cornwall Gardens Trust.

I am most grateful for the help I have received from Rodger Baines, the Executive Director of NCCPG; Professor Grenville Lucas and Rosie Peddle, the Vice-Chairmen; also Richard Nicolle. Birmingham Botanic Garden, The Duchy of Cornwall Nursery, The Linnaean Society and the RHS Lindley Library have all most kindly responded to the many queries I have asked of them.

A novice author could not have been shown greater encouragement and support than that which I have received from Christopher Sinclair-Stevenson, my agent, and from Tim Heald and Penny Byrne. My writing hand has been steadfastly guided by the publishing and editorial team at Headline Books – Heather Holden-Brown and Celia Kent in particular – and my freelance editors Jenny Dereham and Jane Selley – thank you very much. Mel Watson researched the illustrations, finding great treasures. I thank them all for helping me through the process of publishing and after all their efforts any errors which remain lurking in the text are entirely mine.

To ease my way through the actual chore of living while all this has been going on, I have been exceptionally fortunate in the unselfish and loving support of my family, particularly my husband Michael who, being a much better gardener than me, has subsumed his first year of retirement from the Law into drowning in a sea of Latin names and plant dates. Without him, this book would not have been written. I thank him.

Maggie Campbell-Culver
Fowey, Cornwall and St Thois, Brittany, 2001

Notes for the Reader

T HE SUBJECT OF THIS BOOK, looking at some of the hundreds of plants introduced into the United Kingdom over the past one thousand years, is so huge that it is inevitable that more has had to be left out than that which is included. To bring some structure into the helter-skelter onrush of horticultural arrivals I have made each century a separate chapter with, at its beginning, a list of events – who was on the throne at the time, national and international events which had a cataclysmic effect on the incoming flow of plants (the Black Death or the First World War, for instance) or the all-important explorations of the world. Originally, when writing the book, I used these lists as *aides-mémoire* to see what was happening in the big wide world around the relatively minuscule worlds of the plant-hunters themselves. Having explored thousands of plants with their dates of introduction, I now realise the crucial importance of this historical link: the date of entry of a particular plant can reflect very quickly and sometimes dramatically the consequences of trade, war, exploration, technical advances and indeed the whole gamut of human endeavour.

Choosing plants to write about for each century opened up a cornucopia of thrilling stories; the choice was so great that there are inevitable omissions as well. The nearer to our own time we came, the more difficult was the choice. The name of each plant introduction written about in the text is highlighted in **bold**, the same name appearing in **bold** in the List of Plant Introductions at the end of each chapter. However, where occasionally a plant is discussed out of its century, it will appear in the List of Plant Introductions for its own century, with a cross-reference back to the page of description.

To assist the reader to buy plants he or she has read about, I have tried to use the most up-to-date plant name, with earlier names being given as synonyms in brackets. Inevitably, the earliest plant introductions have no specific date within the Plant Lists; it is enough to know they arrived within a given hundred years. They are, therefore, listed alphabetically. It is from the 16th century onwards that the Plant Lists become totally chronological. Within the Lists, I have also included plants which, because of space, have not been written about in the text and therefore are not highlighted. To a number of these I have added a few extra facts. Native plants written about in the text are not highlighted in bold, nor have they been placed in the Plant Lists.

Inevitably with the millions of horticultural arrivals which have occurred during the millennium, the book could have become one (very) long list of plant names and dates. However, besides plant omissions, finding a plant's date of entry is rarely an exact science and I must ask the readers' indulgence if a discovery is made of an earlier or later arrival date of a favourite plant.

I now hold a long and growing list setting out the chronological arrival of plants from abroad and would welcome correspondence on the matter either via Headline or email to maggie.campbell-culver@virgin.net. For readers who would like to learn more about the plants they buy and grow, I recommend they contact the National Council for the Conservation of Plants and Gardens, Stable Courtyard, RHS Wisley, Woking, Surrey, GU 23 6QB; www.nccpg.org.uk

Foreword

In Arcadia

After the retreat of the last Ice Age Britain had the distinction of having the smallest natural flora of any country in the world, yet by the start of the twenty-first century it contained the widest range of any nation on earth.

This came about through circumstances which, on their own, might not have been significant but, in combination, were to create a very British renaissance, a literal flowering. It was inspired by, on the one hand, the competitive aspirations of the aristocratic families who vied with each other to create ever more impressive gardens, pleasure grounds and parks and, on the other, the imperial drive to turn the map of the world red. This created the conditions for the development of the Great British Botanic Garden Tradition. Although cleverly disguised as a scientific institution, it was actually the nerve centre for a quite breathtaking programme of economic development, as crops were transported from one country to another to find the ideal conditions for exploitation. The stories of the mutiny on the *Bounty* and the stealing of rubber from Brazil have passed into folklore, but this was a serious business.

From the eighteenth century onwards the plant hunters were not only searching for plants in their native surroundings but were also seeking plants that could be introduced into domestic cultivation in Britain; hence, many of the expeditions were at least part financed by the great nurseries of the time, for whom a successful introduction meant big money on the back of its novelty value. Maggie Campbell-Culver's beautiful and important book describes the main actors on this stage and the roles they played, but focuses rightly on the plants themselves. These plants have shaped our culture and, to a certain degree, our destiny.

This work of scholarship comes at a time when we are waking up to the true significance of Britain's horticultural tradition. Bodies such as the National Trust and the Royal Horticultural Society have long catered for the seemingly limitless interest of the British public in gardens and gardening. The millions who attend the world-famous Chelsea and Hampton Court Flower Shows are testament to the depth of interest shown in plants.

Increasingly, the scientific community is coming to reassess the importance of plants in medicine and well being. In engineering laboratories across the world, scientists are looking with fresh eyes at plants as perhaps holding the secrets to the harnessing of solar power and new composite materials, investigations which may change the face of the world within a generation. Whatever our political or ethical beliefs, the advances in DNA research and the consequent interest in genetic modification and the potential for customising nature to man's needs is with us. Not for two hundred years have plants been so central to what is happening today.

The great private and botanic gardens of this country are suddenly finding that the collections that have been brought together here from around the globe have an importance far surpassing their ornamental or economic value. In a scenario that would have been unbelievable as little as twenty years ago, scientists are realising that the genetic material of some of the world's most important plants is under extreme threat in their native territories. For example, there is now a greater range of species rhododendrons in the UK than in Nepal, Bhutan and Sikkim, their countries of origin. Similar pressures exist in South America where, the Edinburgh Royal Botanic Gardens claim, some of the genetic material from the temperate rainforests is under massive threat from paper companies of the Far East which operate a rapacious deforestation regime in Chile. Organisations like the National Trust and the RHS are suddenly finding themselves with a role they never imagined having to play. This is not about education; this is about survival, pure and simple.

This lovely book comes at just the right time to create a context for the exciting and challenging years ahead. As the Americans often say: 'If you don't know where you've been, you won't know where you're going.'

Congratulations, Maggie, from a friend and admirer,

Tim Smit
Eden Project, Cornwall

Introduction

An illustration *c.* 1460, of a gardening scene taken from Pliny the Elder's manuscript, *Historia Naturalis*, written in AD 77.

HORTICULTURALLY SPEAKING, Britain would have been considered a Third World country at the end of the last great Ice Age, some eight thousand years ago, when its indigenous flora emerged as a frozen reflection of all that was left from its previous existence. A poor and mean landscape full of waterlogged sedges and ferns is what would have greeted any visitors. A small number of plants did somehow survive the onslaught of the millions of years of ice to emerge triumphant into the post-glacial period. These include the bilberry (*Vaccinum myrtillus)*, the exotic Droseras or sundew plants which trap insects in order to survive, some Alchemillas (distant cousins of *Alchemilla mollis*) and the rare and beautiful Teesdale Violet, *Viola rupestris.* There were in all probability no more than two hundred or so species in total, certainly little growing that could have been turned into a sustaining meal or a posy of flowers.

Nowadays, of course, it is a quite different story. At every turn in our daily lives we are almost submerged with horticultural delights. Walking up and down our garden path, strolling through the pedestrianised centre of a town or in the local park, or even turning into a motorway service area, we take for granted that there will be a selection of flower beds, shrubberies, tubs, planters, pots and hanging baskets filled with flowers, trees and shrubs for us to enjoy. It is the same when one goes for a walk in the countryside: we are greeted with a profusion of plants, from trees and shrubs to flowers, grasses and even weeds (which are only wild flowers growing in the wrong place), as though they are displaying themselves just for us.

Where on earth have they all come from, bearing in mind that we began with such a meanly filled wheelbarrow of native plants? The answer is, of course, that Britain, during the course of about two thousand years, has played host to the biggest planting party in the world. We have become horticultural hedonists as we surround ourselves with plants from every country in the world.

Look around your garden and you will see before you hostas from China and Japan, the scrambling nasturtium which arrived from South America, heather from South Africa, a rhododendron first collected from the Himalayas. California gave us the tall Monterey Pine, while from New Zealand came the hebes; the Mediterranean's contribution was lavender, and even Siberia sent us a gypsophila. In the front garden where the bedding plants are patterned out each May, it is the same story: the bright Busy Lizzies came originally from India, gloxinias and verbenas both originated in

This dainty violet *Viola rupestris* is one of the few surviving plants to have reappeared following the departure of the glaciers.

Central and South America, as did most of the begonias. The two tubs on either side of the front door hold a reminder of earlier times with a pair of topiary box twirls.

Left to themselves, plants are usually very slow colonisers, and yet they are good at adapting to new environments – which is fortunate considering the journeys of many thousands of miles some later introductions underwent. Once here, they then had to adapt to a new terrain and a new climate. But Britain, having an equable climate – not too much of anything – became a ready and eager host. Where plants come from and when they came here is fundamental to the development of our landscape and indeed our whole countryside. It is part of the biodiversity of the land; plant introductions mirror historical events, exploration and scientific discovery, and provide a reason why gardens have developed the way they have, and why our landscape looks as it does.

The ripples of climate change during this long post-glacial period – and indeed up to the present day – have had a profound effect and influence on both the survival and increase of our native and cultivated plants. We are in what climatologists refer to as 'the Sub-Atlantic period', which began about 500 BC and which is both wetter and colder than the

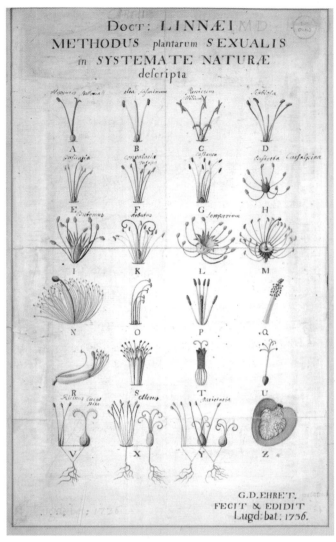

Carl Linneaus, whose *Systema Naturae* was published in 1735, began his series of encyclopaedic works, mainly on plants, in which he set out to instil order and system into the classification and naming of the whole organic world.

previous era. However, within these long phases, there are spells of slightly different weather – the ripples. For instance, from AD 1000 to 1300, the south of England was consistently warmer than the rest of Britain, so the vines that had been developed by the Romans during their occupancy thrived. By 1350, however, it was noticeable that the planting of vineyards, their cultivation and the drinking of English wine were being overtaken by wine imported from France. Even before then, Henry II (1133–89) much preferred wines from

the Gascony, Poitou and Auxerre regions of France.

A further climatic 'ripple' of a mini Ice Age is believed to have begun about 1550 and lasted to around 1840, with a forty-year warm period from 1900 to 1940. Today we are living through a similar period (or ripple) but one which climatologists tell us is man-made, caused by an increase in carbon dioxide thus leading eventually to overheating; in other words, the 'greenhouse effect'. History alone will tell whether this is the sole cause of the present climate change or whether it is linked to the 'ripple effect'.

With all this apparent climatic activity, it seems amazing that in the comparatively short space of a thousand years Britain has become such a truly 'green and pleasant land', with gardens stuffed full of other countries' plants. As we went out and colonised the world, so plants were brought home to colonise our land.

In looking at the thousands and thousands of plant introductions that have occurred over the past millennium, it is interesting to note how a pattern emerges. There is both a historical timescale and a dominant geographical area attached to the plants that arrived (although, of course, by no means is either category exclusive). This correlation is well illustrated by the following ten periods:

Period I, 1000–1560, the longest period, is dominated by plants arriving from continental Europe, due in part to the Norman invasion, the Crusades and monastic influences.

Period II stretches from *c.*1560 to 1620, when plants from the Near East, West Asia and the Balkans began arriving.

Period III is the Tradescant period,1620–62, and encompasses the older John Tradescant's travels to Russia and the North African coast, and his son's three journeys across the Atlantic to North America's east coast.

Period IV, 1660–1720, is dominated by the introduction of shrubs and trees from the eastern seaboard of North America.

Period V is from 1680 to 1774, when the dominant area for plant introductions was southern Africa, especially the Cape of Good Hope.

Period VI, 1772–1820, is the Antipodean period, which begins with the return of Joseph Banks and Captain Cook from their voyage of exploration to the southern hemisphere on HMS *Endeavour*.

Period VII, 1820–60, covers mainly plants introduced from California and the west coast of North America.

Period VIII, 1840–70, is when plants came mainly from South America with plant hunters collecting on behalf of commercial nurseries.

Period IX, 1840–90, takes in expeditions to India and the Himalayas.

Period X, 1890–1930, is dominated by plants arriving from the newly opened China and Japan.

It is endlessly surprising how often the date of arrival of a plant reflects the exploration of a new country, and, conversely, how the closing of a country's borders – or a war – halts the flow.

The first centuries following the Roman occupation of Britain (AD 43–410) show the slow development of our indigenous plants for food and medicine. The few plants which are known to have found a new home here during this period all came from continental Europe. But it was the spread of the new religion of Islam, with the caliphs' territorial conquests of the whole of the southern shores of the Mediterranean, culminating in their invasion of the Iberian peninsula in 711 (where they remained for nearly eight hundred years), which brought the sophistication of the Near Eastern plants to the attention of northern Europeans.

Even the canny invader William the Conqueror can be thought to have contributed to the total of our plant list. When he came to build what we now know as the Norman castles, he preferred to use material with which he was already acquainted rather than stone from unknown English quarries, so the walls of castles like Dover and Sherborne were built from stone imported from the continent. Incidental to its main purpose, the stone itself carried the seeds of a double invasion – seeds of two plants we think of as being most quintessentially English. The first was the pink (now known as the Wild Carnation), *Dianthus caryophyllus,* and the second was *Cheiranthus cheiri*, now called *Erysimum cheiri*, the wallflower. Both were seen blooming on the stone walls of Caen in France, so the pretty little delicate pink which is used in the breeding of nearly all our border pinks is the result of 1066 and all that.

Major events like the Crusades to the Holy Land during the eleventh, twelfth and thirteenth centuries and the Turks descending on Constantinople in 1453 and Belgrade during 1521 seem to have played a significant role in the gentle pursuit of botany. In fact, any movement of people (whether armies, courtiers, pilgrims or, later, tourists) can influence the migration of plants, either accidentally or deliberately. We know that following the Crusades, a number of plants gained a home here. Whether they were brought back as a deliberate trophy of the pilgrims' travels or arrived incognito caught up in the dust and dirt of travel we shall probably never know, but they were different enough from our native flora to be noticed and therefore worth nurturing.

However, plants do not always take kindly to their new home. For instance, another Dianthus, this one called the Wild Pink, which originally came from eastern and central Europe, was undecided about whether it liked it here or not. The plant crossed the Channel several times before it finally felt at home enough to settle here permanently in the 1560s. John Gerard is credited with naming the frilly little flower *Dianthus plumarius*, and although it has gone native in the southern part of Britain it is still being grown in our gardens. So it is with a number of other plants which teeter on the brink of British citizenship.

On the other hand, some find life in Britain so conducive to their well-being that they become positively thuggish in their garden behaviour and one wishes they could be deported. Plants which come (reluctantly) to mind are the Japanese Knotweed (*Fallopia japonica*), which brought its aggressive manners here during the early 1800s, and its relative *F. baldshuanica* (the Russian Vine or aptly named Mile-a-minute), which arrived from Tajikistan in 1883. They are both movers and shakers of all in their path, including tarmac and concrete.

Nearly everyone with an enquiring mind usually has an orderly mind too, for 'order is the source of peace', as John Caie (1811–79), head gardener at Inverary Castle to the 8th Duke of Argyll, stated as he wrote about the desired balance and proportion of nineteenth-century flower beds. In both the pre-Christian era and the early centuries of the first millennium, the uncoordinated and unclassified natural world must have presented an enormous challenge. Many great thinkers in the Greek and Roman world have left us amazing natural history and botanical documents recording their thoughts and observations; *Enquiry into Plants* written by Theophrastus (*c.* 372–286 BC) is the earliest surviving European treatise on the subject. The survey lists, naturally enough, the local Mediterranean flora, but also includes plants from further afield which Theophrastus probably never saw growing but which could have been drawn for him or described to him by a sharp-eyed and inquisitive early traveller. These might have been plant enthusiasts accompanying Alexander the Great and his army on the campaigns he undertook through the Middle East and Asia against the Persians.

Pliny the Elder, born at Verona in AD 23/24, was a pas-

De Marmore antiquo, olim in Ædibus Cardinalis
Maximi Romæ, nunc in Museo R. Mead. M.D.   J. Richardson f. 1730.

Theophrastus is reputed to have been the author of over 227 works, of which *Enquiry into Plants* is the earliest surviving European botanical treatise.

sionate observer of the natural world. He was, apparently, an all-action man, never wasting a minute of his time, and holding a variety of high government offices including the governorship of Spain and command of the Roman fleet. It was while he was stationed with the fleet near Pompeii in AD 79 that he went with some friends and his nephew (whom he had adopted as his son) to observe Vesuvius erupting and, in trying to save his friends, himself perished, overcome by the fumes. He left us his life's work, a prodigious encyclopedia in thirty-seven volumes called *Naturalis Historia* which, when it was finished, he described as being '20,000 matters of importance, drawn from 100 selected authors'.

A third writer from the early days of the first millennium, then undoubtedly the greatest influence in medical and herbal studies, was born during the first century AD in Asia Minor. Pedanius Dioscorides was a physician in the Roman army who wrote a book of plant descriptions called *De*

Materia Medica which had over four hundred colour illustrations; although it only exists in copy form – the first made in *c.* 512 and now in Vienna – it is an astounding document. The influence of his writings was immense. Indeed, some of the names he gave to plants we still use today; for instance, *Asphodelus ramosus*, the Asphodel, *Smyrnium olusatrum*, the plant which now grows wild in our hedgerows and we know as Alexanders (see p.28), and the Biarum genus. However, more importantly, he was honoured by the name *Dioscoreaceae* being given to a whole family of about 220 climbing herbs or shrubs in 9 genera. One of the genera is the *Dioscorea*, which includes the yam species. Remarkably, some of his pharmacological insights have relevance today.

In the ancient and medieval worlds a system called the 'Doctrine of Signatures' became the formalised method by which doctors treated ailments and illnesses, using products from the botanical kingdom. In fact, the system was practised for so long that monasteries were still using it until the birth of modern science in the sixteenth and seventeenth centuries.

The Doctrine consisted of the simple idea of relating parts of a plant to parts of the human body; the assumption was that because they looked alike, the plant would in some way be able to cure the ills of the offending body part. So if one received a head wound, the use of the walnut was recommended as a cure because of the resemblance of the nut to the skull and human brain; in this case, walnut oil was smeared over the wound. The very name of *Pulmonaria* (Lungwort) gives the medical indication of the plant's use. The leaves are spotted and rather rough, and look 'lung-like' so were thought to be helpful in respiratory problems. Any plant with a heart-shaped leaf was considered helpful in alleviating matters relating to the heart, and so on. Although the botanical world does provide a tremendous spread of cures for both the body and the spirit, this extraordinary dictate went almost unquestioned for centuries, with the information being copied, printed and plagiarised, and used by herbalists and medicine men all over Europe. Indeed, in all these early horticultural writings, there is a heady mixture of herbal myth and legend as well as some remedies so disgusting and downright dangerous that it is a matter of wonderment that anyone actually survived to tell the tale. The alarming Doctrine of Signatures was eventually overtaken by naturalists whose systematic observation of plant life led to a more straightforward approach rather than to the unquestioning mumbo-jumbo of earlier writers.

The Greek doctor Pedanius Dioscorides was the author of *De Materia Medica*, which contained information about plants and the ailments they could cure.

This simpler yet more observant approach happened during the fifteenth and sixteenth centuries, when the Renaissance took an interest in the natural world. Now there were men who really observed and studied plants rather than just using them for old times' sake. One such person was Leonard Fuchs (1501–66), for thirty-one years Professor of Medicine at the then newly established Protestant University of Tubingen. The fuchsia was named after him as a tribute to the work he did as a naturalist; however, like so many learned men after whom plants were named, he never knew the plant, since he died long before its arrival in Europe at the beginning of the eighteenth century. Fuchs befriended the English cleric William Turner (1508–68), who has been called the 'Father of British Botany' on account of his having made the first record of British native flora to be written both in Latin and in the vernacular. This was published in 1538 under the title *Libellus de Re Herbaria Novus*. In 1551 he began publishing *A New Herball*, which came out (after various vicissitudes) in three parts, being completed in 1568, the year of his death. His publications have become seminal in the study of the development of early English botany, as also has the *Herball or Historie of Plantes* written in 1597 by John Gerard (1545–1612), and *Paradisi in Sole Paradisus Terrestris* by John Parkinson (1567–1650), which was published in 1629. Each of these books helped lay the foundation for the horticultural enterprises which were to follow.

Although these good men were striving hard to bring order into the plant world and classify the European flora, more exotic material from both the Tropics and the New World was beginning to appear at about the same time. As shipbuilding design became more sophisticated, the Atlantic Ocean proved no barrier to inquisitiveness, and there were at least a further fourteen crossings in the thirty years following Columbus's exploratory voyage in 1492. Seeds, bulbs, corms and tubers were the easiest and surest way of bringing new plants across the seas, and their arrival added an urgency to the attempts to bring about an understanding of the flora, and especially of the medicinal use of plants. The idea of forming botanical collections for study seemed essential, and the first two botanic gardens were begun in Italy, one at Pisa about 1543 and the second at Padua about two years later. They should really be called physic gardens, for their main purpose was to teach medical students to recognise medicinal herbs. They were a modern development from the monastic garden, which had been based on the earlier system of the Doctrine of Signatures.

It was Henry Danvers (1573–1645), later created 1st Earl of Danby, who presented Britain with its first botanic garden, at Oxford. Danvers, who is described as a horticultural patron and innovator – he grew over a thousand fruit trees at his home, Wimbledon House in Surrey – was 'minded to become a benefactor to the University, determined to begin and finish a place whereby learning, especially the faculty of medicine, might be improved'. While under suspicion of murder – he was pardoned in 1598 – he had travelled widely on the Continent and had no doubt visited some of the six or seven established botanic gardens there, all attached to universities. In 1621, after inheriting a large fortune, he decided to take a lease on five acres of Oxford's water meadows. Here, on St James's Day (25 July), with a great deal of pomp, music and speeches, the first British botanic garden was inaugurated. A further twenty-five were created over the next three hundred years.

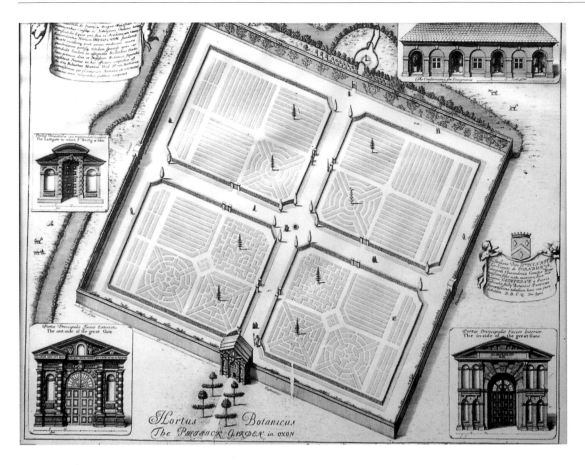

The first of the formal British Botanic Gardens was created at Oxford in 1621. It was followed in 1670 by the Edinburgh Botanic Garden. Three years later the Society of Apothecaries began planting the Chelsea Physic Garden, and in 1762 Cambridge Botanic Garden came into being.

Horticulture and botany became entwined in 1753 when a system of classification was established in the publication by Carl Linné, or as we know him Linnaeus (1707–78), of *Species Plantarum,* a book which for probably the first time brought order and discipline to the botanical world. Travellers and explorers, and eventually plant-hunters, could now scour the world looking for new plants, and when found, these could be placed in their correct family. That at least was the theory, but nothing stands still, and since Linnaeus, the art of botanical naming (nomenclature) has been constantly changing as botanists delve deeper and deeper into a plant's very nature – much to the delight of plant label manufacturers and the frustration of the gardening world.

The word 'nomenclature' has had an interesting life. It started in the Roman world when a *nomenclator* was a servant – normally a slave – whose daily duty was to inform his master of those who had come to pay their respects, usually seeking an influential patron. By 1599, the name was given to a master of ceremonies announcing guests at a banquet, but by 1644 it had been transferred to an inventor of names specialising in the classification of natural objects including plants.

There is no doubt that the last quarter of the second millennium saw the greatest number of new plants being established in Britain from all over the world. However, it was the botanical opening-up of the Far East, especially Japan and China, during the nineteenth century which led to the greatest of the 'flower ages' in the haphazard history of foreign plant introductions to Europe. The Far East is without doubt the horticultural powerhouse of the world, and thousands of flowers, shrubs and trees which now are available from any garden centre in Britain emanated from those countries and have had the most profound effect on our gardening landscape. This development, combined with a significant innovation made around 1835 – the invention of the portable mini-greenhouse (the Wardian Case) by Nathaniel Ward – led to the great era of plant-hunters who were able to take advantage of easier and safer travel in pursuit of their horticultural treasures. Later, still better and speedier travel also helped to heighten the survival rate of discoveries. The first plant to arrive by air was *Primula sonchifolia*, sent from Burma in 1930.

The evidence shows that the scramble for foreign plants was at its busiest and most enthralling when travel and communication was slow, sometimes dangerous and always difficult. During recent decades, when ease of travel is taken for granted and it seems as if the whole world is on the move – or at least on the telephone, fax or worldwide web – the welcoming of new horticultural treasures has slowed down. The timing has coincided with the awareness of the fragility of the natural world and the rise of the conservation movement. The garnering of plants from the four corners of the earth must be considered as coming to an end. After all, a country's natural habit is part of its unique identity and thus of its heritage, which in an age of globalisation, individuals and communities are keen to retain.

While we are infinitely richer in flowers, vegetables and fruit than when we began the second millennium, it is the hybridising and breeding of new and sturdier varieties which has made such an impact during the past hundred years. As the flow of new plants into Britain has slowed from its apogee at the beginning of the nineteenth century, so has interest in the 'designer plant' – a horticultural gem created to fulfil a specific purpose – gained momentum. For instance, *Rhododendron* 'Frosted Orange', *R.* 'Peggy Ann' and *R.* 'Ben Morrison' are three bi-coloured azaleas bred during the latter part of the twentieth century which are compact 1–2 m (3–6 ft) shrubs, hardy to temperatures down to 18°C (0°F). *Lavandula stoechas* 'Roxlea Park' is a strong cerise shade – not a colour one normally associates with lavender. It was first noted in New Zealand and has a very graceful habit. The achievement of breeders and horticulturists in producing a range of plants which fit in with the changing lifestyles and requirements of the twentieth and twenty-first centuries is truly remarkable. Whether it is in making a plant sturdier, smaller or drought-resistant, or in increasing the flowering time, producing a variegated version or a different-coloured bloom for us to be tempted by at the garden centre, botanical expertise seems to have reached new heights. Yet manipulating a plant's size, shape and colour is nothing new, as witnessed by the development of the chrysanthemum, the camellia and the cherry in the Far East long before AD 1000. In addition, the art of *Pen-tsai (bonsai),* the dwarfing of trees, and *Pen-ching*, the making of miniature landscapes, existed in China from the seventh century onwards.

Humans are inevitably the key to the vast majority of the horticultural imports that have taken place around the globe.

But one of the most recent botanical finds in Britain, in the last decade of the twentieth century, and the cause of much excitement, was an orchid indigenous to the Mediterranean which was thought to have been swept here as seed on wind-borne sand from the Sahara. The orchid was identified as *Serapia parviflora* which is a native of the Mediterranean region, particularly of Portugal. Alas, it did not survive, for within the space of two years it had been picked and dug up by collectors to the point of extinction.

Cosmos atrosanguineus is a member of the daisy family, and was first brought from Mexico to Britain in 1835 and grown by William Thompson of Ipswich (founder of the seed merchants Thompson & Morgan). Later, and quite inadvertently, it was transported from South America to South Africa at the time of the Boer War, when hay containing the seeds was sent as fodder for the British Army's horses. Wherever the horse fodder went, nature took its course, and the road verges later bloomed with the new import. In Britain the survival of *C. atrosanguineus* has been more problematical, and despite its hot-chocolate looks and its perfume – and being given an Award of Merit in 1938 by the RHS – it very nearly disappeared from our gardens in the ensuing years, and is also now very rare in its native land. Its naturalisation in South Africa was therefore most fortuitous in maintaining the plant's existence.

Of the thousands and thousands of foreign flora which have arrived here only a tiny percentage have managed to survive, and yet they have so enriched our horticultural experience that Britain is at the very forefront of the gardening world. It is surprising, too, how quickly some of them become established. Think of Japanese anemones, hostas, *Garrya elliptica*, alstroemeria, ceanothus and many more – all sailed into Britain during the past two hundred or so years, and yet are now part of our gardening scene, just as are the pink and the wallflower, which arrived with the Normans. I suppose the truth is you can't keep a good plant from growing – wherever it is.

Knowing where plants come from, the date they were introduced into Britain, the reason why they are named as they are can tell us an enormous amount about our island's history, and give added enjoyment to our own gardening, too. It is in fitting together the jigsaw, and thus bringing into sharp focus the plants which we grow in our own gardens and the stories they can tell, that we gain a fascinating extra dimension to our knowledge of and pleasure in gardening.

Setting the Scene

The Sweet Chestnut *Castanea sativa* is believed to be a Roman introduction and estimates of the age of the Tortworth Sweet Chestnut in Gloucestershire are given as over one thousand years.

Significant dates

43   Romans under the Emperor Claudius invade Britain

60   Revolt of the Iceni under Boudicca; London ravaged by fire

64   Great fire of Rome; Nero begins to persecute the Christians; death of St Peter

117   Hadrian becomes Emperor

118   Invention of the wheelbarrow by the Chinese

122   Hadrian's Wall started; completed 139

161   Marcus Aurelius becomes Emperor

286   Aurelius Carausius attempts to separate Britain from the Roman Empire

324   Byzantium becomes capital of the Roman Empire (renamed Constantinople)

410   Roman legions leave Britain

446   Saxons begin arriving in Britain

529   St Benedict (*c.* 480–547) founds monastery at Monte Cassino

563   St Columba founds monastery on Iona

597   St Augustine (d. *c.* 604) sent by Pope Gregory the Great to convert England; becomes first Archbishop of Canterbury

604   First Church of St Paul in London

664   Synod of Whitby; Roman practice imposed

711   Moslem invasion of Spain

780   Offa's Dyke begun

800   Charlemagne crowned Emperor of the Western Roman Empire

865   First major Viking invasion of England

871   London occupied by the Danes

960   St Dunstan becomes Archbishop of Canterbury

973   Coronation of King Edgar in form still used in twentieth century

978   Ethelred the Unready (978–1016)

991   English treaty with Normans

995   *The Glossary to Grammatica Latino-Saxonica* compiled by Aelfric, Abbot of Eynsham

WITH THE BEST WILL IN THE WORLD, it has to be admitted that Britain was a late developer as far as the art of gardening was concerned. All the floristic evidence suggests that the first 500 years of the second millennium, from AD 1000–1500, were spent clinging on to the detritus left over from the Roman occupation (AD 43–*c.* 410) and the wreckage of the Saxon and Viking raids.

And yet . . . and yet . . . there must surely have been more to it than that. The overwhelming desire for food, shelter and warmth was – and still is – a driving force. We like to think it was we – twenty-first-century sophisticates – who discovered the art of DIY. Not a bit of it; we are mere dabblers in the field compared to our ancestors of a thousand years ago. Then it required everyone's concentrated efforts to cultivate and grow crops, a real community involvement in survival. So it is no wonder that there are only a few scraps of vellum for us to pore over, and artefacts more precious than gold for us to see. Almost everyone's energy was required to keep body and soul together, and the laborious scribing to record such a seemingly mundane and obvious activity must have appeared a somewhat pointless exercise, particularly when almost the entire population was actively involved and knew by example what to do. Who would have recorded the activity or then read about it? Only monks and a few scholars were able to read or write. It was patently self-evident what was required to survive, and although it is a proven fact that people died at an earlier age than say in the eighteenth century (or now), it was usually due to either an accident or disease – often brought about by bad diet – and rarely 'just' from starvation.

And yet . . . and yet . . . in pre-Conquest Britain there are profound signs that an artistic and sophisticated world existed beyond the field and simmering pot, and hints and clues exist as to what was being planted and grown. The English vernacular (Anglo-Saxon) had become the first European language (as opposed to Latin) to develop its own literature. In the monastic establishments around the country – but mainly in the southern half of Britain, in the Viking-free zone – paintings, carvings, wonderful illuminated service books, intricate embroidered vestments and hangings, silver and gold bowls and chalices were being created, finer by far than those produced by the great abbeys of continental Europe at the time. Music too was advanced, with its delicate wide polyphonic range, again led by the monasteries.

Although Christianity had first come to Britain during the Roman period, it had nearly choked to death with the subsequent invasions of the pagan hordes from Germany and the Low Countries, of the Angles, Saxons and Jutes. The Celtic fringes of the land more or less retained their Christian faith, however, and in 597, nearly two hundred years after the Romans had retired to their Italian villas, a Benedictine missionary named Augustine travelled to Britain, and it was he who eventually founded the first English monastery, at Canterbury, and became its first archbishop. Although the Christian faith was slow to take root, and had many lapses back into paganism, by the time of the first Viking invasion in 865, monks and priests were dominating the development of the arts and education.

It is to the learning of Latin that we owe the first list of plants in Old English. It was written by Aelfric, the celebrated Abbot of Eynsham, in 995, and was called *The Glossary to Grammatica Latino-Saxonica*, a Latin–Anglo-Saxon vocabulary. It is in this ancient document that an Old English phrase has survived to come down to us like a clarion call, and that is the wonderful Saxon description of *luffendlic stede*, which means a 'lovely place'. What a ring *luffendlic stede* has to it, and what refinement of thought lies behind the words of a nation just emerging from the sterile arctic winter of the Dark Ages. In the same vocabulary is the word *wyrttun*, meaning 'garden'.

In his long list of words, Aelfric names about two hundred plants, including vegetables and herbs, in Anglo-Saxon and Latin, which were both in general use at the time. That list, combined with the plants introduced by the Romans some four hundred years earlier, gives us at least some idea of what would have been growing at the beginning of the second millennium and before the whirlwind of the next invasion. Whereas the plants which were cultivated during the Roman invasion were all thought of as being imported (and, one suspects, part of the Roman agenda for impressing on the natives how sophisticated was their way of life), the list which the learned Aelfric drew up made no such distinction. It contained words that were familiar, words that were in common

Abbot Aelfric's *Glossary to Grammatica Latin-Saxonica* was written in 995. The vocabulary included the names of plants and trees and has enabled us to have an insight into that early botanical world. It shows the part list of herbs (*mentha, origanum, lilium, viola*) and then begins the Names of Trees – *Nomina Arborum* (*buxus, malus, prunus, abies* and so on), all names which can be identified today.

use, that were probably part of lessons taught in the monastery schools. Whether the plants were indigenous or Roman or had arrived from anywhere else was of no consequence.

As there is no clear-cut list of 'Roman' plants, so there is no clear design available to us of a Romano-British garden. Fishbourne Palace near Chichester in Sussex is the nearest we come to seeing that the style was rather stilted and formal. The gardens here had topiary hedges made of box (Buxus, see p.29) and small edged beds, straight paths and rectangular lawns, the whole made as if for outdoor entertaining on the grand scale. The design displayed none of the exuberance and dash shown in the Romans' indoor decoration and mosaics; perhaps the vagaries of our climate restrained them.

'The climate is unpleasant, with frequent rain and mist, but it does not suffer from extreme cold. The soil is fertile and is suitable for all crops except the vine, olive and other plants requiring warmer climes.' Thus Publius Tacitus (c. 55–120) was able to write with confidence to his masters in Rome about the British weather. Being married to the daughter of Agricola, conqueror of Britain in AD 77, he had to be sure to get his facts right.

This description of Tacitus set the scene for the Romanisation of the land, both domestically and in the wider landscape. The sudden and relatively large influx of the invading army, with its urgent need for cereal, meat and vegetables, must have caused a few supply-and-demand problems to begin with, and a priority must have been the rapid increase in food production.

Whatever Tacitus recommended, top of the Roman wants list would surely have been *Vitis vinifera,* the grape vine. This was a native of Europe and west Asia, although it probably was not planted in Britain, except ornamentally, until after about 280. Italy held the monopoly for cultivating vineyards and producing grapes for wine, and the Roman governors kept strictly to that rule for many years. It was relaxed by Emperor Probus (235–82), who is supposed to have visited England to plant the first vine (reputedly at The Vyne in Hampshire). He thought the army had too much time on its hands, and set them to carry out public works (a not unknown solution today), which included the making of vineyards throughout the Empire. Unfortunately, because the army felt such labour was too much below its dignity – digging and all that – they mutinied on Probus's arrival to

The formal garden at Fishbourne had been created as a level platform and to encourage plants to grow, bedding trenches were filled with marled loam covered with top soil. The pattern of the excavated trenches can be clearly seen in the photograph.

oversee the draining of marshes at Sirmium (now Mitrovica in Kosovo) and murdered him.

However, the monopoly was broken and, despite the climatic problems, the growing of the vine to produce wine entered the psyche of Britain. In spite of that encouragement, by the time of the Domesday survey of 1086, some eight hundred years later, there were only a pathetic thirty-eight vineyards recorded, twelve of which were monastic; the largest of these was the one attached to Bisham Abbey in Berkshire, which was about 4.85 hectares (about 12 acres). English viticulture has always been a fringe activity, due not to lack of interest but to the lack of sun and warmth, so that even today the harvesting of the grapes is a worrying balancing act between ripeness and weather. The growing of vines for wine never ceased entirely but gradually became an aristocratic pastime, that is until a revival began in the 1950s. Now there are some 420 commercial vineyards producing wine from about 1,000 hectares (nearly 2,500 acres).

In early Christian art, the vine also goes under the name of the 'Root of Jesse' and is one of the emblems of Christ. It is a symbol of fruitfulness, and at Ely there is preserved a beautiful thirteenth-century psalter with a Tree of Jesse tracing the genealogy of Christ. Today, there are about sixty-five ornamental Vitis species tendrilling their way around the northern hemisphere, particularly in North America and the

This illustration from Rabano Mauro's 1028 manuscript shows a heavily fruited vine clearly ready for harvesting.

Far East, with most of them producing inedible grapes but wonderful foliage, including a number which set gardens ablaze with their brilliant autumnal colours.

Another known Roman introduction was the Spanish or Sweet Chestnut, *Castanea sativa*. This is now known to be a native of southern Europe, North Africa and south-west Asia, although it was long thought to be a native here. It certainly behaves like one, being one of the few trees that leads a double life, with both a formal and an informal existence. Much later, in the seventeenth and eighteenth centuries it was successfully planted both as an avenue tree in parklands and landscape gardens, as at Mount Edgcumbe in Cornwall, and as a woodland tree in copses and forests. It was also found to take to coppicing very well and it is good to see the practice continuing even now; fence posts made from the wood are very sturdy and long-lasting. In Tortworth in Gloucestershire, beside the parish church, there is an ancient relic of a chestnut, believed to be about a thousand years old.

Because of the complications of its growth and its gigantic size, the writer Thomas Pakenham called it a 'riot of a tree' or a 'great waterfall of a tree'. It seems to bear its years and fame with vigour, for even over three hundred years ago its bulk and great age were remarked on by John Evelyn (1620–1706) in his famous tree book *Sylva*, written in 1664, and again by Peter Collinson (1694–1768) in the following century; these were two of the most important and far-seeing 'tree' men Britain has ever produced.

The Spanish Chestnut even had an Anglo-Saxon name: *cyst* or *cisten*, with the nuts themselves being called *castaynes*. The tree is large and so was rarely cultivated within small-holdings or gardens but was prolific in woodlands and small copses. The Romans ate the nuts, either raw or roasted (as we still do today), and made flour from them. It was also said that the powdered nuts were good for curing piles, although whether taken internally or externally was not made clear.

A tree of real consequence, without which the Romans

would have felt bereft, was *Juglans regia*, the walnut (the name which itself has evolved from the Anglo-Saxon word *wealh*, meaning 'far away, foreign'). The tree – which was known as the Royal Tree of Jupiter and commemorates the ancient notion that the god himself was partial to the nuts – is part of a very small but distinguished family, the *Juglandaceae*. This family, which comprises about 15 species, is widely spread across eastern Europe, northern Asia and North and South America. In fact the noblest walnut of all, and consequently a very valuable timber tree (the nuts in this case being of little use), is one from the east side of the United States: *Juglans nigra*, the Black Walnut. It arrived in Britain in 1656, centuries after the Romans had departed. Because it was said to attract lightning, it was hardly ever used in its homeland, especially not for the building of boats or ships for fear that lightning would strike and cause a fire and the vessel sink.

It was only slowly realised that the walnut tree introduced by the Romans, *J. regia*, was a good timber tree, and therefore to be greatly valued, apart from its yearly harvest of *walnottreis*. A great bed made from walnut wood and decorated in gold and purple was built for one of the royal rooms in the Palace of Whitehall in the 1630s – perhaps being one of the many beds that King Charles I was said to have slept in.

It was nearly always a tree that was planted close to habitation, either in orchards or plantation. It has the reputation for not being attractive to insects, so livestock are often found sheltering in its shade; horses sponged with a solution of the soaked leaves are said to be relieved from harrying insects too. It was sometimes planted as part of a hedgerow – in fact, it still is on the Continent, particularly in France. Because the tree is so long-lived, or at least very slow to mature, it is considered a most unlucky event if it is blown over or dies other than being felled by man; some used to believe that it presaged a death in the family. (The fact that the timber is worth a lot of money might have something to do with it as well.)

The walnut is grown not only for its timber but for its harvest of nuts, and subsequently for the oil produced from the nuts, which makes a very fine dressing for salads. Folklore tells that the oil, rubbed into the scalp, is also supposedly a cure for baldness. Personally, I think it should be saved for salads. The immature nuts (known as 'wet walnuts') may be pickled and eaten with cold meats, and the mature nuts are welcomed in November and December. Today they arrive in our supermarkets often from China or North America. In

Roman times they were distributed at weddings as an encouragement for fertility during the marriage.

Several other dual-purpose fruit trees thought to have been brought to this garrison outpost of the Roman Empire were the fig, the medlar, the sweet cherry and the plum. All produced fruit to eat, while the wood of the walnut and the cherry was used to make furniture – as it still is. Just as in a later age, when Britain was busy anglicising large parts of the world and the first British settlers were planting trees and shrubs with which they were familiar, so these imported fruit trees helped to make the Roman centurion feel at home. In

Common European Walnut.
Juglans regia

The earliest recorded evidence of the walnut tree being cultivated is in 1498 when the inhabitants of the Yorkshire village of Willstrop quarrelled with a landowner. They vented their anger by breaking through a hedge and 'hewid and kit doonn 100 walnottreis'.

The fig tree said to be planted by Archbishop Thomas Becket at his Palace at West Tarring near Worthing in Sussex.

both cases, the indigenous population was amazed at the sophistication of the alien food.

The fig (*Ficus carica*) has survived to the present day with its Latin name unchanged, and even now, after some eighteen hundred years of British domestication, it still gives off an aura of hot Mediterranean sun. To fruit well it needs a sheltered warm spot, which, since its place of origin appears to be Asia Minor, makes sense. It has been cultivated from time immemorial and was known to be familiar to the Romans; Pliny describes over 20 cultivars. No doubt in the warmer southern half of Britain it could have been planted in villa gardens. That at least is the traditional view, although no written evidence survives. Another tradition is that an abbot from Fécamp in Normandy was the first person to introduce the fig tree to Britain, this time specifically to Sompting Abbots in Sussex; yet again, there is a story that Thomas Becket planted one of the fig trees at the Old Palace

of the Archbishops of Canterbury at West Tarring (also in Sussex). The area was certainly known for growing figs, and even now there are large healthy trees growing against old houses, producing a good crop each year.

The earliest written reference to a fig tree is in 1525, when Reginald Pole (1500–58), later Archbishop of Canterbury, returned from his studies in Italy bearing several trees for planting at Lambeth Palace in London. They were apparently of the 'white Marseilles kind', and one grew to the enormous size of at least 16 m (about 50 ft) in height. It did not succumb until the severe winter of 1814. In 1648, a fig tree was brought back from Aleppo in Syria by a Dr Pocock and planted at Christ Church, Oxford; this one lived until at least 1833, despite having been severely burned during a fire in 1803.

There are about 800 species in the genus, found in nearly all the subtropical and tropical forests of the world. Some are free-standing trees (such as *Ficus elastica*, the India Rubber Plant, see p.199), some are climbers and some are stranglers. The fig we know from the ubiquitous hotel buffet breakfast is one called the Kadota, so named because of its similarity to the shape of a specific Greek pot named a *kados*. As becomes the tree's many biblical connections, one that grows against the wall of the church at St Newlyn East in Cornwall is said to have sprouted from the staff of St Newlina, who was believed to have been a Christian princess martyred for her beliefs. From the New Testament comes the story from St Mark's Gospel (11:13): 'And seeing a fig tree afar off having leaves, he came, if haply he might find anything thereon: and when he came to it, he found nothing but leaves; for the time of figs was not yet.' This appears to be the inspiration behind the naming of Palm Sunday as 'Fig Sunday', when traditionally figs were eaten. At least it is a tradition in the village of North Marston in Buckinghamshire, and one supposes that in this case the figs were dried. In the Lake District it is Good Friday which is associated with figs, this time in the form of a drink, ale, sugar and ground ginger being mixed with the stewed and strained figs.

The refinement of the Roman civilisation failed to usurp the name local people gave to the medlar when it arrived: 'openarse', a descriptive name if one looks at the fruit. It was called this, or alternatively *cul de chien* ('dog's arse'), throughout Europe. The Latin name is *Mespilus germanica*. The medlar is indigenous across Europe and, despite its vulgar vernacular name, is a most decoratively shaped tree with charming white flowers which looks good planted either in a garden or an orchard. It is grown for its fruit, which is picked in October or early November when it is still not ripe; laid in straw, it softens over the next few weeks and can be eaten raw or made into a jelly. One hesitates to mention it, but the jelly was recommended by doctors as late as 1907 as a cure for stomach looseness.

The cherry native to Britain is *Prunus avium*, which grows wild throughout Europe and western Asia. It is the species most used to develop the cultivated sweet cherry, and was one of the names on Abbot Aelfric's list. The varieties have never ousted our own native cherry, which by the sixteenth century was being called the Gean (meaning juicy) or Mazzard Cherry. Pliny knew both the sweet cherry varieties and *Prunus cerasus*, the Sour Cherry; the latter, a native of south-west Asia, being more of a bush than a tree. This species was the main progenitor of all the cooking and bot-

The Medlar, an original Roman introduction, probably continued here during the Dark Ages, but as a rarity. By the thirteenth century it was one fruit tree usually to be found in a monastic orchard.

tling varieties of cherry, including the Morello Cherry (*P. cerasus austera*).

The plum, *Prunus domestica*, was also known by Pliny, as well as by Ovid and Virgil (all of whom described the different varieties of plums there were). In addition to our native plum, other natives to survive are the damson (*P. damascena*), the bullace, (*P. insititia*) and the sloe (*P. spinosa*).

The simmering pot of stewing vegetables and herbs that the Romans added to during their stay was pretty meagre, but the leek, *Allium ampeloprasum* var. *porrum*, was one of them – after all, it was known to have been eaten by the Emperor Nero (AD 15–68) to the exclusion of almost all else when he was training his voice for his singing recitals. Conversely, because it was thought that apples were considered bad for the voice, he refrained from eating them.

The leek's close relation, the garlic, *Allium sativum*, was probably only ever found growing wild in the exotic hotspot of the Kirghiz Desert of Central Asia, but was grown and eaten with enthusiasm almost all over the known world. However, it never seemed to have been completely popular in Britain; it was recorded by the apothecary Nicholas Culpeper (1616–54) in his renowned *Complete Herbal* of 1653 that 'Its heat is very vehement; and all vehement hot things send up but ill-savoured vapours to the brain.' Four hundred years later, in 1956, the *RHS Dictionary of Gardening* states that 'most people find it sufficient to rub the dish with one

of the small divisions of the Garlic bulb', and then adds as an afterthought, and rather sanctimoniously, 'Its flavour is too pungent to our taste.' It may have been too strong for British tastes then, but it certainly is not now. Its popularity in cooking has grown to such an extent in the past fifty years or so that garlic festivals are now held, and it is grown on the Isle of Wight as a commercial crop. The storybooks promise that attacks by vampires are sure to be held at bay by the wearing of the corm (as is whooping cough and yellow jaundice); more recently, it has been scientifically shown that garlic can also protect against some forms of heart disease. New life has been breathed into an old remedy.

The addictive flavour of *Asparagus officinalis* seems not to have been left at home by the Romans either. Named by Theophrastus in his *Enquiry into Plants* which he wrote around 320 BC, there are some 100 species, all native to the Old World. It came originally from the salt steppes of eastern Europe and from the Mediterranean coast, and never seems to have gone out of favour (unlike the leek, which was hardly touched during the eighteenth and nineteenth centuries). John Gerard (1545–1612) in his writings rather poetically mentions that the edible asparagus spear could be the 'size of the largest Swan's quill'. The cabbage (*Brassica oleracea*) was most likely another import from middle Europe, although it may have arrived here a little earlier than with the Romans.

The Romans are believed to have also brought with them, perhaps as an antidote to all the onions and garlic, the dill plant, *Anethum graveolens*. Originally from Asia, this fennel lookalike is now naturalised over the whole of Europe. It was first called *Peucedanum graveolens* – which was the name Hippocrates gave it and is the Greek word for 'parsnip'. Its

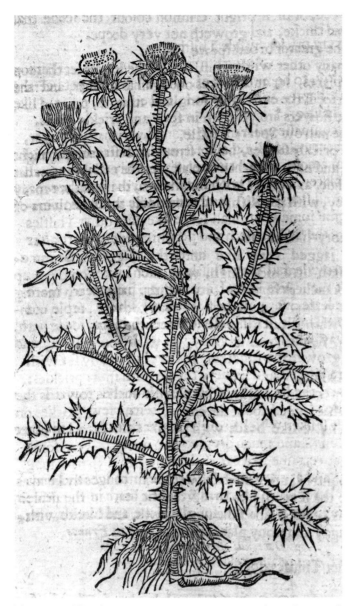

The stately Thistle *Onopordum acanthium* is a worthy architectural plant in the herbaceous border, despite its vulgar vernacular name of donkey fart.

was a 'gallant expeller of wind and provoker of terms'. During the Roman occupation, the herb was supplied in great quantity to the forts in this country, and one can only assume the Roman soldiers required it as an expellent rather than a sleeping draught. It was used as a spicy cooking herb as well, its leaves to flavour fish (especially gravlax) and its seeds to flavour vinegar (for pickled cucumber). A sweet and spiced dill cake was baked and eaten as a *digestive*.

Lovage (***Levisticum officinale***) was believed to have been a native of south-west Asia but is now thought to have originated in the Mediterranean. Together with sage (see p.81), it was included on Aelfric's list. It was an ale-flavouring herb and was also used to make a tea-like cordial. Among other things, it would have been cultivated for its aromatic leaves, which, when bruised and stewed in lard and laid hot on spots or boils, helped to make them better. Nowadays, its lemony-celery taste makes it a useful flavouring in soups.

Al-kharstif is an Arabic word and a pointer to the fact that the artichoke originated from the North African coast. It was the Italians who took the Globe Artichoke to their hearts (or stomachs) and developed it for gourmet eating. ***Cynara scolymus*** is its Latin name (*scolymus* is the word for the Spanish oyster plant, the roots of which you can eat). There are about 10 species in all, including that giant of the vegetable world, the Cardoon, ***Cynara cardunculus***, which arrived in 1658. They all belong within the thistle family (*Asteraceae/Compositae*).

The artichoke itself was once considered an aphrodisiac, and as the cookery writer Jane Grigson commented, it is 'the vegetable expression of civilised living'. No wonder the Romans wanted to show it off to the natives of their uncivilised northern province. It is presumed that once the Romans had departed for warmer climes, the artichoke went too, like so much else they introduced, but it was later found gracing the tables (and gardens) of the sixteenth century. John Parkinson (1567–1650), who wrote his book *Paradisi in Sole Paradisus Terrestris* in 1629, was full of praise for the vegetable and described how it should be prepared and eaten. The Globe Artichoke is one of those plant introductions from abroad which, however long it has been part of the British scene, never seems to lose its feeling of 'foreignness'.

There are a number of plants of a similar type whose foreign chromosomes have never quite settled in or taken on the mantle of their adopted British home. However, one member of the thistle family which has made itself at home here is what has come to be known as the Scotch Thistle,

vernacular name dill could be derived from the Anglo-Saxon word *dillan*, 'to lull'; a concoction made from the leaves said to have been drunk for its soothing properties and mild soporific effect seems to bear this out. For at least a thousand years teaspoons of dill, or gripe, water have been fed to babies to soothe their colic. By the seventeenth century, the adult version included white wine and, Culpeper reported,

nothing to do with Scotland. (The thistle usually referred to as the emblem of Scotland is a native, *Cirsium vulgare*, the Spear Thistle.)

Alexander the Great (356–23 BC), King of Macedon and conqueror of almost all the then known world, had much to answer for, but something for which he was not responsible, despite its being named in his honour, was the Alexanders, a vegetable that came into Britain at about the time of the Roman occupation. *Smyrnium olusatrum* was a sort of precursor to celery (*Apium graveolens*) and every bit of it could be eaten. The root was parsnip-like, the stems of the leaves tasted like celery and the little flower buds were used raw in salad, as were the leaves, with the ripe seeds being ground up to make a pepper. It must have been a very welcome plant, since it comes into flower and leaf very early in the year, during the 'hungry gap' period. The Alexandrine name derives from the leaf shield which, when pulled back, resembles the shape of the great hero's helmet; it is also sometimes called Horse Parsley. It seems now to have quite reverted to the wild, which is a pity because as well as being a most useful plant, it is also tall and handsome. It is most usually found in coastal areas, specifically the Isle of Wight, North Wales and Norfolk, and I see no reason why it should not be used now in much the same ways as the Romans did.

While the Romans were here, the fruit trees and vegetables they brought with them inevitably played a more important role than any flowers they might have introduced. That attitude only reflected what had gone before and what was to continue for hundreds more years – plants were dismissed or only given a cursory glance if they did not feed you or purge you. Their decorative and aesthetic qualities by and large remained unremarked behind a veil of hunger and ague.

However, there were a few flowers for which we can be grateful and which still grace our gardens today. One which must have played a dual role is the pretty ruched-petalled mauve or pink poppy whose antecedents are unknown but whose seed heads are so beloved of today's flower arrangers. It is *Papaver somniferum*, the Opium Poppy, the source of the laudanum which once brought relief to generations of families. (One source reports that 50,000 lb – approximately 22,800 kg – of 'loddy' was manufactured during only a few years in the eighteenth century.) An import from the southern Mediterranean and North Africa was the Stinking Gladwyn, or *Iris foetidissima* to give it its Latin tag; its 'gladwyn' name is a relic from the Anglo-Saxon world. John

Alexanders *Smyrnium olusatrum* (shown centre right) usually grows as a weed on waste ground, but certainly until the seventeenth century it was planted and eaten as a vegetable.

Onopordum acanthium. The first half of its botanical name is a combination of two Greek words: *onos*, which means 'ass', and *porde*, meaning 'to fart'. This thistle's early vernacular name of 'Donkey Fart' seems a rather surprising one even for a thistle, but it presumably reflected the result (or, indeed, the end result) of the donkey's digestive habits. Its current vernacular name of Scotch Thistle is equally strange, since it has

Gerard believed it to have a 'purging qualitie' from a drink made from the root.

The Romans seem to have been responsible for two plants which were said to cure sore eyes. The first is the Common Mallow, *Malva sylvestris*, whose territory stretched from North Africa into Europe. Children long ago discovered the dubious pleasure of eating the plant's ripe seeds, as they are no doubt still doing. It was found that the boiled leaves could be made into a salve to soothe aching joints and rheumatism as well as cure styes and soreness around the eye.

The other arrival could do the same, except in this case the poultice was conversely placed on the good eye so that the sore eye could be remedied. If that seems a mite strange, so will the fact that every bit of the plant is very poisonous, hence its common name of Hemlock or Poison Hemlock. It was the poison used in the drink given to those condemned to die for having offended against the Athenian state, and was also the poison that Socrates took – a handsome plant but not a happy one. Gerard says that the whole plant has 'an ill-favoured scent' and today it appears not to be commercially available. It is a member of the *Umbelliferae* family, and the Latin name, *Conium maculatum*, derives very closely from the Greek *koneion*, thus giving a clue to its long use.

Another introduction from the *Umbelliferae* family which was perhaps slightly more welcome was *Aegopodium podagraria*. Its derivation from the Greek of *aix*, meaning 'goat', and *podion*, 'a little foot', describes why this plant was in such demand – it was supposed to cure gout, as painful an affliction then as it is now. One would think therefore that this would be a most useful plant, but no, because it is, of course, Ground Elder and definitely not welcome in gardens on account of its voracious colonising habit. Its other name is Bishop's Weed (do bishops suffer a lot of gout?). In California, apparently, it is nowhere near as aggressive and is thus rather more welcome in gardens there.

One shrub which the Romans used frequently during their time in Britain was *Buxus sempervirens*, a native shrub throughout Europe (including Britain), Turkey and North Africa. It would have been quite familiar to the Romans, who used it in their own Italian gardens. Evidence shows that they were keen to clip and shape trees, not only box, into shapes, mainly architectural and decorative. The phrase for this was *nemora tonsillia*, the fashion for which developed around the end of the first century. According to Pliny, an ornamental garden was an *opus topiarium* and a gardener who worked in one was referred to as a *topiarius*. The word 'topiary' as we now understand it was being used in our vocabulary by 1592, when interest was shown in the art of clipping trees and shrubs into many different and sometimes extraordinary shapes. Later, in the eighteenth century, the phrase 'hortulan architecture' was used for a short time to describe the same process.

Box has a funereal side to it, with sprigs of the shrub having been discovered in several excavated Roman coffins. A similar practice of throwing box into the grave was reported in a number of instances, particularly in the north of England, right through to the nineteenth century. Apart from the recorded belief that box was one of the many shrubs considered to be unlucky if brought into the house, no reason appears to have been discovered why it should have been used in this way.

The tree itself, in its natural form, is very graceful and can reach a height of approximately 4.5 m (15 ft). A considerable number of varieties have been developed from *B. sempervirens*. Perhaps the most important, and one of the oldest – it was listed by John Gerard in his *Herball* – is *B. sempervirens* 'Suffruticosa'; the Edging Box, as it was often called. It is thought to have originated in Holland, though no one is quite sure, and to have arrived in England by the seventeenth century, when it was considered a 'mervaillous fine ornament'. There are some 70 Buxus species, all of which are evergreen and which come from Europe, Asia, Africa and Central America. *B. balearica*, with its bright glossy leaves, was introduced from southern Europe in 1780 and is still available, as is *B. microphylla* (the Small-Leaved Box), which came from Japan in 1860. This species has never been found growing naturally in the wild, and nowadays is considered to be of (Japanese) garden origin. The leaves of all the species contain the extract buxine, which is a narcotic and a sedative and was used medically well into the twentieth century.

The stem wood and the root of box are dense and closely grained. As a consequence, the wood is very heavy and sinks like a stone in water, but it is this very heaviness that gives it its great strength and beauty. It has been turned, carved and engraved into flutes, pegs for stringed instruments, combs, rolling pins, chessmen, nutcrackers, sacred statues and cabinets. The cross-cut stem wood, because of its hardness and smoothness, was an excellent surface for wood engravings from which prints were made. Quite regularly, some six hundred tons per annum were brought to England from Turkey for this purpose alone – box wood was always sold by weight, and demand nearly always outstripped supply.

"CINDERELLA"
STANLEY BAXTER.
THIRD COSTUME

FLATTEN HAT MORE

TOMATOES

ANTHONY HOLLAND

Theatre designer Anthony Holland used the colourful carrot as the basis for this costume. The carrot was part of the Roman diet, and accompanied the Moors when they arrived in Spain early in the eighth century. Much later it gained a reputation as an aphrodisiac.

Following the Emperor Honorius's proclamation that the people of Britain were 'to look to their own safety', the Roman legions withdrew to defend their motherland in AD 410. No sooner had the people of Britain seen the back of the Romans and four hundred years of Roman domination, however, than they were faced with attacks from invading barbarians. The slow disintegration of the Roman way of life, and the constant harrying of the hordes from northern

Europe ushered in the Dark Ages, which in turn eventually gave way to the most dominating invasion of all, the Norman Conquest. So little is known of gardens or plants during those gloomy six hundred years that it can be thought of as a horticultural 'black hole'.

Yet life was not held in suspended animation for all that time. What we do know from the scant evidence that has survived is that there appears to have been little distinction made between working in the gardens, the orchard or on the farmland. Once again the provision of food from the cultivated land was of overwhelming importance; shelter and the wherewithal for warmth, too. We learn little from the early religious settlers from the Continent, but they must have brought with them plants or seeds to cultivate in the monastery gardens. In due course the travellers would have gone in both directions, thus beginning the first exchange of plants between countries. It is just that so little information has survived that we are left to imagine and conjecture what we can.

At the very end of this period, however, we do learn a little about the Anglo-Saxon idea of a lovely place, the 'luffendlic stede' of Abbot Aelfric's vocabulary, written in 995. The two hundred named plants on the list probably comprised a mixture of native and introduced plants. While some of the names are unidentifiable today, they would have all been words in common use at the time. Even if a plant was not actually being grown in Britain, it might well have been brought in from abroad in a dried form, for herbal or medicinal use.

A familiar vegetable to us, and one on Aelfric's list, was *Daucus carota*, which combines both the Latin and the Greek words for carrot. Although there is no disputing what it is, the vegetable then would not have been the orange-coloured one we know today, but a regal purple-coloured root which came originally from Afghanistan and trickled, courtesy of the Moors, into Spain, Holland and then Britain. The modern carrot is a product of Holland and reached this country sometime in the Middle Ages. As with nearly all vegetables, it not only tastes good but does you good as well, helping to make your hair curl and enabling you to see better in the dark – that is, if you eat enough. The game of slicing off the feathery carrot top and placing it in water to watch the leaves unfurl and grow is not something invented by mothers in the twentieth century to introduce their children or grandchildren to the wonders of nature; it has been around for at least a hundred and fifty years. In fact the idea

HELLEBORUS niger flore albo etiam interdum valde rubente et True Black Hellebore or Christmas rose

The Christmas or Lenten Rose *Helleborus niger* is one of about twenty species which all come from Asia and Europe.

may have been used much earlier, since highly fashionable ladies in the seventeenth century were given to wearing the carrot topknot as decoration; strategically placed, it took the place of feathers.

Other vegetables included in Aelfric's vocabulary were the native celery (*Apium graveolens*), which took over in the simmering pot from the Alexanders, and the cucumber (***Cucumis sativus***). The latter may well have been brought here during the Roman period from southern Europe; the seeds were used against all the 'hot' complaints and the skin for a cool balm to soothe a fevered brow. Even today, cucumber pads are used to relieve eye tension, and the vegetable is used in the preparation of many herbal skin products.

Also from southern Europe but now naturalised in Britain came Sweet Fennel, ***Foeniculum vulgare***, a plant which has both health and slimming qualities. (Greek athletes took it to build up their stamina and give them courage – the Greek name for the plant is *marathon*.) Its leaves are often used in the cooking of fish, whilst in Italy the young, tender stems are peeled and eaten raw in salads. In Britain in the Middle Ages the seeds were often chewed to help alleviate pangs of hunger. The Puritans even had a special name for them: 'Meetin' Seeds', since they were frequently found to be a useful antidote to the very long sermons they had to endure. The bulbous root fennel which is eaten as a vegetable is a varietal form known as *F. vulgare* var. *azoricum*, or Florence Fennel. In Italy it is sometimes eaten as a dessert, which, in the middle part of the seventeenth century, is how the 3rd Earl of Peterborough enjoyed it. He was one of the first people to grow fennel in England, planting the seed he collected during his travels in Italy in his well-known garden at Parsons Green in London; see also p.136.

Another root vegetable which also used to be eaten as a dessert, this time as a sweetmeat with honey, was the parsnip, *Pastinaca sativa* – a member of the *Umbelliferae* family, which stretches from Britain to Siberia. It was known both here and in France as the 'pasternak', a corruption of the original Latin name of *pastinaca*. Not very surprisingly, however, the French did not actually care for them; indeed, by the sixteenth century neither did we. If they were grown at all, they were used as animal fodder. Whatever their fate here, they had travelled to the West Indies by 1564 and reached Virginia in 1609. Today Americans cook them with brown sugar and fruit juice to serve with turkey – sounds delicious. They later regained their standing in Britain, becoming one of the vegetables roasted with a joint of beef, and along with the carrot, turnip and even the swede are now part of the winter menu.

Abbot Aelfric's list also included woad (*Isatis tinctoria*), a native hedgerow plant cultivated for the manufacture of the famous blue dye which was produced from the crushed and then fermented leaves. The dye was used both for Celtic body-painting and for dyeing woven cloth.

Flowers too, nearly all of them native, had their place in Aelfric's intriguing list. What is not known is whether they were singled out because they were cultivated as part of a 'lovely place', or were just viewed as commonplace. Our two native water lilies, *Nuphar lutea* and *Nymphaea alba*, with, respectively, yellow and white blooms, are there, as is a peony species, *Paeonia officinalis*, which is indigenous anywhere between France and Albania. It is not known when the plant arrived in England, but it was certainly well known in medieval times, since evidence shows that in 1299 the infirmarian of Durham Priory bought 3 lb (about 1.5 kg) of the

The biomes of the Eden Project near St Austell in Cornwall (photographed early in 2001) are a dramatic example of the marriage of technology, botany and horticulture, enabling us to see plants from all corners of the globe growing in their 'natural' habitat.

seeds. These were used medicinally to help stave off nightmares and for pain relief, but of course some could as easily have been planted in the garden. About the same time, *P. mascula*, a species from southern Europe, was brought to Britain, possibly by the Augustinians when they settled on the little island of Steep Holm in the Bristol Channel. The island has provided a more lasting refuge for the plant than for the monks, whose gardening activities ensure a yearly echo of their past life.

The peony has a very long and honourable history – in Greece it was named to honour Paeon the Healer, and it was one of the earliest flowers to be recorded there in cultivation. It was grown not only for its beauty and its godly association, but also because its roots could be used as a tonic.

As we have seen, the Romans were acquainted with and grew the Opium Poppy, but Aelfric made no mention of it in his list. Instead, he listed the native Field Poppy, *Papaver rhoeas*, which is widespread throughout Europe. This lovely poppy with its paper-thin petals had to wait nearly a thousand years to be taken seriously as a garden flower. In 1880 the Revd William Wilks noticed, in the corner of a cornfield, one poppy edged with white. He sowed the seed, and then the resulting seed, and so on and on . . . until, after ten years' work, he launched on to the gardening world the 'Shirley Poppy', naming the beautiful full-skirted, strikingly coloured and marked blooms to commemorate the parish of Shirley, near Croydon, where he was the vicar. In the 1890s, when the poppy was first put on public display at an RHS Flower Show, it received a first-class certificate. William Wilks became Secretary of the Royal Horticultural Society and was responsible for the beginnings of the society's annual flower show, which was then held in the grounds of Chiswick House. In 1913, when it outgrew that site, it moved to Chelsea, where it has famously been ever since.

Another native which Aelfric noted was *Verbascum thapsus*, or Great Mullein, which must have been a familiar plant. With its tall and quite sturdy single bloom, it has attracted a number of vernacular names, among them Aaron's Rod, Jacob's Staff, Virgin Mary's Candle, and Torches. The last perhaps reflected the fact that Roman soldiers were supposed to have dipped the dried flowers into oil and ignited them, thus creating flares to light the way. Although it was given its Latin name by Pliny, the first recorded use in England of its proper name (apart from on Aelfric's list) was not until 1562. 'Mullein' appears to come from an Old French word, but whatever the name it is a delightful plant, with its burgeoning rosette of soft silver leaves and its ramrod-straight yellow flower. All these early garden flowers, whether native or foreign, had to earn their keep, and Mullein was said to be able to cure bronchitis and asthma if the dried leaves were smoked.

Another native on the list was the low-creeping perennial *Oxalis acetosella*, the Wood Sorrel, which is also the true

shamrock. In Morayshire it is known as 'sookie sooricks', while south of the border it is sometimes called 'Bread and Cheese', as the flower and leaves make a tasty snack. Also listed was a spurge (Euphorbia), a number of which are indigenous. The one native verbena mentioned was *Verbena officinalis*; this plant is part of a very large worldwide family of about 250 species and, true to form, was seen as efficacious. So amazing were its powers believed to be that no fewer than thirty different complaints were said to be cured by it. It must have been a great comfort to have the plant growing near your back door; a sort of chemist's shop on call. Not only could it give practical help, but it had sacred and druidical qualities as well; people believed it attracted lovers to each other on the one hand, and kept witches at bay on the other. An altogether amazing plant.

There were a few floral incomers included on the list. One was *Melissa officinalis*, the Bee or Lemon Balm, a member of the *Labiatae* family. Again, this was a dual-purpose plant because it not only looked good but did you good as well. The leaves, infused in water, made a tea which both soothed and helped cure colic. Culpeper was quite positive about its curative properties and wrote that 'It causeth heart and mind to become merry.' Originally a native of southern Europe, it settled here so well that it became a garden escapee and is now found growing wild and presumably merrily too. Another incomer listed, this time from south-east Europe, which also has acquired native tendencies since its arrival is the wonder plant which gives relief to migraine sufferers, the comfortably named Feverfew. Its Latin name, *Tanacetum parthenium*, reflects the medieval Latin word for our own native tansy, *T. vulgare*. They are both in the same large *Compositae* family.

Helleborus niger, the Christmas Rose, came from central and eastern Europe and may have arrived in the Romans' baggage trains. It is not known whether it faded out after the Romans left, or remained to become a well-known plant from medieval times. Culpeper says it grew 'in gardens but . . . may be found in the woods in Northamptonshire'. He also gives a series of common names for it – Setter-Wort or Grass, Bear's Foot, Christmas Herb and Christmas Flower, but curiously not, as we know it today, the Christmas Rose. With so many common names and with it naturalising itself, it would appear to have been a long-established garden plant by the early part of the seventeenth century and one not likely to have needed any further introduction.

A reintroduction of the Christmas Rose was, however,

recorded in 1629. This plant is a real beauty, with its sturdy yet fragile blooms shining 'with a blazing white innocence', as twentieth-century plant-hunter Reginald Farrer put it. However, within its seductive beauty it holds an ancient and dark secret, for its black roots (hence its second name, *niger*) were for centuries known as a cure for mania, insanity and melancholy. So powerful was it that the Greeks prayed to Apollo for safe keeping whenever they had the occasion to disturb its roots, while Gerard, in the seventeenth century, believed that a 'Purgation of Hellebore is good for mad and furious men'. Like garlic, the Christmas Rose had the powerful aura of warding off evil spirits, and a sensible precaution was to plant it as near as possible to the entrance of the home so that assorted devils could not enter.

At the other end of the terror scale was the Periwinkle, also on Aelfric's list. *Vinca minor,* with its sparkling small twirls of blue, arrived from Russia and the Caucasus, and is the flower from which originates the old custom of a bride wearing something blue on her wedding day. Legend has it that the Periwinkle is supposed to promote love and fertility for newly married couples.

Britain in pre-AD 1000 was, then, in gardening and horticultural terms, a country whose place was at the back of the shrubbery (just supposing there was one), so humble were our gardens, and so poorly were the native plants we had to play with. We were far behind the sophisticated sunnier courtyards surrounding the Mediterranean where the influence of the Arabic and Near Eastern worlds was beginning to unfold and where everything grew in seemingly sumptuous splendour.

How is it then that during these past ten centuries the situation has completely reversed? It is Britain whose gardens are now the envy of the world and whose horticultural expertise is so much admired. Where else but in Britain would so much excitement have been generated over the largest greenhouse ever created anywhere, filled 'with the world' and named Eden? Anywhere else it would have been thought of as either slightly ridiculous or arrogant; here it seems entirely logical. The Eden Project was the vision of one man, Tim Smit, who, in the 1990s, had already re-established the legendary Heligan Garden in Cornwall. He has now gone on to create a remarkable 'global garden' in a disused china-clay quarry, also in Cornwall. Its aim is to 'promote the understanding and practise the responsible management of the vital relationship between plants, people and resources, leading towards a sustainable future'. It surely reflects something deep in our psyche. Does it all go back to our deprived 'plantery' poverty, and have we over the past millennium been making up for lost time in our search for the floral delights of the world? It is without doubt one of the major wonders of our age that we can look back across a thousand years to the furthest edge of the darkest shrubbery and realise all these centuries later that our gardens are now flooded with the delight of blooms from around the world.

It is this story of how these plant introductions came about and the influence they have had on our gardens that, like so many plant collectors before us, we are poised to explore.

Plant Introductions in the period before AD 1000

Acanthus mollis Bear's Breeches. S. Europe. A Roman introduction. The flowers were displayed ornamentally. Possible reintroduction (see p.117).

Acanthus spinosus Spiny Bear's Breeches. Italy to W. Turkey. A Roman introduction. First written evidence occurs in Parkinson's list of 1629.

Acer pseudoplatanus Sycamore. Europe, S. W. Asia.

Aegopodium podagraria Ground Elder. Europe.

Allium ampeloprasum var. *porrum* Wildleek. S. Europe, W. Asia.

Allium cepa Onion. Asia. One of the oldest cultivated vegetables, later recorded as being cultivated in India, China, Arabia, Greece and Egypt.

Allium sativum Garlic. Central Asia.

Anethum graveolens (syn. *Peucedanum graveolens*) Dill. S.W. Asia.

Anthriscus cerefolium Common Chervil. Europe, W. Asia. A Roman introduction.

Asarum europaeum Asarabacca, Wild Ginger. W. Europe. Now naturalised in Britain.

Asparagus officinalis Asparagus. E. Europe.

Borago officinalis Borage. Europe. Said to have been introduced by the Romans, but now considered to be a native.

Brassica oleracea Cabbage. Mediterranean to temperate Asia.

Calendula officinalis Pot Marigold. S. Europe. Reintroduced (see p.59).

Cannabis sativa Hemp. Asia, possibly India. Reintroduced (see p.81).

Carum carvi Caraway. Europe to W. Asia. Possibly introduced by the Romans but now grows wild. The seeds are used in cooking.

Castanea sativa Spanish Chestnut, Sweet Chestnut. S. Europe, N. Africa, S.W. Asia.

Conium maculatum Hemlock, Poison Hemlock. Europe, Asia, N. Africa.

Coriandrum sativum Coriander. E. Mediterranean. A Roman introduction and on Aelfric's list. Sometimes known as Chinese Parsley.

Cucumis sativus Cucumber. Possibly from N. Africa or S. India.

Cumium cyminum Cumin. E. Mediterranean. The seeds are used in curries and to flavour bread.

Cynara scolymus Globe Artichoke. Probably N. Africa. Reintroduced (see p.117).

Daucus carota Carrot. Europe to Asia.

Ficus carica Common Fig. W. Asia, E. Mediterranean.

Foeniculum vulgare Fennel. S. Europe.

Helleborus niger Christmas Rose. C. and E. Europe. Reintroduced (see p.145).

Iris foetidissima Stinking Gladwyn, Stinking Iris. S. and W. Europe, N. Africa, Azores, Canary Islands.

Iris germanica Garden Iris, Common German Flag. Mediterranean but now naturalised along river banks in Britain.

Juglans regia Walnut. S. E. Europe.

Lactuca sativa Lettuce. Europe. Name probably derived from *L. scariola*. Many varieties cultivated by the Romans.

Lagenaria vulgaris Bottle Gourd, Calabash Cucumber. Asia, Africa. Reintroduced (see p.119).

Lepidium sativum Garden Cress. W. Asia. The cress of 'mustard and cress', and included on Aelfric's list.

Levisticum officinale Lovage. Mediterranean.

Lilium candidum Madonna Lily. Balkans, Turkey, W. Asia. Reintroduced (see p.49).

Malva sylvestris Common Mallow. N. Europe, N. Africa, S.W. Asia.

Mandragora officinarum Mandrake, Devil's Apple, Love Apple. W. Balkans, N. Italy, W. Turkey, Greece. See also p.82.

Melissa officinalis Balm, Lemon Balm. S. Europe.

Mentha spp Mint. Although several species, including Pennyroyal, are native, some of the stronger-flavoured species arrived from S. Europe with the Romans; these include *M. spicata* (Spearmint). The herb also appeared on Aelfric's list.

Mespilus germanica Medlar. S.E. Europe, S.W. Asia.

Onopordium acanthium Scotch Thistle, Cotton Thistle. W. Europe to W. and C. Asia.

Paeonia officinalis Common Peony. France to Albania.

Papaver somniferum Opium Poppy. Greece, Near East.

Petroselinum crispum Parsley. C. S. Europe. Introduced by the Romans, who not only used it in cooking but also apparently put it into their all-important fish-holding tanks to keep the fish healthy. The herb appeared on Aelfric's list.

Pimpinella anisum Anise. Greece and the Middle East. Introduced by the Romans. The seeds were used to flavour cakes, but now it is most commonly found in a variety of alcoholic drinks including anisette, arrack, ouzo and pastis.

Prunus cerasus Sour Cherry. S.W. Asia.

Prunus persica Peach. China. Probably brought here by the Romans. Reintroduced (see p.61).

Rubia tinctorum Dyer's Madder. Mediterranean, W. Asia. Introduced by the Romans. A red and orange dye extracted from the roots was an important feature of the medieval wool trade.

Rumex scutatus French Sorrel. Europe. A leaf vegetable used like spinach, and introduced by the Romans.

Ruta graveolens Rue, Herb-of-Grace. Mediterranean. A herb introduced by the Romans but little grown after their departure. Reintroduced (see p.118).

Salvia officinalis Sage. Mediterranean; see p.81.

Satureja hortensis Summer Savory. S. Europe. A strongly flavoured annual herb whose crushed leaves were used to alleviate wasp stings. It was introduced by the Romans, and appeared on Aelfric's list. Reintroduced (see p.118).

Satureja montana Winter Savory. S. Europe, N. Africa. A perennial introduced by the Romans and, as its relation above, reintroduced later (see p.118).

Sinapis alba White Mustard. Mediterranean. Introduced by the Romans. Thought to be the mustard tree of the Bible (Matthew 13: 32).

Smyrnium olusatrum Alexanders. N. Africa.

Tanacetum balsamita (syn. *Balsamita major*, *Chrysanthemum balsamita*) Alecost, Costmary. Europe to C. Asia. Related to our native tansy (*T. vulgare*). The leaves give a spicy flavour to beer.

Tanacetum parthenium (syn. *Chrysanthemum parthenium*) Feverfew. Europe, Caucasus.

Thymus vulgaris Garden Thyme. W. Mediterranean to S. Italy. Introduced by the Romans and stronger than our own native Beckland Thyme, *T. serpyllum*, which was included in Aelfric's list. See also p.65.

Urtica pilulifera Roman Nettle. Mediterranean; see p.54.

Vicia faba Broad Bean, Horse Bean. Origin unknown; see p.42.

Vinca minor Lesser Periwinkle. Europe, S. Russia, N. Caucasus.

Vitis vinifera Grape Vine. Mediterranean.

Normanisation
1000–1099

Significant dates

To ENTER THE SECOND MILLENNIUM AD is like entering a brightly lit city, so different does it seem from the preceding five hundred years. Suddenly the miasma has cleared, the Dark Ages are behind us and the path ahead is familiar. Before the Normans landed on the beach at Hastings in 1066, the native Britons still had to contend with the Danes, who had maintained their frenzied raids from 980 until 1016, when Cnut (he of the not-turning-back-the-tide at Bosham in Sussex) ascended the English throne.

Almost the first thing to happen of any significance in the plant world occurred about AD 1000, and was an event of which very few people in England would have been aware. It took place in Italy at the Abbey of Monte Cassino, the home and burial place of St Benedict. Here many of the ancient classical and Arabic documents were brought for the monks to translate and copy, and then distribute. It is the remnants of these documents which are so invaluable to us today. The event of such importance to the botanical world was the very first translation of an ancient Latin text of medicinal and herbal information into the vernacular Anglo-Saxon. The document originally had been written around AD 400 in Latin, and was believed to have been the work of an other-

Left: **By the fifteenth century the great houses and castles of the nobility had their own medicinal herb gardens, following the earlier example of the monastic (and Roman and Greek) herbary.**

wise unknown man named Apuleius Platonicus. We only know of his manuscript by hearsay, since no copy has survived. What has survived, however, is an illustrated herbarium made about fifty years after the translation (*c.* 1050), which combines the work of both Apuleius and Dioscorides, the latter being the physician who wrote the highly influential *De Materia Medica* (see p.14).

The *Herbarium* is in the British Library in London and consists of 140 folios (pages). A further smaller group of 44 folios survives as a later copy – done at some time during the last thirty years of the eleventh century – and is in the Bodleian Library in Oxford. According to the eminent scholar and writer Wilfrid Blunt, these documents laid the foundations for first a distinctive Anglo-Saxon style and later an Anglo-Norman style of herbarium. The illustrations and drawings of the plants are wonderfully esoteric: some are clearly painted from life, others have muddled features, whilst a few are entirely mythical. However, like the vocabulary of Aelfric, the herbarium is an indication of the amount of information that was being discovered and promulgated at the time.

New information about herbs, and remedies associated with them, was now being shared, as copied documents were distributed to monasteries and abbeys. The monks were able to learn about the newest treatment for all manner of ailments and diseases, and, importantly, which were the correct plants to be sown, grown and harvested. Ideas were shared, and information about horticulture, growing crops and the keeping of beasts (especially sheep) flowed between these established intellectual sites. The value of such a development cannot be underestimated.

When the military generals of one country want to invade another, or stage an internal coup, one of the first essentials is to secure both the access routes and the lines of communication. This is precisely what William of Normandy did in 1066. Commanders were posted to the coastal ports, and most of the existing Anglo-Saxon bishops and abbots were replaced with Normans. (Today it is airports that are secured and radio and television reporters who are replaced.) There then followed, very swiftly, the 'Normanisation' of the country. William's friends and acquaintances were placed in the most powerful positions throughout the land – not only as a reward for services given but so a close eye could be kept on the locals. The Norman lords embarked on an extensive building programme, initially as an act of aggression but later for defensive purposes. The early constructions were of wood, and so well organised was the invasion that several timber forts were preconstructed and brought over from Normandy in the form of early flat-packs, the pegs to hold the wood together packed into barrels. The forts were then able to be erected speedily on the chosen sites. Berkhamsted, Windsor, Lewes and Arundel were all constructed in this manner; all four were later rebuilt in stone.

By 1086 – a mere twenty years after William's arrival in England – a full inventory of most of England was being undertaken, to find out what and how much each landholder held in land and livestock, and what it was worth. This was so the King's treasurers could work out what they could extract in the way of taxes. This was the Domesday Book, so-called because, like the Day of Judgment, there could be no appeal against it. The feudal system, the ownership of all land by the King, had arrived. The invading Danes earlier had not sought to impose such a stringent system, and the resident Saxons had looked after their land as a community.

Apart from human invaders, however, there were others against which no barriers were raised. One such invader was the Wild Carnation or Clove Pink, *Dianthus caryophyllus*, and the other was the wallflower, known then as *Cheiranthus cheiri*. Pliny wrote of the discovery of the Dianthus, which was found over 2,000 years ago in southern Spain, and both it and the wallflower have been associated with our gardens for at least a thousand years. As mentioned in the Introduction, it is thought that the seeds of both plants may have been carried to this country with the building stone

Dianthus caryophyllus **from John Parkinson's book** *Paradisi in Sole Paradisus Terrestris*, 1626.

brought over from Caen for the construction of William's new castles. Certainly, the Clove Pink is still often associated with Norman buildings and ruins, and it is recorded that it was growing on the walls of William the Conqueror's castle at Falaise well into the nineteenth century.

In the case of the Dianthus – or, as the Greek philospher Theophrastus first named it, *diosanthos* (divine flower) – the plant is a native of the Mediterranean. As becomes its long history, it acquired a number of different names: gillyflower (or gilloflower, as John Parkinson called it in 1629), Chaucer's 'gilofre' and Shakespeare's 'gillyvore' all derived from the Arabic word for clove, *quaranful*, which refers to the aromatic smell of the leaf and hence its Latin name of *caryophyllus*; Sops-in-Wine was another name (because the flower was dunked into ale or wine to flavour it); and then, finally, carnation or, as it first appeared, 'incarnacyon'. That word appears to be derived from the Latin *carnatio*, meaning 'fleshy' or 'flesh-coloured', but was not associated with the flower until 1535.

The pink, or pynkes, was one of the names used to describe **D. plumarius**, the Common Pink, from eastern Europe, which was the plant used in due course as the main parent in the cultivation of the numerous pinks and carnations. There are three native pinks: Cheddar, Deptford and Maiden. The word 'pink' was rarely used to describe them, although it had a meaning of 'twinkling' or 'small' which seems quite appropriate. It was not until the 1720s that 'pink' came to denote the colour, and since pinks were just that, it means it was not an invented word, more an invented colour.

As was noted in the Introduction, **Dianthus plumarius** had to make several attempts to establish itself here permanently – which it did finally around 1560. Both *D. caryophyllus* and *D. plumarius* have gone native and can now be found listed in books on wild flowers. Numerous varieties of pinks, many with wonderful names like Lustie Gallant, Ruffling Robin, Master Tuggie's Princess (all three gillyflowers and grown by John Parkinson), Whole Podder and The Bleeding Swain, have been bred and then lost to cultivation.

So fascinating and complex a perfumed trail have they led us through to the present day, there appears to be no rhyme nor reason why, along with the Madonna Lily and the rose, they have attained a cultural persona quite apart from the role they play in our gardens. One of the reasons could be that, just like 'nursery food', they have a high comfort factor.

Their story, like that of the rose, is complicated and long and crosses several countries, but one of the most interesting strands involves the very British town of Slough in Berkshire and concerns a husband and wife who, in the mid-1870s, ran the Slough workhouse. The husband was an enthusiastic gardener who developed a love of growing 'florist flowers', the pink in particular. This gardener's surname was Sinkins, and it is the heavily perfumed double-fringed white that he named after his wife which is perhaps the most famous pink of all. Slough later fed on the fame of the flower, since on the borough's coat of arms is depicted a swan (for the River Thames) with a 'Mrs Sinkins' in its beak.

Charles Turner, known as the 'king of florists', owned the Royal Nurseries at Slough, and John Sinkins agreed to sell the stock of his new dianthus to him on one condition: that it was named in honour of his wife, Mrs Sinkins.

We have travelled a long way from the turmoil of the Norman invasion to the sophistication of Victorian England, but we must go back to the wallflower we left suspended in the dust of the stone imported from Caen, and trace its long and honourable history. The *Cheiranthus cheiri* came originally from southern Europe and although its meaning is obscure, the name was retained until the 1980s, when, with the entire Cheiranthus genus, the wallflower joined the Erysimum genus and is now **Erysimum cheiri**. The first of those names comes from the Greek *erysimon*, meaning 'blis-

FLORIST FLOWERS

Originally 'florists' were people who were involved in the growing of flowers and who developed and exhibited a number of pot-grown plants, including the tulip, Auricula, Ranunculus and, of course, the pink. The meaning of the word today is quite different: a florist is someone who is involved with cut flowers, not with the growing of them.

Probably the Chinese were the first to specialise in specific plants, for as early as the fourth century the chrysanthemum was being collected and developed there, and the Japanese were not far behind with their cherry trees and camel-

Stands to display auriculas became 'auricula theatres' and the one shown here is at Calke Abbey in Derbyshire.

lias. In Britain it was not until the 1630s that the first plant became a speciality – and that was the tulip; see p.139. By the eighteenth century there were eight different flowers which were considered to be florist flowers and worthy of their own specialist societies and shows. Apart from the tulip, there were societies for the Primula, Auricula, Ranunculus, hyacinth, anemone and both parts of the Dianthus group, the pink and the carnation. All were grown in pots, and hundreds of different varieties were cherished and nurtured, and then prepared to be exhibited and displayed before their peers. However, they were grown to increasingly strict rules regarding height, form, colour and so on.

Later, in the nineteenth century, the initial eight genera were joined by what came to be called 'fancies'. These included Dahlias, pansies, viola and Gladioli. The chrysanthemum too came into its own as a 'fancy', so much so that in 1846 the Stoke Newington Chrysanthemum Society was formed, becoming, in due course, the National Chrysanthemum Society.

tercress', and was originally used by Theophrastus. He must have been a most observant man, since the plant can indeed suffer from a fungal outbreak of something which is now called White Blister. There were about 10 species when the wallflower was Cheiranthus, but when it moved to Erysimum, becoming at the same time a member of the *Brassicaceae* family, it embraced about another 70 species.

The wallflower, just like the pink, had a number of country names, including Heartsease, Cheiry, Bee-Flower and Yealowe Violet. One of its other names was Chevisaunce, or Cherisaunce, which was an old word for 'comfort' (although by 1600 it had come to mean 'achievement'). John Gerard (1543–1612) wrote that 'the wallflower groweth upon bricke and stone walls, in the corners of churches everywhere, as also among rubbish and other such stony places'. In 1613, Gervase Markham wrote about the wallflower in his book *The English Husbandman*, saying that it should be grown especially in the bee garden 'for it is wonderous sweet and affordeth much honey'. Later, in 1770, the Revd William Hanbury thought the name 'wall flower' to be particularly inept because 'many old walls in one part or another exhibit in bloom a not very inconsiderable share of the flowering tribe'; perhaps, as the writer Alice Coats suggests, we should call it by its old name of 'comfort'.

The sweetness of the wallflower has always been admired, and over the centuries they were often added to nosegays and, according to John Parkinson (1567–1650), used to 'deck up houses'. Most of the European species probably made their way into British gardens during the seventeenth and eighteenth centuries. During the seventeenth century, there was much talk about a rare white 'wallflower'. Sir Thomas Hanmer (1612–78) remarked in 1659 that the plant was 'tender'; later, the great Philip Miller (1691–1771), *Hortulanorum princeps* (or head) of the Chelsea Physic Garden, thought it was not a wallflower at all, but a variety of stock (*Matthiola incana*, see p.68). Even today, no one is quite sure what plant it was. The eminent medievalist John Harvey (1911–97) suggested that it could have been a form of *Arabis albida* (now *A. caucasica*), the Rock Cress, but that seemingly was not introduced until 1798.

It appears that none of the North American species of wallflower, like *Erysimum asperum*, the Western Wallflower, which is indigenous to Texas and California, arrived until the nineteenth century. The Siberian Wallflower (*E. x allionii*), despite its somewhat exotic name and colour –

In the twelfth century the petals of wallflowers were pounded into a conserve to relieve the aches and pains of gout.

bright orange – was actually raised by John Marshall, an English nurseryman, in 1847, and the fizzy double yellow-flowered beauty *E. x kewense* 'Harpur Crewe' was a remainder of a very old variety rescued by the Revd Henry Harpur Crewe in the nineteenth century.

The Domesday Book might be a record of the 'Great Inquisition of the lands of England, their extent, value, ownership and liabilities', but it is almost silent regarding gardens – or garths, as they were more often called, a word usually referring to a space beside or enclosed by a building, hence the enduring tradition of building houses with back gardens. The survey did list fishponds, orchards and vineyards, noting that most of the latter were newly planted which presumably means they were being made by the Norman cognoscenti on their recently acquired lands. Woodlands, copses and forests were all examined in detail, and it was while the survey was taking place in 1086 that the Old Windsor Forest was enclosed for the new king's pleasure. A few years prior to that, in 1079, another important area had been enclosed for the King, the *Nova Foresta* in Hampshire.

While the Domesday Book gives scant details about gardens and the plants that were being grown, there is no doubt that nearly all the horticultural knowledge we have relating to this time comes from the monasteries (particularly those of the Benedictine order), many of which were newly built or under construction during this first century of the new millennium. Each abbey displayed the same principles of layout; they were self-contained and self-sufficient settlements designed to fit in with both the spiritual and physical work of the monks. It was this and their work in the community which led them to be such a civilising influence on the surrounding population. The essence of the contemplative life was exemplified by the cloister, usually attached to the outside of the nave of the church (which thus provided it with some shelter from the north). This was a quiet and meditative area, usually consisting of a lawn (or at least turf) with paths made in a cruciform pattern, sometimes with a cross or fountain at its centre.

Quieter than the cloister, and used for praying and contemplation, was an area known as paradise, an enclosed and sometimes circular garden often sited to the north of the

Gervase Markham (c. 1568–1637) was born at Cottam near Newark in Nottinghamshire. He was a writer on agricultural and horticultural subjects, his three books being *The English Husbandman* (1613), *Cheap and Good Husbandry* (1614) and *Farewell to Husbandry* (1620). In them he gives detailed and practical instructions on how to grow plants, specifically with the new designs then emerging. For the colouring of the intricate patterns of the 'knots', he recommends a number of different materials which could be used. One such was yellow clay or yellow sand (and if neither was available, then he helpfully suggests 'Flanders Tile which is to be bought of every Iron-munger or Chandelor'). White was procured from beaten chalk or well-burnt plaster; lime was not to be used, as it soon decayed. Purest coal-dust 'well cleansed and sifted' would give you black; red should be made of broken useless bricks 'beaten to dust', and blue could be obtained by mixing chalk and coal dust together 'till the black have brought the white to a perfect blueness'. He offered the further nugget of information that to protect fish from being stolen, 'sticke stakes slant-wise by everie side of the Pond, that will keepe theeves from robbing them'. It all sounds very much like a twenty-first-century garden make-over.

Revd William Hanbury 1725–78 was born in Warwickshire. In 1753 he became rector of Church Langton in Leicestershire, where, two years previously, he had begun sowing seeds from 'distant countries particularly North America'. So extensive were his plantings that by 1758 they were valued at £10,000, and in his *Essay on Planting* he put forward the idea of creating a trust for the benefit of the parish and its church. The trust prospered during the eighty-six years of its life, keeping the building in good order, providing an organ (and paying for an organist), employing a village schoolmaster, and founding a public library, a picture gallery and, finally, a choral college in Oxford.

church, near the monks' cemetery. The name derived from the old Persian word *pairidaeza* meaning 'enclosure', and it was not only monasteries but some medieval churches as well which found such areas beneficial. They were always looked after by the sacristan, the 'obedientiary' or responsible official who was charged with the running of the church, and because he had to find flowers for decoration, he also oversaw the planting of them. These paradise blooms were used to decorate statues, side altars and shrines, and on special feast days, no doubt with some extra help, the whole church.

The Rule of St Benedict meant a well-ordered life for the monks. Each day, apart from the all-important times set aside for singing the Divine Office and private prayer, the brothers had to find time to build and repair the abbey buildings, look after the livestock and work in the gardens and fields. The abbey had to produce much of the food which the community needed.

The vast number of vegetables required to keep the brothers and the local population fed must have meant a never-ending struggle. The kitchen gardens would be planted mostly with Alliums – leeks, onions and garlic – which, together with the all-important herbs, would have helped to flavour the endless mush of starchy beans and peas known as a pottage. This was the staple daily diet. Both the pea and the 'bene', which was an early relation of the modern broad bean, *Vicia faba*, had been grown since ancient times; there was the advantage that they were easy to grow and to harvest, and took little space to store, an essential attribute. Leeks were so fundamental to the good life that kitchen gardens of all descriptions were often called leac-garths, in the same way as we sometimes cultivate our 'cabbage patch'.

Everything that was in season would be added to the monastery cooking pot, to make the porry, or broth, including 'colewort' or kale (*Brassica olearacea*) and the leaves off the tops of what we would think of as root vegetables, such as parsnips, turnips and beet. Then, to relieve the blandness, nettles, orach (*Atriplex hortensis*, probably a native), plantain and sorrel, cumin and fennel seeds (the latter were often chewed to keep hunger pangs at bay), linseed (*Linum usitatissimum*, the native flax plant) and barley were all added to the 'mess of pottage'. Finally, salt, vinegar and mustard would be employed to give it extra spice. The whole of the medieval world enjoyed strong flavours, so the hotter and stronger the herbs, seeds and flavourings, the better the food was enjoyed. It was only from the fourteenth century, when cooking in general was improving and more meat was being eaten, particularly by the monks – who, by then, were ceasing to be such strict vegetarians – that the heat began to go out of everyday food.

As well as the leac-garth, there was usually a small garden attached to the infirmary planted with medicinal plants and herbs, set within tidy box hedges. This garden was the responsibility of the infirmarian. If the abbey were big enough there would be a vineyard and an orchard, too; apples in particular were grown for the brewing of cider, although ale was the main drink. Barley was the cereal grown for making bread, and was also used, malted, for ale. Together with

FOREST LAW

The word 'forest' has come to mean a very large area of trees, but the original word had nothing to do with trees. The phrase 'forest law' derived from the enclosure of land, by the King to use for 'the chase', with its own Crown-appointed officials. The hunting and game laws, together with the wood and timber regulations, evolved over the years and were administered quite separately from the law of the Kingdom. Included under these vast reserved areas were whole towns and villages and, although difficult to assess, scholars believe that at one time, during the reign of Henry II (1133–89), as much as a third of the country was under 'forest law' control.

The whole focus of the laws was the preservation of the beasts of the forest, deer – red, roe and fallow – and wild boar were the main animals protected for hunting by the Crown, but smaller game too were subject to rules and regulations. Because the laws were deemed so important (at least to the King and his court), the forest clauses were taken out of the Magna Carta in 1217 and augmented to form a completely separate charter, known as the Charter of the Forest.

Gradually over the centuries, manorial and other rights took over or were sold, and the only vestigial remains of forest law now is the control of grazing for domesticated animals exercised in the New Forest by the verderers and their court.

oats, these field-grown crops were usually cultivated outside the abbey grounds.

Most households, too, would have grown the same vegetables and cooking herbs, so the simmering pots of domestic kitchens and the monastic world would have been little different. Similarly, the domestic gardens and garths would have contained some of the same healing herbs as were grown in the infirmary gardens of the monasteries.

A plant which the Italians knew of, *Artemisa abrotanum*, and which was on the list that Abbot Aelfric made, may well have settled in Britain sometime during this century. It comes from southern Europe and has always been known as Southernwood – which is what its second name means; it is also known as Lad's Love and Old Man. Its yellow blooms have a heavy scent and the Romans believed that if a sprig of the shrub were placed under the pillow, its magical properties would act as a powerful aphrodisiac. There are about 300 species in the genus, most of them found in the northern hemisphere.

There are numerous other plants, berries, fruits and roots which were used in the daily task of keeping body and soul

together, but none appears to be specifically associated with having entered England during this particular century.

There is still only very scant knowledge of flower gardens and what was grown in them at this time, but the century marks the beginning of England's long love affair with the rose. This flower, together with what we now know as the Madonna Lily, played a significant role in our Christian development.

By the start of the second millennium, the rose was already very well known, having been growing throughout Europe for centuries. Both the Greek and the Roman civilisations were cultivating it, and in its heartland in the Middle East, the Persians called it 'a messenger of the garden of souls'. The flower is so instantly recognisable and is so much bound up in our vision of the Garden of Eden, paradise gardens, the cult of Our Lady and courtly love during this period that, like the Dianthus, it has become part of our Englishness. As Alice Coats remarks: 'The Rose is not a family whose history has been neglected.' The rose also happens to have a very complicated story as well.

Although the five native roses (briars) must have often been grown in the garths or picked from the countryside (for their flowers and their hips), they were not significant in the development of the garden rose. This springs from another five, the 'Five Ancestors' as John Gerard succinctly calls them. The first is the beautifully perfumed *Rosa gallica* (now *R. gallica* var. *officinalis*), a native from France and southern Europe to western Asia and cultivated from time immemorial. Known as the Apothecary's Rose, its petals were used in potpourri and conserves. Then there is *R. moschata*, the Musk Rose (and earlier known as the Holy Rose of Abyssinia), which is the white climbing rose from West Asia and is believed to have arrived here in 1590. The third of Gerard's roses is *R. damascena*, which was thought to have originated in Damascus, hence its name, but which grew all over the Middle East. It was introduced here in 1573 and was used to make attar of roses. *R. x alba*, although very old, is thought to be a cross between *R. canina* and *R. damascena*. Called the Great White Rose, it may have been a native of southern France but, again, it has been cultivated for so long that no one is quite sure. The last of the five is the Common Moss Rose, *R. x centifolia* 'Muscosa', the old Cabbage Rose, which has been described as having up to sixty petals and as being the most fragrant in world.

From very early days, the rose was a flower which was

Rosa damascena versicolor is known to have been in existence before 1550. It got its vernacular name because the shrub carries the two symbolic colours of the opposing houses of York and Lancaster.

Rosa gallica var. *officinalis* is known by so many old names that its story seems to cover the whole history of the Middle Ages. It has the luscious sobriquet of Crimson Damask Rose.

highly thought of and any association with it was considered most desirable. Its development as a royal emblem began with Eleanor of Provence, who, in 1236, married Henry III in Canterbury Cathedral and took as her emblem the *Rosa x alba*, the White Rose. The story of the war between this rose and *Rosa gallica* var. *officinalis*, the red rose which her son Edmund brought from France, is told on p.88.

Down the centuries, the rose in all its forms has continued to attract attention. It is England's chosen flower – even though the majority of species, about 100, come from Asia. Of the rest, about 20 come from America and some 30 from Europe and north-west Africa. The flower is recognised as a beacon of love and peace all over the world, so much so that on each St Valentine's Day about 380 million roses are sold. In 1986, to sniffy disapproval from some quarters, the red rose was adopted (some say hijacked) as the Labour Party's emblem. At the same time, and making much more sense, a 'Tudor Rose' Award was instituted at the yearly Royal Horticultural Society's Hampton Court Flower Show.

Romsey Abbey in Hampshire, founded in 967 and housing Benedictine nuns, is attributed with growing the first

English garden rose on record. It was around 1092, the story goes, when the King, William II – or William Rufus as he was known – with his attendant courtiers clattered into the courtyard of the abbey 'as if to look at the roses and other flowering herbs'. The politically sensitive heiress of the Saxon dynasty, the young Edith of Scotland, was receiving her education at the convent at the time, and the obviously worldly-wise abbess thought the roses were just an excuse. The twelve-year-old girl was dressed as a nun and escaped their attention. There is no record as to whether the King enjoyed seeing the roses, or indeed even if they were in bloom at the time.

During the coming centuries we shall meet the rose several times as it develops its different characteristics and comes to be England's garden flower *par excellence*. However, we end the eleventh century as we began it, looking well beyond these shores. We go further this time, however, than Italy and its transcribing monks, to Jerusalem and to events which would profoundly influence our English gardening world for the next two hundred years: the capture of Jerusalem in 1099; the subsequent creation of the Christian kingdom of

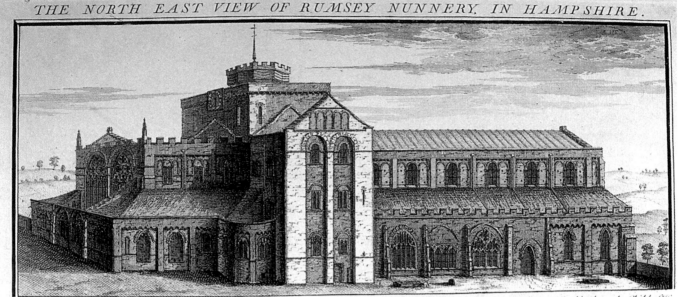

THE NORTH EAST VIEW OF RUMSEY NUNNERY, IN HAMPSHIRE.

THIS NUNNERY was founded by K. Edgar for Nunns of the Benedictine Order, & dedicated to y.̇ B.V. Mary, In this the only Child of K. Stephen was profess'd & became Abbess, but was privately convey'd thence & Married by Humphrey of Alface, Son of an Earl of Flanders, who had two Daughters by her. But Ecclesiastical Authority prevailing, she was forc'd to retire again to this Monastery according to her first Vow. —— The Present Proprietors are the Dean & Chapter of Winchester.

Romsey Abbey in Hampshire was a tenth-century Anglo-Saxon foundation whose tradition of growing roses in the garden is now continued at nearby Mottisfont Abbey, where the National Collection of Old-fashioned Roses is housed.

Jerusalem (whose boundaries stretched from the Sinai Desert in the south to the northern edge of Syria); and the founding of two orders of chivalry. The first of these was the Knights of the Hospital of St John of Jerusalem – known as the Hospitallers and created in 1070; and the second was the Templars, the Knights of the Temple, who were founded in 1118, between the First and Second Crusades. Both were military-religious orders created specifically to ensure that Jerusalem, the Holy City, remained Christian as opposed to Islamic. The Hospitallers in particular also tended both spiritually and physically to the pilgrims and Crusaders who made the long journey to the Holy Land to pray and fight for the Christian kingdom.

There were no frontiers or passports in those days; Britain was part of Europe. The pilgrims returned with the sights, smells and sounds of southern Europe; their baggage was not laden with tourist trifles, but with souvenirs of seeds – of flowers, shrubs and herbs – and maybe even a sapling or two. While most horticultural trophies must have died on the long trek home, some established themselves to live a generation or two, while a few – a very few – have lived long enough to grow into the third millennium.

Plant Introductions in the period 1000–1099

c. 1000 *Artemisia abrotanum* Southernwood, Love's Lad, Old Man. S. Europe. See also p.117.

c. 1066 *Dianthus caryophyllus* Wild Carnation, Clove Pink. S. Europe. See also p.204.

Erysimum cheiri (syn. *Cheiranthus cheiri*) Wallflower. S. Europe.

A *Wilbarewe of Plants*
1100–1199

The Abbey of Strata Florida in Dyfed, Wales, a Cistercian foundation, is typical of the locations favoured by that Order which sought seclusion in deserted and uncultivated lands.

Significant dates

1100   Henry I (1100–35)

1120   Loss of the *White Ship* off Barfleur; Henry I's son William drowned

1135   Stephen (1135–54); civil war in England

1147   Second Crusade

1154   Henry II (1154–89)

1167   Oxford University founded

1170   Murder of Thomas Becket in Canterbury Cathedral

       St Dominic born (d. 1221), founder of the Dominican Order

       Arrival of the wheelbarrow in England

1182   St Francis of Assisi born (d. 1226), founder of the Franciscan Order

1189   Richard I (1189–99)

       Third Crusade

1199   John (1199–1216)

IT IS CLEAR THAT TWELFTH-CENTURY GAR-DENING owes much to the powerful influence of Christianity in the twin forms of the continuing growth of new monastic orders settling in Britain, and the Crusades. Both brought with them new ideas and methods. One could say that both the practical and the spiritual sides of horticulture and gardening were digging their roots deep into our soil and our minds. It was the time, too, when the symbolism and practices of half-forgotten pagan ideas became entwined with the new and experimental plants that were being grown, and godly and knightly love was being codified through flowers. The idea of courtly love emanating from Aquitaine and Provence was a product of the realisation of the beauty of the natural world. It was brought to England by the arrival of Eleanor of Aquitaine, Henry II's highly cultured wife, when he ascended to the throne in 1154.

In August of 1100, William Rufus was killed in the New Forest, not far from the rose-covered Hampshire abbey he had visited only a few years before when trying to glimpse Princess Edith, heiress to the Saxon royal line. His brother Henry (who was hunting with him) immediately rode to Winchester, seized the Treasury, took it to London and within three days of his elder brother's death had himself crowned as Henry I at Westminster. Three months later, in November, in an equally decisive move, he took as his bride that same Edith, who, in an early example of political correctness, changed her name on marriage to the more Norman Matilda.

During the whole of this century, and in stark contrast to the world of the court, monastic life flowered. The premier order were the Benedictines, who had arrived here in 597 with St Augustine and by now had some well-established monasteries, such as Worcester, Abingdon, Glastonbury, Peterborough and St Albans. In 1107, just over five hundred years after St Augustine had arrived in Canterbury, Queen Matilda endowed Holy Trinity Priory, Aldgate, London, as the first English home of the Augustinian canons. The Rule of St Augustine had been developed first in France from the Benedictines, but it was stricter and harsher, based on prayer, poverty, chastity and work. Records of the Augustinians' many gardens (and a vineyard) at Aldgate show that they flourished for over two hundred years in the City. By 1350 the Austin canons, as they came to be called, had over two hundred abbeys and priories, with one of the most celebrated being at Walsingham in Norfolk.

The White Monks, or Cistercians (named for where they were founded, Citeaux, near Dijon, in France in 1098), kept a very strict observance of the Rule of St Benedict. They arrived in Britain in 1128 at Waverley in Surrey, and turned out, quite inadvertently, to be the creators of some of the most beautiful scenery in the country, for the order sought out the remotest and most desolate sites on which to practise their own particular style of monastic life. They settled mainly in the north and west of England. The north at this time had hardly recovered from the Danish whirlwind of the previous century, but many of the foundations there subsequently became some of the wealthiest. The Cistercians were the great agrarian order, and through their shepherding skills (required for wool, and for parchment made from the sheepskins, but not for meat, since the monks were vegetarians) they moulded the moorland, valleys and high plateaux around their many abbeys.

By the end of this century there would be over a hundred English Cistercian establishments, the majority built in the most inhospitable places – places which we now consider to be some of the most beautiful and stunning imaginable. A great deal of Christian courage must have been needed to begin building in such a seeming wilderness, 'remote from the comings and goings of people'. It is difficult to imagine today what it must have been like, although if one is able to visit the remaining ruins of these abbeys at a quiet time, it is possible to reap a little of their spirituality.

The first two Cistercian abbeys to be built were in Yorkshire, at Rievaulx in 1132 and in 1135 at St Mary at the Springs, later to be called Fountains Abbey. In the eighteenth century, the latter's ruined remains helped to create the most sublime and romantic vista along the little river Skell to the gardens at Studley Royal. Other foundations followed, the majority in settings of great serenity, such as Hailes Abbey, just below the Cotswold escarpment in Gloucestershire, and Buckfast, on the edge of Dartmoor in Devon. In Wales there was Tintern Abbey, spectacularly set on the banks of the river Wye, and Strata Florida beside the river Flur in what is now Dyfed. In Scotland among the various foundations there was Melrose on the river Tweed in the Borders, and Kinloss on the Moray Firth (here, Thomas Crystall, known for his gardening skills, was one-time abbot). Beaulieu Abbey in Hampshire was founded at the beginning of the next century in 1204, carved out of the New Forest.

The Poor Brothers of God of the Charterhouse, the Carthusians, were the most austere of the monastic orders. They also chose quiet and remote sites on which to build their monasteries. Their spiritual needs were seemingly fulfilled by living as a community but in individual cells as hermits. Whereas other orders prayed and worked together, and together administered the running of the monastery, the Carthusians lived and ate and worked a tiny patch of garden each in his own prayerful isolation, while the administration was carried out by lay members of the order. Their English foundation, at Witham, on the Somerset/Wiltshire border, was established in 1178 by Henry II as a penance, part of the fall-out from the murder of Thomas Becket eight years previously.

Only six Charterhouses were ever established in England. Hinton in Somerset was one, and another was Mount Grace in Yorkshire; eventually there was one in London, at Spitalcroft. We know something of the contents of this last garden, because at the time of its dissolution in 1538, Henry VIII's gardeners from Hampton Court (itself built on land originally belonging to the Hospitallers) were sent to clear the little plots. They left with bay trees, cypresses, yew and 'grafts and other such like things as was mete for his Grace in the said garden'. Three baskets of herbs went as well as ninety-one fruit trees and 'for Master Leyton: a bundle of roses'.

The introduction and growth of these new monastic orders from the twelfth century and for the next four hundred years is important in the matter of plants and garden skills. Almost all the orders were founded on the Continent, either in France or Italy, where their 'mother houses' were. They knew no national boundaries, and travel and communication between the various houses was swift and frequent. It must have been somewhat similar to the life led by European Union officials today, except that because of the universality of Latin – every court and official and Christian community conversed in it – communication was easier then than it is now. Therefore, even if the abbeys, priories and convents were in lonely and remote places, they were not intellectually isolated. The exchange of ideas and knowledge as well as, inevitably, plants, seeds and saplings all flowed between them.

So it is to an Augustinian abbot that we turn for some gardening information during this century. His name was Alexander Neckham (1157–1217), and he was a foster-brother of Richard I. He was an academic and teacher attending both the University of Paris and the newly formed Oxford University, and just like Aelfric did earlier, he deplored the sloppy way in which Latin was written and spoken. Later he became Abbot of Cirencester, from where he wrote his famous prose work *De Naturis Rerum*, which listed herbs, plants and trees to be grown in the garden. (It is interesting to note that in the maritime section he alludes to a mariner's compass as well.) His other work is the poem *De Laudibus Divinae Sapientiae*, part of which again includes plants for the garden, with another division devoted to trees and crops. His two works were written before 1200, as by then, they were being widely circulated. Other authors frequently referred to his manuscripts and they seem to have been in cir-

culation for generations. There are still over twenty manuscript copies of his writings in existence.

De Naturis Rerum elaborates what should be grown in a
'noble garden', and includes our native violet of which there
are several varieties, the wild daffodil (*Narcissus pseudonarcis-
sus*), the marigold, peony and poppy, as well as a long list of
herbs including parsley, fennel, mint and rue. But top of the
list are the rose and the lily – those two cult flowers again.
Both had by this time a long pedigree of cultivation,
although not necessarily in this country, and as we have seen,
the rose had the most complex legends of Christian and
courtly worlds woven about her.

The lily was *Lilium candidum*, the White Lily. (It came to
be called the Madonna Lily only in the nineteenth century,
to distinguish it from other white lilies which were then
being introduced.) It had been cultivated from earliest times,
and across the whole of Europe it came to be regarded as a
symbol of purity associated particularly with Our Lady. In
Britain we have to thank the Venerable Bede for transferring
the flower from its original pagan beginnings to the
Christian symbolism of the Virgin. The earliest English rep-
resentation of the flower appears to be in the ninth-century
miniature of Queen Etheldreda (the founder of Ely
Cathedral), who is shown holding the lily. This miniature is
in an illuminated manuscript known as a benedictional (a
book of blessings) painted by the monks of Winchester for
their bishop, the great St Ethelwold (*c.* 908–984). Nearly all
the images of the Virgin Mary and the Annunciation which
were painted during the first half of the second millennium
include the lily somewhere in the painting. The flower is
dominant in the depiction of the Annunciation painted in
the fourteenth century by Simone Martini (1284–1344) and
now in the Uffizi Gallery in Florence, and continues to be so
into the nineteenth century, with the famous painting *Ecce
Ancilla Domini!*, also of the Annunciation, by Dante Gabriel
Rossetti (1828–82).

And what of this almost holy plant? One hesitates to use
the word 'divine' in such a context, but it does have star
appeal: its breeding and regal quality oozes from all of its six
satin-white petals, its wonderful golden stamens and its fragrant honeyed perfume. Because the bulb has been cultivated for so long, no one now is really quite sure from where it
originated, although most think its wild ancestors came from
the Balkans, Turkey and Asia Minor. It was brought to
Britain by the Romans during their occupation in the first
half of the first millennium.

Lilium candidium, the Madonna lily. The plant's association with
the Virgin Mary is shown in *The Annunciation* painted in 1850
by Dante Gabriel Rossetti at about the time he formed the
Pre-Raphaelite Brotherhood with the painters Millais and
Holman Hunt.

One cannot, of course, live by a lily alone or, in this case, by mooning over its beauty and associations, but as long as one was not intimidated by its weighty character the bulb could be stewed and eaten. Europeans, however, were not alone in eating lily bulbs: *L. tigrinum* (now *L. lancifolium*), the Tiger Lily, and brought from China in 1804 and *L. auratum*, the Golden-Rayed Lily, collected from Japan in 1862, were devoured for over a thousand years in the Far East. *L. pardalinum*, which grows in California and arrived here in the year 1875, caught the eye (or the belly) of cooks there in the eighteenth century. However, as the writer Alice Coats remarks, 'I would as soon turn cannibal as devour a lily.' One knows how she feels.

The genus, of approximately 100 species, is in its own *Liliaceae* family, and slowly but steadily the indigenous lilies from Turkey, Asia Minor and eastern Europe found a domestic home here. They are all striking plants with a bold demeanour, and as well as their culinary and curative qualities, they also make a strong addition to the flower bed. References to lilies in garden books of the sixteenth and seventeenth centuries became quite frequent, a sure indication of their popularity in gardens. They were, as were all bulbs, one of the easiest of items to transport – in a pocket, pouch or saddle bag – and so made a worthwhile souvenir to bring home after travelling abroad.

The Scarlet Turkscap Lily, or Scarlet Turn-Again Gentleman, is *L. chalcedonicum* (its synonym is *L. heldreichii*, after Theodore von Heldreich, a nineteenth-century German botanist who settled in Greece). Chalcedon is close to Constantinople (or Byzantium as it was then called, now Istanbul) on the Asian shores of the Bosphorus, and near the lily's indigenous home of Greece, so the former name seems more appropriate. By the time John Parkinson (1567–1650) talked of it in 1629, he referred to the fact that 'the red martagon of Constantinople is become so common everywhere'. The reference here to 'martagon' (a special form of turban) has always been the cause of some confusion, since another of these cottage garden lilies (which has similar-shaped 'turban' petals and the same rather unpleasant smell, but is much pinker in colour) is *L. martagon*, the Martagon Lily, or, as it is now known, the Common Turkscap Lily. The two are closely related, but this latter lily is a native of continental Europe and Mongolia, and had arrived in England by 1596. On the other hand, *L. chalcedonicum* had supposedly arrived in England some thirty years before John Gerard (1543–1612) mentioned it in his *Herball* of 1597; it is known that it grew

wild in Palestine. It is perhaps therefore not too unreasonable to suggest that they could both have been introduced much earlier than recorded, not only by soldiers returning from the Crusades, but also as a result of the upheaval of the Turkish invasion of the Byzantine Empire. Constantinople fell in 1453, and some of the fleeing survivors of the nobility undoubtedly would have carried with them not only their gold and silver but their equally precious lily bulbs to wherever they found refuge in Europe. It is a different story today; one rarely sees *L. chalcedonicum* in gardens, and it is difficult to find stock.

Like the rose, the lily family is an international favourite, and species and varieties have kept arriving from almost all over the world. They are one of those plants that, however long they stay in Britain enjoying our English climate, retain their foreignness and have a look of the harem about them – perhaps because so many of them originated in Turkey and Asia.

There appear to be no definite 'first-timers' making their debut in Alexander Neckham's list of 1199, but one possible newcomer was Hyssop, *Hyssopus officinalis*, a partially evergreen bushy aromatic herb native to south Europe. It was simmered in the cooking pot, and later, oil was extracted from it and used in the making of eau-de-Cologne. The flower is particularly attractive to bees, and hyssop honey is considered both tasty and efficacious in helping sufferers of rheumatism.

Hyssop is one of the 130 herbs required in the making of the brandy-based liqueur Chartreuse, which is made at the Grande Chartreuse in Grenoble, France, the mother house of the Carthusian order – a seeming dichotomy of indulgence for one of the most rigorous and austere of medieval orders, even though the liqueur is 'not consumed on the premises'. Hyssop is found growing wild on at least one abbey ruin today, that of Beaulieu in Hampshire. These could be the seeded remnants of the plants grown by the monks as a cooking herb, or for attracting bees to the garden, or for cutting as a 'strewing herb'. This last described any cut aromatic flowers, herbs, leaves or small branches which were 'strewn' on the ground to overcome the rather unsavoury smells which tended to linger about. The word 'hyssop' is of Hebrew origin, *azob*, meaning 'a holy herb', but it is not the

Right: Hyssopus officinalis is contained in a genus of only five species, in the Labiatae/Lamiaceae family, all aromatic herbaceous perennials.

Hyssopus officinalis

hyssop of the Bible. That is an altogether different plant, which needs to be in a greenhouse in order to survive in England – *Capparis spinosa*, the Common Caper.

Teucrium chamaedrys is another plant whose first entry was listed by Alexander Neckham. We know it is a plant that Dioscorides used, and the name is supposedly derived from Teucer, the King of Troy, who is said to have been the first to employ its medicinal properties; it helps promote the healing of wounds. There are approximately 300 species in the genus, which belongs in the huge *Labiatae* family. It was brought to Britain from southern Europe and is called the Wall Germander; it promptly jumped over the garden wall into the surrounding countryside, and is one of those incomers which felt so at home that it was soon taken for a native. It probably is not the germander recommended by John Parkinson as a useful edging plant for knot gardens and parterres; that is more likely to have been *Teucrium polium*, which came from the Mediterranean and western Asia about the same time. This one has grey leaves, and Parkinson thought it could also be used as a 'strewing herb for the house, being pretty and sweet'. Even this may not be the correct plant, as it is a bit of a softy as far as the winter climate goes, and requires a little mollycoddling. Needless to say, this plant has remained sedately in our gardens, where it often disports itself decoratively on the rockery. Cats find it attractive. Confusion has always surrounded the Germander family.

'Dragons' is a word with medieval connotations, and was the English way of speaking of *Dracunculus vulgaris* (syn. *Arum dracunculus)*. This plant, one of only 3 species in the *Araceae* family, was recorded by Alexander Neckham in *De Naturis Rerum*, and came from the Mediterranean region. Whoever returned with this plant in their baggage could not have been very popular, since it is a real stinker, particularly in the late spring, when the purple-blotched spathes – leaves – are unfolding, giving the plant a vulgarly suggestive look. Its olfactory distinction means that even today it is not much grown, despite the fact that Pliny and Dioscorides were of the view that a piece of the root should be always carried as a deterrent against serpents.

Anything that would guard against fire, thunder, lightning and pestilence would seem to be worth cultivating, and a particular plant on Neckham's list was thought to do just that. Indeed, its reputation had been established all over continental Europe long before, by the first Holy Roman Emperor, Charlemagne (768–814), who issued an edict that all homes on his estates were to grow them – on the roof. It is the Houseleek, or, as it was known in the twelfth century, *Jovis barba*. Culpeper called it 'sengren' (or 'sengreen', meaning

Nicholas Culpeper encouraged his readers to look into the bottom of the spathes (leaves) of *Dracunculus vulgaris* and, as he says, to '…see how like a snake they look'.

The Houseleek *Sempervivum tectorum* was believed to have at least sixteen medical uses, and it was a tradition for the builder of a new house in Kent to 'plant' some on the roof before the owners moved in.

'evergreen'), and advocated it be grown on roofs as a protection against disaster. This succulent consists of a mat-forming cushion of rosettes with no resemblance to the vegetable. Today it is known as **Sempervivum tectorum**, and you may still sometimes see it growing on roofs, casting its pre-fire brigade spell on the home below.

The Houseleek belongs in the *Crassulaceae* family, in which there are about 49 species, coming mainly from the mountains of Europe and Asia. Other names for it are Devil's Beard, and Welcome-Home-Husband-Be-It-Never-So-Late – perhaps that is why it should be encouraged to grow on the roof, to keep an eye on domestic comings and goings rather than to ward off lightning bolts. The literal translation of the Latin name is 'always grow on the roof', and the Irish name for it, *Buachaill a' tighe,* means 'warden of the house'. The juice, when squeezed from the fleshy leaves, was believed to have a healing quality for sore eyes, although in Cornwall it was the feet it helped in curing corns, while in Suffolk it was warts and ringworm.

Another plant which appears on Neckham's list originally travelled from southern Europe, but no date of arrival is chronicled and it may well have been growing here for so long and been so familiar that no one bothered to record it. This is the marigold, which seems the most English of flowers. The Romans certainly knew of it, using the flowers for decoration, as a dye and as an infusion to be drunk to relieve sleeplessness and nervous tension. Whether it survived from then through to the Middle Ages or was a reintroduction will probably never be known. Neckham, in his manuscript, refers to it as *calendula* (its old Latin name, meaning the 'first day of the month'). He is careful to distinguish it from the native Blue Chicory, *Cichorium intybus* – the specific Latin-based epithet for Wild Chicory – which he calls *solsequium,* as opposed to the Orange Pot Marigold, which he calls *solsequium nostrum quod calendula dicitur.* The word *solsequium* is again a Latin word, and relates to midsummer, so is a reminder of their joint flowering time rather than the colour of their leaves. The vernacular name for the marigold was 'Gold' or 'St Marygold', and it was considered so useful and homely a flower that by the seventeenth century it had been taken across the Atlantic and was a resident of North America.

The marigold's botanical name is **Calendula officinalis**, which reflects its almost year-round flowering. It is one of those plants which knows its place in the scheme of things and is so cheerful that you just have to smile when you look

Calendula officinalis **was grown for its efficacy in making a soothing ointment from the leaves, for use in cooking and to decorate the church, because the plant could bloom in every month of the year.**

into its sunshine face. As the garden guru William Robinson (1838–1935) once remarked: 'Few plants are more colourful or less fussy as to soil and situation.' Its contribution to the well-being of the human race is not confined to its character, however, for it is also a most useful culinary and medicinal herb. Its petals were once sprinkled into the pottages simmering over the hob to add a touch of sharpness, and were also used in egg dishes (a few discerning cooks still use them today). Later it helped retain or increase the yellow colour of butter and cheeses. A salve was also made from the pounded leaves to relieve stings and irritations; today, Calendula ointment is available from chemists and health food stores. There is no better antidote for the relief of the native stinging nettle (*Urtica dioica*): snatch a few marigold leaves and rub them hard so the juice comes out on to the area of the offending nettle sting. It seems more efficacious than using the native dock leaf, *Rumex* sp.

THE NETTLE

Despite the antisocial habits of nettles, they are rather useful plants – although Pliny clearly did not think so, saying there was nothing more hateful than the nettle. Not only can the fibrous stems be woven into cloth, but the young growth of fresh leaves may be eaten; when cooked, they taste much like spinach. The plant also produces a green dyestuff which came into its own during the Second World War when, on the orders of the County Herb Committees, forty tons of nettles were gathered and

An illustration from Elizabeth Blackwell's *A Curious Herbal*.

laid out to dry on cricket pitches, tennis courts and croquet lawns. It seems to have been a secret mission for the war effort, since no one was told where the dried weeds were going to or what they were going to be used for. It became the received wisdom that the extracted juice was going to be used to dye camouflage material.

At Common Moor near Liskeard in Cornwall, 1 May was designated Stinging Nettle Day; this was when a leaf had to be picked, rolled up in a small dock leaf and eaten, supposedly to keep you from harm throughout the year. Both in south Devon and in Cromer in Norfolk it was 2 May which was celebrated, slightly differently, as Sting Nettle Day, seemingly with much more fun. Games were played which involved young girls being chased by lads waving the nasty stingers. In Nottinghamshire on 29 May, Oak and Nettle Day was celebrated; this also involved the use of the stinger, this time for chastisement if you were not wearing your oak leaf to commemorate the restoration of Charles II to the throne in 1660.

There is an imported species of nettle which is fairly rare and, according to reports, much more vicious in its stinging capacity – *Urtica pilulifera*, the Roman Nettle. It was apparently dumped on us by the Roman invaders, and legend says the soldiers would deliberately beat their bodies with the irritant fronds to help keep themselves warm and stimulate blood flow. It is also said that the leaves of this nettle mixed with chicken mash are good for poultry, since they will clean their blood and improve their laying. All nettles have a reputation for relieving the aches and pains of rheumatism.

One of the fruit trees on Alexander Neckham's list was probably grown here during the Roman occupation, since it was also on Aelfric's list of 995, but it is only during this century that, for the first time, there is evidence of a planted tree. It was the mulberry, and it is referred to in the scene-of-crime documents written up following the murder of the Archbishop of Canterbury, Thomas Becket, in 1170. The whole of the Christian world was profoundly shocked by the Archbishop's violent death, and in a show of remorse, Henry II demanded a detailed account of what had happened. It appears that four knights (one of whom later committed suicide) had travelled from France – where the court was spending Christmas – to the Archbishop's Palace at Canterbury. There, under what has come to be confusingly called a 'large sycamore tree', they divested themselves of their cloaks and put on their armour.

The eminent medievalist John Harvey explains that the name in use for the mulberry tree at that time was *Celsus major*, but that its ancient Latin name and synonym was *Sicomorus* or 'sycomor' – a corruption of the word 'sycopsis', which comes from the Greek words *sykon*, meaning 'fig', and *opsis*, meaning 'appearance'. It would seem that because there are similarities between the fig and the mulberry (all those seeds and, when ripe, the same colour), medieval botanists considered them to be of the same family (which, indeed, they are, both being in the *Moraceae* family) and thus gave them similar names. This would have been reinforced by the existence of the *Ficus sycomorus*, the Mulberry Fig Tree – or, as it was also known, the Sycomore Fig, a native of Egypt and Syria. This is the sycamore of the Bible, but is too tender to grow in Britain. From the above, it is clear how confusingly named the mulberry was. Thus, when later writers or researchers perhaps without any botanical knowledge saw the word 'sycomor', they quite naturally referred to the tree as a sycamore. (The latter is a quite different tree – *Acer pseudoplatanus* – a native of central and southern Europe and Asia, but grown in Britain for so long it is almost considered a native.)

The only thing that is clear is the confusion – which still exists today in a faint echo of that earlier muddle, for in the Royal Horticultural Society's *Dictionary of Gardening* (1956) the mulberry is shown with the vernacular name of the Sycamine Tree. There is even more confusion since the Greek word *moron* and the Latin *morum* were also used by early classical writers for both the mulberry and the blackberry – now *Rubus*. It is a similar story in the kitchen, where

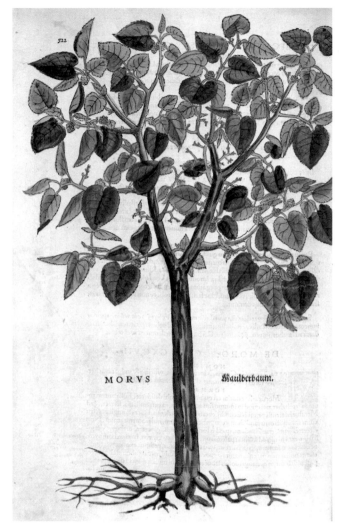

A mulberry tree was planted in Shakespeare's garden at Stratford-upon-Avon in 1609, of which descendants are still growing at the Botanic Gardens, Kew.

recipes often indicate that the two fruits are interchangeable.

The Black Mulberry (*Morus nigra*), whose original home is believed to be western Asia, is supposedly a symbol of prudence since not a leaf stirs until the very last hint of frost has gone, and, rather like the Gingko tree (see p.166), it is cautious and slow in its growth, waiting some twelve or fifteen years before fruiting.

Another early recorded planting of the mulberry, this time in London, was noted in the new garden of the Drapers Hall (home of one of the City livery companies), which was built in 1364. The tree lived for an astonishing 605 years, only

dying some thirty years short of the third millennium. According to Teresa McLean, writing in 1981, cuttings were taken from the original tree and planted 'in the little plot of garden that remains next to the Drapers Hall, off Throgmorton Street'. Also in the City, in Basinghall Avenue, in the garden of the Hall of the Girdlers' Company (which was chartered in 1449), there is a mulberry tree descended from one which was there in 1750.

While we are in London, there is a document about City life which throws some light on gardens and on the sophisticated way of life in the second half of the twelfth century. In about 1180, William Fitz Stephen, a Londoner, talked 'of the Gardens planted. Everywhere without the Houses of the Suburbs, the Citizens have Gardens and Orchards planted with Trees, large, beautiful, and one joining to another.' The King had two residences in London, one at the Tower and one in Westminster Palace, each with grand and kingly gardens to go with them. Numerous fruit trees and beds of soft fruit seem to have been planted in both gardens – pears, cherries, peaches, apples, raspberries and strawberries (a royal drink was concocted from the two latter fruits for Henry II in 1186).

The cultivation and growing of fruit was one of the great enthusiasms during the whole of the Middle Ages, orchards of apples and pears being recorded in all the most noble and monastic households. There were even special fruit-picking ladders, described by Thomas Tusser (*c.* 1520–80) as in couplet in his remarkable book *Hundredth Pointes of Good Husbandrie*: 'Light ladder and long, doth tree least wrong', an adage that holds good today. One of the reasons for the popularity of fruit must surely have been its sweetness, both for bees to make honey and for humans to consume, but much of the harvest was turned into juice: apples in particular were made into cider, and pears into the not so popular perry, while the native gooseberry, raspberry, crab apple, bullace (damson), sloe and strawberry were literally pressed into service when their fruit ripened. Fruit, nuts and vegetables were all to be seen in the London markets at this time.

By now apple trees had been domesticated for a long time; they really were the 'first fruit' of the world, and the first to be domesticated was known appropriately as the 'Paradise Apple'. It is *Malus pumila* – from the Greek *melon*, meaning 'with a fleshy exterior' (referring to the seeds being covered), and *pumila*, meaning 'dwarf'; confusingly, Linnaeus called it *Pyrus malus*. This is the wild crab apple, found growing in Europe, Asia and the Himalayas, and the first to be named.

THE STORY OF TWO APPLES

The Ribston Pippin, an apple allegedly grown from pips sent from Rouen in the 1690s, became the most widely grown apple during the nineteenth century. The original tree is said to have perished as late as the 1930s, but not before it became one of the parents of Cox's Orange Pippin. This was named for Richard Cox, who, in 1830, is believed to have planted two pips from the Ribston Pippin: one became the Cox's Orange and the other Cox's Pomona, and all three apples are still grown today. Ribston, in Yorkshire, has another claim to fame; see p.59.

Granny Smith and her son in Tasmania.

One apple introduced into Britain during the twentieth century, and now well known to everyone, is believed to have begun life in Tasmania as a crab apple. A few young trees were given as a gift to a woman who had emigrated with her family to Australia from Sussex in 1838. Because the fruit was so successful as a cooker she planted some of the pips; many years later her two sons-in-law, who were both professional apple growers, began to cultivate her apple in a different area, one that was cooler and drier, and where the fruit could develop into full maturity. The *Agricultural Gazette* of the time reported the new initiative. The original grower was known affectionately to her family and friends as Granny Smith – and so the Australian apple was named 'Granny Smith', and is one of the best-known apples today.

In all there are about 35 species of Malus, nearly all of which have been found growing in those areas. They form part of the *Rosaceae* family. Our native crab apple, *Malus sylvestris,* is presumed to have been partially domesticated long before the twelfth century, as the Anglo-Saxons certainly enjoyed their cider. The monks brewed it ceaselessly – there was little else

to drink apart from ale or wine – so every monastery and manor house usually had its own brewing house. Eventually the Pearmain apple (so called because it is long and pear-shaped), which is both a cider and a dessert apple, and Bitter-sweets, an old cider apple, came to be acknowledged as the best for making the juice. The Pearmain is still being grown and eaten today, but the Bitter-sweet appears to have been lost in the mists of time. John Evelyn (1620–1706), in 'Pomona' (part of his book *Sylva*), discourses on varieties of cider apples but makes no mention of Bitter-sweets.

The crab apple itself was popular over the centuries for hedging, and in the early eighteenth century Sir Robert Walpole of Houghton Hall in Cambridgeshire used 5,000 'crabbe settes', which cost him £1, for just such a purpose. (In 1718, when he was hedging the parish boundary separating Chippenham from Snailwell, he used 36,000 hawthorn and 280 elm saplings, which were planted at a cost of £5 18s 8d, or £5.94 in today's money.) One of the earliest dual-purpose apples, good for both cooking and eating and still available today (but interestingly only listed as a cooker), was the Costard, a word which means 'ribbed'. This apple was so popular for so long that the name has come down to us in the word 'costermonger'. In 1292, 136 kg (300 lb) of them were delivered to the court when Edward I was visiting Berwick Castle.

Apples have always excelled in the English climate, and until comparatively recently at least three-quarters of the fruit grown and eaten here was home bred. Domestically they have been grown for so long in our orchards that they have been taken for granted, and both their early history and the arrival of different species often went unrecorded. The story is different now: there is a revival of sorts going on, and we have been made aware of the hundreds of different varieties that still exist, some very local ones indeed. There used to be, until relatively recently, up to 3,000 varieties recorded, over 600 of which, in the 1950s, were growing in the RHS's garden at Wisley, and this did not even include any cider varieties.

Pears, like apples, were an obsession of the medieval world, just as they had been of the Roman world. Pliny enumerated 39 varieties in his *Natural History*, but they all stemmed from the common European and British native *Pyrus communis,* like the apple a member of the *Rosaceae* family. It was to be found in Britain's copses and woodlands and would surely have been cultivated from Anglo-Saxon times. A handsome pyramid of a tree, it grows large and is long-lived.

One of the reasons it became so popular was probably because of the ease with which it could be trained to grow against a wall or fence: the extra warmth and protection that this provides was and still is helpful in producing dessert pears of good quality. Its clusters of white flowers are attractive in the spring, and if you bury your nose in the blossom there is a faint whiff of fish, which presumably is attractive to the little two-winged fly which helps with its pollination. Although some pears are self-fertile, it must have soon been discovered that the flowering times of different varieties had to be carefully considered. It would have been no use planting a very early flowerer (like a Doyenné d'Eté, for instance) alongside one that flowered almost three months later (say, the Doyenné du Comice): the results would have been very unproductive. (Both these delicious pears were raised in France in the 1850s by the President of the Comice Horticole at Angers.)

It is in the twelfth century that the pear story really begins. In the Cistercian abbey gardens at Warden in Bedfordshire was grown a pear that had been sent especially from Burgundy. The variety proved very popular, since it was late-maturing and could be stewed and eaten in the winter when other fruit was scarce. The name persisted, and Wardens came to be synonymous with cooking pears, at least in the West Country, up until the eighteenth century. Alexander Neckham was rather dismissive of them, calling them 'hard cold fruits'. This was echoed by John Evelyn in 1669; writing about pears grown especially to make perry, he found Wardens 'so tart and harsh that there is nothing more safe from plunder, when even a swine will not take them in his mouth'.

The juicy dessert pears with their glorious golden skins were developed mainly in the warmth of Italy and France. In the 1500s, no fewer than 232 varieties were listed in an Italian manuscript; 209 of them were served to the Grand Duke Cosmo III during the course of one year – which works out at four varieties a week. Perhaps he was of the same opinion as William Robinson, the great British gardening guru of the nineteenth century, who reckoned that there were only ten minutes in a pear's life when it was worth eating! In Britain, in the Horticultural Society's fruit garden at Chiswick, 677 varieties were being grown in 1831. By 1858 there was a British Pomological Society, and that was the year that the pear 'Beurré Hardy', which holds an RHS Award of Garden Merit, was first exhibited.

In 1885 a National Pear Conference was held at Chiswick

at which a 'New Seedling Pear' was introduced; this was described with 'skin dark green and russet; flesh salmon coloured melting, juicy and rich. Tree robust and hardy'; how could anyone resist it? The RHS were obviously impressed, and immediately granted it an Award of Garden Merit. Because it was the only new pear that year to receive such an award it was called simply 'Conference'. How low we have now sunk in the pomona world: not so long ago estimates were given of about a thousand varieties of *Pyrus communis*, and yet now often the only pear we can buy to eat is a tough-skinned Conference.

The Crusades, or *Croisées,* as they would have been called until the sixteenth century, stretched across the whole of the twelfth century, having begun in 1095 with the call to arms by Pope Urban II for the whole of Christendom to expel the Muslim occupiers from the Holy Land. They eventually petered out in the first half of the thirteenth century, having failed to achieve any of the grand and knightly promises that were originally made. What these rumbustious journeys did do, however, was to create opportunistic trade and technology links between Asia Minor, Turkey, Egypt and Europe. For instance, the techniques of wind power were well understood in the east but had not been developed in Europe, so the craft of windmill building, invented by the Arabs for grinding grain, gradually breezed its way across Europe, followed later by water-raising windmills. Something equally as practical may well have come to England in 1170 as a result of contact with the Crusader world: this was the 'wilbarewe', which was believed to have been invented in China as early as the second century.

The first prototypes must have been fairly unstable, as the one small wheel was placed in the middle of the load, the handles being set either at the front (when the whole contraption was known as a Wooden Ox) or at the rear (when it became the Gliding Horse). It is surprising that this useful implement was not invented by the Romans: they made do with baskets for moving material around their gardens, and sledges pulled by either humans or animals on their farms. By the time the wheelbarrow reached Britain, experiments had been made with a two- and a three-wheeler version, but with the handles settling into the now familiar pushing position.

Most of the travelling to what was then called Palestine and is now Israel and part of Jordan was obviously done by walking or riding across Europe. In an age when the word 'holiday' meant 'holy day' and the phrase 'package holiday'

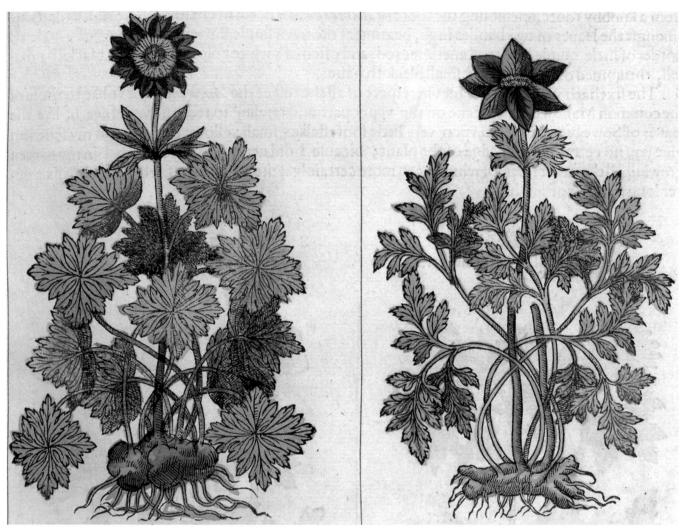

Anemone coronaria, the red-petalled beauty, has come to be associated by scholars with the biblical lines from Luke 12:27: 'Consider the lilies of the field, and how they grow, they toil not, neither do they spin. And yet I say unto you, that even Solomon in all his glory was not arrayed like one of these.'

was unknown, one is never quite sure whether it was the sheer excitement of an adventure or Christian fervour for the Holy Land which drew so many to make the pilgrimage. Probably a little of both. There can be no doubt, though, that with so many people tramping their way backwards and forwards through eastern and central Europe, Turkey or down through Italy to board a boat for the eastern Mediterranean and the Holy Land, the impact of new scenery, climate, food and plants must have been overwhelming, and have given the returning traveller (assuming he survived) a cache of stories to tell for the rest of his life. There is no doubt also that with

so much human (and animal) movement backwards and forwards across Europe, the inadvertent introduction of plants from one country to the next must have happened frequently, as often as horticultural souvenirs from a strange and wonderful land were brought back deliberately.

There is one Crusader 'maybe' for this century, and a rather apposite one it is, bearing in mind the introduction of the windmill, and that is the Windflower, *Anemone coronaria*, which blew its way into Britain. It was a native of the Holy Land, where it grew in great abundance and was called the Blood Drops of Christ. It grew prolifically in Italy too.

There is a story that Umberto, Bishop of Pisa, when returning to Italy from a Crusade, had his ship ballasted with soil rather than the usual sand. On arrival, the soil was unloaded and carted off to Pisa, to the Campo Santo, where in due course and like a miracle these beautiful flowers appeared (rather as the red poppy does on newly dug or disturbed ground).

Theophrastus, in his *Enquiry into Plants,* named it the Anemone or, even more charmingly, the Daughter of the Wind. It is a member of the *Ranunculaceae* family and is in a genus of about 120 species, spread mainly across the northern hemisphere. The simple, sprightly, saucer-shaped flower gave *A. coronaria* the cottage name of Poppy Anemone.

Many authorities believe the Windflower to be the New Testament 'lily of the field', and as he tramped across Europe, its bloom must have reminded the returning pilgrim of all he had seen and experienced in the Holy Land. The seed he undoubtedly collected would have been enthusiastically distributed when he got home as a lasting and memorable souvenir from Palestine. It is more than likely that the 'Great Double Windflower of Bithynia', as John Gerard called it, made its way to Britain in this manner. It is known that the Templars grew the Windflower at their remote priory at Ribston near Knaresborough in Yorkshire. The horticultural reputation of this northern priory and its links with Crusader plants extended beyond the flower border and encompassed fruit and trees also. The knights were obviously keen and experimental gardeners, having the reputation for growing one of the first mulberry trees in England; they also seem have been involved with the early planting of the Oriental Plane *Platanus orientalis.* (See p.56 for Ribston's other claim to fame.)

However it arrived here, *A. coronaria* is a flower that has contributed a great deal to the breeding and development of the modern anemone. In the wild it is variable in colour – red, white or violet – but there is a cultivated blue variety. *A. coronaria* has also given rise to two different groups of varieties, one consisting of single-bloomed flowers developed around the French town of Caen in the eighteenth century and called the de Caen Group, and the other of double or semi-double blooms cultivated in the 1880s in Ireland and called the St Brigid Group.

In the sixteenth century a close relative arrived from southern Europe and Asia Minor, *A. pavonina,* the Peacock Anemone, which was known to be grown by John Gerard. Indeed, by the sixteenth and seventeenth centuries a veritable gale of new varieties were being grown and were noted not only by Gerard (in his *Catalogue* of 1596), but by John Parkinson in 1629, and in the *Garden Book* written in 1659 by Sir Thomas Hanmer (1612–78). In the seventeenth century, the anemone was one of the flowers taken up by the florists in their endless search for the perfect bloom; see p.117.

The delightful *A. coronaria* is important in the development of our gardens, as it is possibly the first introduced flower to be grown for its beauty alone. It never seems to have been recommended to purge or salve either inwardly or outwardly, but was just grown for itself – the first of many floral beauties that were to be increasingly seen in our gardens.

Plant Introductions in the period 1100–1199

c. 1100 *Anemone coronaria* Windflower. E. Mediterranean to C. Asia.

Calendula officinalis English Marigold. S. Europe. Reintroduced (see p.110).

Hyssopus officinalis Hyssop. S. Europe.

Lilium candidum Madonna Lily. S. E. Europe to E. Mediterranean. Reintroduction (see p.35).

Morus nigra Black Mulberry. S. W. Asia.

Sempervivum tectorum Common Houseleek. S. Europe.

Teucrium chamaedrys Wall Germander. N. Europe.

The Queen's Mermelada
1200–1299

Eleanor of Castile, wife of Edward 1, brought expert gardeners from Aragon in northern Spain to England in 1289.

Significant dates

1204 Fourth Crusade; sack of Constantinople

1209 Cambridge University founded
Franciscan Order founded
(arrives in England 1224)

1215 Magna Carta signed by King John at
Runnymede on the river Thames

1216 Henry III (1216–72)
Dominican Order founded
(arrives in England 1221)

1258 Thomas Aquinas (1225–74) writes *Summa Theologiae*

1261 Byzantines regain Constantinople

1271 Marco Polo travels to China, continuing until 1295

1272 Edward I (1272–1307)

FRUIT WAS AN ENTHUSIASM and continuing theme in the thirteenth century, just as it had been during the previous fifty years, not only for food and drink but for ornamentation as well. Although the Romans had been acquainted with the peach and probably brought it with them, it is thought to have been reintroduced into Britain in the early years of this century, no doubt brought by returning Crusaders and travellers from southern Europe, where it had been established and the fruit eaten with relish for well over two hundred years. It seems to have arrived first in Spain and, like so many early imports, was thought to have originated in the east, in Persia (Iran), hence its Latin name of *Prunus persica*. It was only in the nineteenth century, when the Far East was opened up for plant exploration, that it was realised the trees grew wild in China and did not originate in the Middle East. The peach had been cultivated in China for thousands of years, so it surely must have been one of the plants which travelled the early Silk Route, eventually finding its way into Europe. It is not difficult to imagine those early travellers filling their saddle bags with succulent fruit at the start of their long and arduous journey, and the stones being later planted.

When the peach arrived in Britain, it was a tree for the most refined of gardens, and for the most royal, a difficult tree to flower and to fruit and ripen. It needed much expertise and ingenuity to help it cope with the vagaries of our British climate. All this was reflected in its scarcity and cost.

In 1252 it was reported by Richard of Wendover (a surgeon who wrote an anatomical treatise) that the death of King John in 1216 supposedly had been brought about by eating too many peaches and drinking too much ale. He died in the October, so he could have been gorging on the fruit for several months, but it would seem an unfortunate and unusual way to die. Since thirty-six years passed between the King's demise and the appearance of the written account, a little artistic licence could very easily have crept into the subsequent telling of the story. In any case, the tragic event certainly seemed to have no effect on the popularity of the peach tree.

The delightfully named *Cydonia oblonga* made an early and fruitful entrance in this century. Its natural home was south-west Asia, although its name links it with the Cretan town of Cydon (now Canea). This is the quince, the lone species in this genus which belongs to the *Rosaceae* family. There is a great deal of confusion between this and the three 'japonica' quinces, all of which belong in the Chaenomeles genus and to which *C. oblonga* is related; they also form part of the *Rosaceae* family. These three shrubs came from China and Japan, the first – *Chaenomeles speciosa* – arriving in Britain in 1784, long after *Cydonia oblonga*. This was followed in 1869 by *C. japonica*, the Japanese or Maule's Quince (named for William Maule, from Bristol, whose nursery promoted it), and finally *C. cathayensis*, found in central China. It is the fruit from this last species which apparently makes the best quince jelly.

They were first placed in the pear (Pyrus) genus, then classed with the Cydonia, then back to the Pyrus. This part of the *Rosaceae* family had, by the early part of the twentieth century, become large and unmanageable, and was broken up into several groups, with the three 'japonica' quinces

The fruit of *Cydonia oblonga* need the protection of a warm south-facing wall.

becoming the Chaenomeles genus – a Greek word referring to the shape and splitting of the fruit. The group has remained together in this genus ever since. The fruits of both the Cydonia and the Chaenomeles have to be cooked to make them edible.

The quince was clearly a luxury fruit. It appears to have come into Britain with Eleanor of Castile, probably in 1254, when she married, at the tender age of ten, Edward, the eldest son of Henry III. She brought with her the Spanish taste for the sweet quince confection called 'mermelada', the original marmalade; the word comes from the Portuguese for quince, *marmelo*. It was not until the eigthteenth century that citrus fruit came to be identified with the making of marmalade. The original mermelada is now known as quince cheese, and is still made today. In a fruiterer's bill of 1292, the cost of a hundred quince was given as 4s – as compared with 3d per hundred for ordinary apples or pears. In the same year, the cost of quince saplings was given as 2s for four, these to be planted in the garden of the royal residence in the Tower of London. Some years earlier, when they were being planted as an orchard in the royal garden at Westminster, the cost was 41s a hundred. Clearly, after allowing for discount for bulk, the cost of a quince sapling in 1292 had not reduced in price, but the vogue for the fruit had become firmly established.

Long before this time, the quince was believed to have been both the golden apple of the Hesperides and the golden apple of Aphrodite. In a love poem of the Arabic world of the tenth century, the fruit was extolled for its beauty and perfume: 'it is yellow in colour, as if it wore a daffodil tunic, and it smells like musk, a penetrating smell. It has the perfume of a loved woman, and the same hardness of heart.' Whatever its early reputation, Culpeper (1616–54) was ready with a practical use for the silky down which comes off the skin of the quince: when it is mixed with wax and spread over the head, 'it brings hair to those who are bald, and keeps it from falling off, if it be ready to shed'.

A plant which remains useful today put in its first appearance very early in the century and was perhaps originally collected by Crusaders or travellers striding across the wild and rocky mountains of south-west Asia and Europe on their way home. It is *Saponaria officinalis*, an upright perennial, perky in its outlook and with red, pink or white flowers. In 1659, Sir Thomas Hanmer (1612–78) thought the 'dowble blush onely esteem'd, which is fulsomely sweet', but added a word of warning that it 'spreads too much for a little garden'. What he did not write about was its soapy qualities. Its English and American names reflect its cleaning qualities and its usefulness, with its country names including Soapwort (*sapo* is the Latin word for 'soap') or Latherwort, Crow Soap, Fuller's Herb, and then Bouncing Bet (on account of its perky nature, perhaps?). The flowers are similar to our own wild Red Campion (*Silene dioica*); indeed, the plant has leapt the garden wall and is now naturalised in the countryside, growing along hedges and streams. Both plants belong in the same family, that of *Caryophyllaceae*. It is the leaves of the Saponaria which give the plant its soapy name. If they are swirled around in water, a considerable lather can be achieved. The Romans used the suds as a water softener, and even in the twentieth century its properties of cleansing softness were a recognised part of the treatment and conservation of old and precious tapestries. Saponaria has another use should you be unfortunate enough to be the recipient of a black eye or a bruise. Apparently a freshly dug root placed on the offending spot will assist in your recovery. This can probably be explained by the saponins which can be extracted from the root (and to a lesser extent from the leaves), which herbalists recommend also as a mild diuretic.

Queen Eleanor of Castile was in the same mould as her forebear, Eleanor of Aquitaine, and very much the modern woman, even to the extent of accompanying her husband on a Crusade. As well as beginning the passion for mermelada in England, she has also been credited with bringing back from her travels in the Near East the idea of the Arabic style of floor covering, the carpet, and she is believed to have intro-

Soapwort or Bouncing Bet *Saponaria officinalis* is part of a genus containing about twenty species.

Hanmer describes as bearing 'a stalke full of very sweet dowble white flowers in May, and requires good earth and to stand in a warme place abroad all wynter, otherwise to be kept in potts'. John Parkinson (1567–1650) knew Sweet Rocket rather strangely as 'Close Sciney' or 'Close Sciences' – no one seems to know why. Its old Latin name derives from the original Greek name which Theophrastus gave it, and by which it is known today, *Hesperis matronalis*. The Latin *matronalis* part of the name refers to its decorative use in the celebrations held on 1 March each year, the Roman Festival of Matrons. There are about 30 species of Hesperis in all, and they have a huge range, spreading from Siberia across into the whole of continental Europe. *H. matronalis* itself is a very variable biennial with flowers ranging in colour from white to pale lilac or purple. The plant is a member of the fragrant *Cruciferae* family and is most sweetly perfumed on warm evenings, which is why it is named after Hesperus, the evening star and the guardian of the love apple on the Isles of the Blest.

Another member of the *Cruciferae* family also thought to have come into our gardens sometime in the thirteenth century is the plant especially designed with children in mind, Silver Pennies or Honesty. The plant has been with us so long and is so instantly recognisable that it has a string of vernacular names, including Sattin-flower, Silverplate, Prick-song-woort, Judas Pence, Shillings, Two-Pennies-in-a-Purse and Money-in-both-Pockets! Just to add to the delights, it is known in France as Herbe aux Lunettes, or Spectacle Plant. However, its Latin name, *Lunaria annua*, recalls the moon, and that is just what the round silvery seed pods look like. It is a very widely spread native of Europe but is believed to have originated in Italy. Both Sweet Rocket and Honesty feel so at home here that they have crossed the divide and gone native.

A spectacular introduction from the Middle East, and trailing the influence of the Crusades with it, is the hollyhock, *Alcea rosea* (syn. *Althaea rosea*), which Queen Eleanor is thought to have brought to England following her travels in the Holy Land with Edward I. The old Saxon name for mallow was *hoc*, hence 'holy-hoc', and its Greek derivative name is from *althaia*, meaning a cure: the pigment found in the flowers is supposed to have healing qualities. The flowers are occasionally recommended in Chinese cookery, and the leaves, so John Evelyn (1620–1706) reported, were sometimes used as a pot-herb. There are about 60 species in the genus and they are all at home in rocky sites widely spread

duced, or at least reintroduced, to Britain the idea of regular bathing and baths. (King John was regarded as effete because he was known to have bathed at least six times in one year.)

Horticulturally, her name has always been linked with the introduction to England of Sweet Rocket (sometimes called Dame's Violet or Summer Lilac). The plant certainly seems to have arrived here during this period, and it is a native of Castile, so perhaps, as a reminder of her homeland, *Viola matronalis flora obsoleto* (as Sweet Rocket was then called) travelled with the young bride when she made her home in northern Europe. However it arrived in Britain, John Gerard (1543–1612) had by 1596 been introduced to 'a very beautiful kinde . . . having very faire double and white flowers'. A double-flowered variety appeared about 1640, which Sir Thomas

There are about 30 species of Hesperis which are natives of Europe, Siberia and China. *Hesperis matronalis*, the Dame's Violet or Sweet Rocket, was a plant grown by the Romans.

across Europe and Asia. *A. rosea* is thought to have originated in western Asia (it grew wild in Palestine as well). On arrival here, it obviously settled in well and was widely grown. It was mentioned in the long narrative poem of practicality called the *Feate of Gardening* by Jon Gardener of Kent. This has the distinction of being the first gardening work written in the English language, although all we have now is a copied survival made around 1440. The 'holy-*hoc*' had other names as well: the Outlandish Rose is one – it does look like a rose – and Rosa Ultramarina, too.

Two further species had arrived by the sixteenth century:

A. ficifolia (now ***Alcea ficifolia***), originating from Siberia although somewhat perversely known as the Antwerp Hollyhock; and ***Althaea cannabina***, from the south of France. Anyone who grows hollyhocks knows that the stalks are stringy – hence its second name, *cannabina*, or hemp. Neither of these species grows as tall as *A. rosea*, which can reach nearly 2.5 metres (about 8 ft). However, *Althaea cannabina* grows to between 1.8 and 2.35 metres (between 6 and 7½ ft) tall, and an attempt was made to grow it as a commercial fibre crop. Apparently 113 hectares (280 acres) were planted on the banks of the river Dee at Flint, and although

Lunaria annua, **Honesty, is one of only three species in this genus which are native across Europe and into W. Asia.**

the experiment was not repeated (one wonders how their height was supported), the plant yielded a useful blue dye.

Hollyhocks became increasingly popular flowers during the nineteenth century, almost rivalling the dahlia in their complicated petal arrangements, and special classes at flower shows were arranged for the cognoscenti to show off their breeding successes. One connoisseur in particular, William Chater, a landscape gardener and nurseryman from Essex, specialised in growing them. He is commemorated today with a beautiful anemone-flowered hollyhock called *A. rosea* 'Chater's Double'. The hollyhocks (Alcea) and mallows (Althaea and Malva), as well as the Lavatera genus are all very closely related genera and are found in the *Malvaceae*, mallow, family.

There are three thymes which grow wild in Britain – *Thymus serpyllum*, *T. pulegioides* and *Acinos arvensis*, or Basil Thyme – and all can be used in cooking. Stronger in flavour than any of them is Garden Thyme, or ***Thymus vulgaris***, which comes from the western Mediterranean and southern Italy and is a most useful plant. Being a native of a warm climate, it is more flavoursome than the native thymes, and is

one definitely to be used in the simmering pot. It would have been an easy traveller in the knapsack or baggage of returning pilgrims. The plant is part of a huge genus, with over 350 species in it, and in trying to codify their differences, botanists have divided the plants into two groups: the carpeting and wild thymes; and the cultivated species. They all belong in the *Labiatae* family.

All thymes are bee-friendly plants, and the native species must have been noted for that trait alone, long before there were cultivated plants growing in the garden. One of the native plants, *T. serpyllum*, is the herb Shakespeare referred to when he wrote, 'I know a bank whereon the wild thyme blows'. It is sometimes known as Shepherd's Thyme, and was believed to bring ill luck if brought into the home. As with many sweet-smelling flowers, it was often thought that the souls of the dead resided in the blooms. It is just as well, therefore, that no such notions are attached to *T. vulgaris*, the pot herb of the bouquet garni.

The Garden Thyme must also have been used as a strewing herb to ward off the plague; certainly when it is crushed or brushed against, it gives off a warm sunshine fragrance that reminds one of summer days. It was one of the herbs used to make up the posies distributed on Maundy Thursday to keep any smell at bay during the ceremony of the washing of the feet, a custom going back to the fifteenth century. Today, the Queen's posy is made up by the Queen's Herballist and always includes the sweet-smelling rosemary. Several of the cultivated thymes were recommended by John Evelyn to 'improve and meliorate the Aer about London'. The Crusaders carried it with them on their journeys, and it was sent on its way with the first settlers to America to give them courage. It was an old belief that not only the plant but even the word itself was a source of strength: the Latin word *thymus* comes from the Greek derivative *thumos*, which means 'courage'.

Legend tells how a certain grey Mediterranean shrub was used as a clothes drier by the Virgin Mary to spread the infant Jesus's washed clothes out to dry. After she had collected the tiny garments from the plant, it was left with the most beautifully fragrant perfume as a reward. The plant is lavender, Lavandula, the name coming from the Latin *lavare*, 'to wash', which reinforces the legend. No one really knows when it first came to Britain. The Greeks used the oil from the leaves in their bathing rituals, as did the Romans. The plant is indigenous to Italy (and around the Mediterranean), so it could have been introduced to Britain

Left: The Lentil *Lens esculenta* has been cultivated from ancient times, the seeds being highly nutritious.

Right: The Chinese or Japanese Lantern *Physalis alkekengi* has been domesticated for a long time, being noted for growing in Roman gardens, even though it has the reputation for being invasive.

by the Romans or have been brought here with the monastic orders in any of the preceding centuries. Oddly enough, it was neither on Aelfric's list nor was it one of the plants which Alexander Neckham (1157–1217) named in his *De Naturis Rerum*. Certainly there is evidence that Lavandula was here by 1265, when it appears in a manuscript of that date. (Now in the British Library, the manuscript was one collected by Robert Harley, Earl of Oxford, in the eighteenth century.)

The lavender we are talking of was *Lavandula spica*, now *L. angustifolia*, which accurately describes the shape of the narrow, rather lance-like leaves. Like thyme, this would have been a strewing herb for the house and was another garden plant which bees found (and still find) irresistible. There are some 25 species of this aromatic evergreen shrub and they all live in hot and dry rocky habitats stretching from the Canary Islands through the Mediterranean into Asia and India. In England we can only manage to grow a few of them; see p.105.

Another member of the *Labiatae* family which chose the thirteenth century to put in an appearance was marjoram, or Joy of the Mountains. This aromatic herb from south-west Europe and Turkey always seems to have been associated with happiness, and in Ancient Greece was often wound into head dresses and garlands for brides and grooms. *Origanum majorana* is the species we call Sweet Marjoram. As becomes such an ancient herb, the first part of its name is from classical Greek, while the *majorana* comes from Arabic. Culpeper believed that our own wild marjoram, *O. vulgare*, kept adders at bay.

Something new for the simmering pot came here during this century, although it had been eaten throughout the Middle East for thousands of years, and that was the lentil, *Lens esculenta*. This pulse has been recognised in cultivation for so long that it is believed to be the mess of pottage that Esau sold to Jacob, as recounted in the Bible. There are few species in the genus, and they are akin to peas in that they

supermarkets today was developed from *P. edulis,* the Cape Gooseberry – so named because it was first grown commercially in South Africa, although the plant originated in South America; in fact, *P. edulis* was known originally as *P. peruviana* var. *edulis.* It is not hardy in this country but in 1878 it was grown as a commercial crop in France. It is a useful winter fruit as it has a long shelf life.

The word *Physalis* comes from the Greek word *physa,* or 'bladder', and refers to the distinctive orange housing around the seed; the plant is also known as Bladder Cherry. The second name, *alkekengi,* has its roots (like the Origanum) in Arabic; although the word was being used as early as the fifteenth century, its meaning was then and still is obscure. In the sixteenth century William Turner (*c.* 1508–68) wrote in his book *A New Herball* that the bladders are '. . . round and rede lyke golde . . . whyche garlande makers use in making of garlandes'. By the nineteenth century, a more sophisticated approach to decoration was apparent when Jane Loudon (1807–58), a prolific writer on gardening subjects, recommended leaving the lanterns in water for six weeks to soak off the orange skin and leave the glowing red seed inside the skeleton of the bladder.

The stock, or gillyflower, a cottage garden plant which is held in great affection even today, is thought to have reached our gardens round about the middle of the century. Its perfume is of cloves and the Elizabethans called it the 'gilliflower' because of its lovely carnation-like scent (see p.38). In France the clove is known as the *clou de girofle* and the flower is named *Giroflée* for the same reason; its Italian equivalent is *Garofolo.* The twentieth-century writer Alice Coats comments that Henry Lyte (*c.* 1529–1607) of Somerset, a friend of John Gerard, described the plant as a 'Garnzie' Violet. One wonders if this is the plant referred to by Gerard as the 'Orange Tawnie Gilliflower' from Poland which was given to him by a certain John Leete, who apparently had trade connections with eastern Europe and who was three times Master of the Worshipful Company of Ironmongers.

The pre-Linnaeus name for the gillyflower was *Leucoia incana,* derived from the Greek word for 'white', and *incana,* meaning 'grey-leaved', which describes it most accurately – more so than its present Latin name of Matthiola (see p.215). This latter name commemorates a sixteenth-century botanist and author who wrote a celebrated life of Dioscorides which, in its day, must have been a bestseller, for there are still copies in existence in Latin, Italian, Czech, German and French.

In the wild, ***Matthiola incana*** grows along the coast of

arc annual climbers with a small pale blue flower. The lentil can be grown in England but needs sandy soil and a warm situation. The word 'lens' which we use in film-making and photography, the optic eye, relates directly to the shape of the lentil seed, and was in use that way as early as 1693.

Many of us in the twenty-first century will be familiar with the next plant, for the fruit of the Physalis is now widely available in our supermarkets. However, the plant that arrived here in the thirteenth century was ***Physalis alkekengi,*** a different species; apparently it did not stay long, for it was noted as a reintroduction in 1548. The plant grows wild from Europe all across Asia to Japan and is known as the Winter Cherry. Like Honesty, with its silvery orbs, this is another plant which entrances children, especially in the winter, when it sports bright orange seed pods which look like Chinese lanterns – another of its common names. Opinion is divided as to the taste of the small, rather sour seed which the lantern (calyx) holds. The fruit that we can buy from our

Marco Polo was a Venetian who travelled extensively in Asia, India and China between 1271 and 1295, and was probably the first European to visit Yunnan, in S.W. China.

south and west Europe and into central and south-west Asia, and in South Africa. Of the 55 species in the *Brassicaceae/Cruciferae* family we in Britain can seemingly grow only about three or four. We have one native stock, *M. sinuata,* which has a white flower and is happiest growing among sand dunes, and even *M. incana* is occasionally found growing outside the confines of the garden, along our southern seashore. Almost every early writer on gardening matters extolled the plant's virtues, including John Gerard, who, in 1597, wrote in his *Herball or Historie of Plantes* of their 'diverse colours . . . for the beauty of their flowers, and pleasant sweete smell'. According to Thomas Hanmer, about fifty years later, the single cream-coloured bloom was the sweetest perfumed of all. He also spoke of 'dowble' flowers in shades of purple and tawny colours 'which are many tymes

finely stript or markt with white', and a double yellow which 'tis a stranger yet in England'.

No trace of the Night-Scented Stock, *M. longipetala* subsp. *bicornis* seems to be recorded prior to the early nineteenth century, but one feels it must surely have been a plant to rouse the interest of early travellers in Greece and Asia, where it grows wild. However, what has been noted is its sensitivity to darkness, when it opens its flowers and releases its delicious perfume. It seems that even an eclipse of the sun will set it off.

It is the versatility of the stock which continues to keep it in the front line of gardening. Horticulturally there are a number of distinct strains which have been bred from the cultivars of *M. incana,* some of which have been crossed with our own native stock. One such group is the Brompton

The Stock *Matthiola incana* grown as cultivars makes a spectacular sweetly scented show planted as spring or summer bedding.

plants from this group will flower within that time. The East Lothian Group is another old, distinct type, one or two of which may still be grown. Nowadays, there are a number of groups which have been bred in different colours and heights, but all are 'dowble' flowered – for sturdiness and vigour; single-flowered stocks tend to sprawl and straggle. This points not only to the continuing popularity of such an old garden plant but also to its ability to cope with modern-day demands.

The thirteenth century saw the glimmerings of flowers being grown for their own sake, a sure sign of the confidence of living in a stable (and enlarging) population with a plentiful supply of food, much of which could be obtained from local markets. These were increasingly to be found, particularly in large towns like Kingston-upon-Thames in Surrey, whose market began in 1242. Life was becoming more civilised as ideas from the Continent and the Middle East were slowly being adopted, particularly by the rich and influential, including such refinements as glass within the window frames, carpets on the floor, tapestries on the wall, and baths. With such luxuries being applied to the interior of homes, it was not unnatural for the garden also to proclaim itself, and for the craft of gardening to be taken up. If one thinks back to the beginnings of this century, that was an almost revolutionary idea. The following century, however, was to bring a baptism of fire to Europe and Britain as the flea-borne plague ravaged the whole of Europe with ghastly consequences. Just as it was beginning to blossom, horticulture once again had to take a back seat as the country fought for survival.

Stock, which received its name in the eighteenth century having been bred by George London and Henry Wise in a nursery on a hundred-acre site in Brompton Park (where the museums of South Kensington now stand). Then there is the Ten-week Stock, which was first mentioned in 1754:

Plant Introductions in the period 1200–1299

c. 1200   *Cydonia oblonga* Quince. S.W. Asia.

    Hesperis matronalis Sweet Rocket, Dame's Violet. S. Europe, Siberia, W. and C. Asia.

    Lens esculenta Lentil. Middle East.

    Lunaria annua Honesty, Silver Pennies. Italy.

    Origanum majorana Sweet Marjoram, Joy of the Mountains. S.W. Europe, Turkey.

    Physalis alkekengi Winter Cherry, Chinese Lantern. C. and S. Europe, W. Asia to Japan. See also p.118.

    Prunus persica Peach. China. See p.76.

    Saponaria officinalis Soapwort. S.W. Asia.

    Thymus vulgaris Garden Thyme. W. Mediterranean, S Italy.

c. 1250   *Matthiola incana* Gillyflower, Stock. S. and W. Europe, Arabian peninsula, Egypt.

1260   *Alcea rosea* (syn. *Althaea rosea*) Hollyhock. W. Asia.

1265   *Lavandula angustifolia* (syn. *L. spica*) Lavender. W. Mediterranean. See p.105.

1299   *Paeonia mascula* (syn. *P. corallina*) S. Europe; see p.32.

A Quartet of Trees
1300–1399

Cupressus sempervirens the Mediterranean Cypress shown here growing in one of its favourite habitats in Tuscany, where it is an integral part of the English image of the Tuscan countryside.

Significant dates

THERE SEEMED NO REASON to doubt that this new century would follow the example of the old, with its worldly pleasures and luxuries continuing the 'upward mobility' of at least some of the country's inhabitants. The continuing rise in population (now thought to have been around five million, having risen slowly from about two million in 1066), with the subsequent growth and organisation of towns and villages, seemed all set to turn Britain into a confident and forward-looking nation. More people meant more houses, needing a constant supply of timber for building, the replanting of trees, the cultivation of land for food, larger flocks of sheep for the production of wool and thus cloth – in other words, an expanding economy.

The cultivation of gardens, too, was a growth industry and gardeners were engrossed in both creative planting and the medicinal and cooking requirements of the home (particularly with the innovation of chimneys now making life easier in the kitchen). The Worshipful Company of Gardeners, founded in London in 1345, enjoyed (and still does) the privilege of presenting bouquets to the Queen Consort (and later, when relevant, to the reigning Queen) and the royal princesses at coronations and royal marriages. Another responsibility of the Gardeners Company at that time (but no longer) was to ensure that streets of the City of London were kept clean.

We know that at least four trees arrived during this century, all of them destined to make a long-term impact on our landscape. The first, **Picea abies** (or *P. excelsa* as it was also known) is the Common or Norway Spruce, which is certainly common and not only in Norway, where it grows wild, but nearly everywhere else in Europe too. Planted in profusion, it now flows like cold lava over our remoter hills and valleys. A redeeming feature, however, in earlier centuries was the enormous value of the timber for building – not only of houses but of ships during the formation of the Royal Navy.

The making of beer from the spruce must have seemed an unexpected bonus for such a dour tree when it first arrived here. The centre of this trade was apparently Danzig (now Gdansk in Poland) where it was called 'black beer', although later the best spruce beer was, by all accounts, made from **Picea mariana,** which was an introduction from the northeast of North America about 1700. The tips of young shoots were boiled down in water to provide an essence which was then added to beer to make spruce beer. The Tsarist Russian Army liked to strengthen their spruce beer further by lacing it with a touch of horseradish, ginger, mustard and a tot of alcoholic spirit.

The spruce took on a new lease of life in the nineteenth century when Prince Albert introduced the German tradition of the 'Christmas' tree to Windsor Castle. In the twentieth century this was the tree that lit up London's Trafalgar Square at Christmas, a gift from the people of Norway in grateful remembrance of Britain's support to their country

Martin Luther, in the years *c.* 1535, seems to have been the first person to use *Picea abies* as a decoration to celebrate the Christmas story, having set up a branch of the tree for his children with candles on it to replicate the starry December sky whence the Christ child descended to earth.

during the Second World War – a gesture which is continued to everyone's pleasure each December.

Because of its great size – it can grow to 30.5 metres (100 ft) – the 'firr', which it was frequently called, is a forest or landscape rather than a garden tree. In the eighteenth century the 'landskip' designer William Sawrey Gilpin (1762–1843) had refined ideas concerning the impact of trees in the countryside, one of his contentions being that belts of trees should never be planted because they looked too strong in the landscape. This was quite the opposite to what Lancelot (Capability) Brown (1716–83) recommended. Gilpin also thought a picturesque scene could be enhanced more with round-headed trees rather than columnar ones, which is perhaps why he preferred the 'firr' to our native Scots Pine, for as the tree matures, the top of it flattens out and thus blends more harmoniously with our round-headed trees.

There is no record of what Gilpin thought of the cypress of the ancients, the almost vertical Italian Cypress, which made its debut in about 1375. *Cupressus sempervirens* is native throughout the Mediterranean area (it is often called the Mediterranean Cypress) and western Asia. Despite its early introduction to Britain, it has an unsettled look, as though it does not quite know what to make of our climate (or of us) and is looking forward to returning to the warmth of the sun and fellowship of the old Greek and Roman worlds. There is no evidence to support the idea that this was another Crusader import, but the dark exclamation mark of its form against the blue sky must have made it seem a desirable inclusion in English gardens. It has certainly provided a fascinating subject for painters over the centuries, particularly in the French and Italian schools. It is often portrayed in the company of that other signature of southern Europe, the contrasting round-headed *Pinus pinea*, the Stone or Umbrella Pine which arrived in our country in the middle of the sixteenth century. (This is the pine from which come the pine-kernels used in cooking.)

Found all over the northern hemisphere, the Cupressus genus encompasses about 24 evergreen species, nearly all of which are conical in shape. One which will be especially familiar in south-west England, where it has been planted in belts as shelter against the salt-laden winds, is the Monterey Cypress, *Cupressus macrocarpa*, brought from the Monterey peninsula in California about 1838. This has a shape not unlike a cedar, except rather more untidy. It has a powerful presence; not a tree to be ignored.

The Monterey Cypress is one of the parents of *x Cupressocyparis leylandii*, the Leyland Cypress, the other being *Chamaecyparis nootkatensis*, also from North America and arriving in this country in 1853. The result of the pairing is known as an inter-generic hybrid or, as it was once described rather more succinctly, a bastard offspring. This reflects the bad press *C. leylandii* presently receives. It is not quite in the triffid class but, rather like a recalcitrant child, if not disciplined the tree will terrorise the neighbourhood. *C. leylandii* must never be allowed to get out of hand as not only is it almost the swiftest-growing offspring of all time, but it grows very densely, shutting out the light and creating a glowering effect around it. Found twice as a seedling at Leighton Hall in Wales, first in 1888 and again in 1911, the cross must also have occurred a third time, for it was discovered thirty years later in a Dorset garden. The tree is named for C. J. Leyland of Leighton Hall and it is most often planted as a screen or barrier to shelter or hide something; if grown as a hedge (there is a difference) and clipped from an early age, it can make the most delightful backdrop, with a certain touch of lightness about it. When grown as a specimen tree in parkland, the Leyland Cypress takes on quite a different character and is considered to be a noble being.

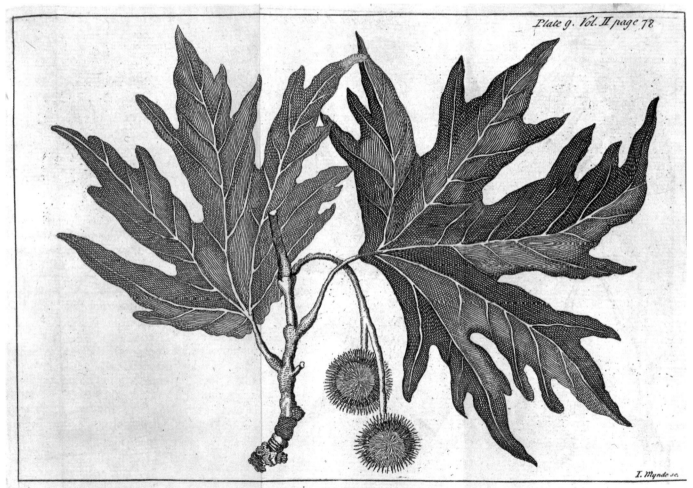

The Plane Tree

The leaf of the Oriental Plane *Platanus orientalis* shows the close resemblance the shape of the leaf has to the outline of the Peloponnese where Athens and Sparta fought a bitter war between 431 and 404 BC, ending with Spartan victory.

Another noble tree of great stature, which frankly knocks the socks off the Leyland Cypress, and which is believed to have been introduced here during the fourteenth century, is *Platanus orientalis*, the Oriental Plane. The history of this tree, however, goes back far beyond the fourteenth century into antiquity. It was, and is, grown widely in the Middle East, as reflected in its original Persian name of *chenar*, the Chennar Tree. Geoffrey Chaucer (*c.* 1345–1400) mentions the plane tree in the Knight's Tale (line 2922) of *The Canterbury Tales*.

John Evelyn (1620–1706) has much to say about this beautiful specimen in *Sylva*, his book on trees, including a snippet of most useful information. Apparently, Cicero and Hortensus would use wine to 'refresh their Platans' in place of water – an interesting use of the wine lake, so do remember that when the next drought occurs. Evelyn also tells us that the tree was so doted on by Xerxes, King of Persia, that, by the river Meander, on his way to Sardis, he 'made a halt and stop'd his prodigious army of seventeen hundred thousand soldiers . . . to admire the pulchritude, and procerity of one of these goodly trees'. It is then reported that he stripped his fellow officers, travelling concubines and much of the army of their jewellery, gold, scarves, bracelets and anything else he presumably considered appropriate to drape through its branches which he then stepped back and admired 'to the concernment of his grand Expedition'. In the eighteenth century, George Frideric Handel (1685–1759) reproduced the story in his opera *Xerxes* (which was not a success, although one of the tunes was recycled into the more famous 'Largo').

The deeply cut shape of the leaf of the Oriental Plane was used as a reminder to schoolboys that the outline made a map of the Peloponnese (neatly combining botany with geography), but what it also does is distinguish it from the London Plane. This is the one tree from this small genus of only 6 species with which everyone is familiar, but its development is still something of a mystery. The Latin name is *Platanus x hispanica* – the cross indicating that it is a hybrid. The two parents are considered to be *P. orientalis* and *P. occidentalis*; the latter is from North America, where the common names for it are the American Sycamore and the Buttonwood, which refers to the seed balls that hang from the bare branches throughout the winter. The tree was first recorded in 1631 in the Lambeth garden of John Tradescant (1570–1638), but the species did not do well in Britain. The first London Plane – whose long Latin name was originally *Platanus inter et occidentalum media* – was known to be growing in the Oxford

A fine example of the handsome White Poplar or Abele *Populus alba* which can reach a height of 40 m (130 ft).

Botanic Garden by 1666, having arrived in the country three years earlier. A theory was put forward in the nineteenth century by Dr Augustine Henry (after whom a form of the tree was named) that the hybrid had occurred at Oxford between the two trees from different continents, but since the New World partner refuses to flower here, it must have taken place elsewhere and the offspring itself introduced here. In 1731, when Philip Miller (1691–1771) planted one in the Physic Garden at Chelsea, he called it the 'Spanish plane'. Since Spain had long grown the Oriental Plane, perhaps that is where the cross with its American cousin *P. occidentalis* took place. Its later name of *P. x hispanica* would seem to reflect this theory.

Another theory was that the London Plane was a 'sport' or variation of the Oriental Plane, but whatever its antecedents, experts agree that the hybrid occurred naturally rather than having been a planned cross. The London Plane is certainly one of our greatest arboricultural assets. With a grandeur and dignity which becomes our capital city, it never looks out of place – unlike some roadside plantings of cherry trees which, after two weeks of glorious pink or white blossom, have a squatness about their shape which seems to diminish a road rather than enhance it. The bark and trunk of the London Plane, which can be viewed without craning one's neck, are full of interest and colour, and the tree itself is no wimp, tolerating severe pruning and city pollution.

The oldest surviving plane trees in London are thought to be those planted in Berkeley Square in 1789, so it seems entirely appropriate that this was the square chosen in the song for nightingales to warble their trilling call. There is a theory that they grew particularly well here because they were planted on a well-matured plague pit. One of the largest, and probably the oldest London Plane of all is in the Bishop's Palace at Ely. In 1960 it was just over 33 metres (111 ft) high and thirty years later, in 1990, when measured by the dendrologist Alan Mitchell, it was recorded as being 1 metre (3 ft) higher. Topping that are two giants at Bryanston School in Dorset which, when measured in the mid-1990s, were, according to the tree historian Thomas Pakenham, both over 46 metres (151 ft) high.

The White Poplar *Populus alba* also made its entry sometime during the fourteenth century, one of the 35 mostly deciduous species of the *Salicaceae* family that are native across the northern temperate zones. This particular species has a very wide geographical spread and can be found growing naturally anywhere between north Africa, Turkey and Siberia, so it will be no surprise that it has become naturalised in our landscape. It joins three other Populus species: our native aspen, *P. tremula*, whose name aptly describes the movement of its leaves; the magnificent but rare Black Poplar, *P. nigra*, originally a native of central southern Europe and Asia but which has been here so long that there is no record of its arrival and it should now be considered a native of this country too; and the robust *P. nigra* ssp *betulifolia*.

How this sub-species was discovered is rather unusual, because even though it was a European native it was first noted on the banks of the Hudson River in New York State early in the nineteenth century by a M. Michaux, who named it *Populus hudsonica*. Examination by botanists confirmed that it was a relation of *P. nigra* and since the tree was not an American native, an early settler must have taken it with him to remind him of home. It was renamed *P. nigra*

The London Plane *Platanus* x *hispanica*, photographed growing in the grounds of the Bishop's Palace in Ely, Cambridgeshire, in 1928.

spp betulifolia (meaning 'with birch-like leaves') and has been known in Britain since the eighteenth century. It was later called the Manchester Poplar; its toleration of industrial pollution meant it came to be used as a street tree in northern Britain and Manchester in particular; see p.178.

The poplar that is most noticeable is the Lombardy Poplar, *P. nigra* var *italica*. It is from the same branch of the family and arrived here in 1758 in the baggage of Lord Rochford on his return from Turin in northern Italy. Like the Italian Cypress, this is another great exclamation mark in the landscape, and it is shown off to its best advantage and looks its most natural when planted in lowland river basins with a distant prospect of hills. Nearly all the Lombardy poplars in cultivation are male and are propagated by cuttings; the female tree is infrequently planted, but one can always recognise it since during early summer great tufts of white down drift on the wind, looking from a distance not unlike cotton-wool balls.

It would be many years, if not generations, before this quartet of trees – the Spruce, the Italian Cypress, the Oriental Plane and the White Poplar – made any impact at all on our countryside. Brought from across Europe and beyond during this century, three of them would reach heights of 30 metres (100 ft) or more while the Cypress would eventually settle for a little less, some 20 metres (70 ft). All four, though, would become trees of stature in our landscape, if not our gardens.

A tree that has no great landscape pretensions but is a small and glorious addition to our gardens is believed to have also arrived during this century. At least, the consensus of expert opinion is that the almond tree made itself known in Britain in the mid-1350s, even though it had been found in various lists before then. It was known on the Continent, particularly in the south, for its edible nuts, but unless planted in a very favoured site here in Britain it is the early blossoming of the flowers – the heralding of spring – which we find so entrancing. With its delicate pink and white petals, it is much prettier than either of its close relatives, the peach and the apricot. Like the poplar, the almond also came from western Asia and north Africa. It is one of the 200 species of the Prunus genus that all belong in the *Rosaceae* family, and its Latin name is now *P. dulcis*, meaning 'sweet plum' or 'cherry', although up until the latter part of the twentieth century it had always been known – rather sensibly – as *Prunus amygdalus*, which was the Greek word for almond.

The peach (*Prunus persica*) had already arrived in Britain from China (see p.61) but an even closer relative of the almond, the apricot, *Prunus armeniaca*, is believed to have arrived later, in 1542. In the same way that the peach was thought to have come from Persia (now Iran) – hence its *persica* name – the apricot too turned out to have been misnamed. It had *armeniaca* added to its *Prunus* name in the belief that the tree came from Armenia but, just like the peach, which had made the journey earlier, the apricot actually travelled from China, probably along the same Silk Route. It later proved its home territory when it was found growing wild in northern China by Dr Bretschneider (1833–1901), a great botanical discoverer. Modern visitors to China's Great Wall can enjoy massed apricot blooms in early spring; here, according to the highly regarded plant-collector Roy Lancaster, 'tens of thousands' of the native apricot have been planted.

The 'pomgarnet tre' is a shrub which, like the almond, arrived sometime in the middle of the century. Growing on the cusp of tolerance in Britain, it has been cultivated here more for its camellia-like blooms than for the production of its round glowing fruit. Pomegranates certainly featured on royal tables in England, since we know that in 1352 one Bartholomew Thomas bought five of them for Queen Philippa at a cost of ten shillings, a considerable amount in those days. Although the plant originated in south-west Asia, the fruit must have been imported from southern Europe, where it would have been grown in orchards. A syrup made from the juice of the seeds, grenadine, has long been used in the cooking of the Middle East and southern Europe; a more modern recommendation is to drink the juice ice cold mixed with gin. The plant's Latin name, *Punica granatum*, is a contraction from its old name of *Punicum* which itself is linked back to the word *Phoinikes*, or Phoenician. Sometimes it is referred to as the Carthaginian Apple. Its second name, *granatum*, aptly describes the crevices and granite rock where it likes growing. The pomegranate was once a symbol of fertility (all those seeds), and transferred smoothly into the Christian world as symbolic of the Oneness of the Universe. There are several Renaissance paintings of the 'Madonna with pomegranate'.

While the Chinese dedicate June in their floral calendar to this particular fruit, a poem written in 1897 by André Gide (1869–1951), 'Les Nourritures Terrestres', accurately describes the fruit and the flower.

> A little sour is the juice of the pomegranate
> like the juice of unripe raspberries.
> Waxlike is the flower
> Coloured as the fruit is coloured.
>
> Close-guarded this item of treasure, beehive partitioned,
> Richness of savour,
> Architecture of pentagons.
> The rind splits; out tumble the seeds,
> In cups of azure some seeds are blood;
> On plates of enamelled bronze, others are drops of gold.

Right: In the sixteenth century Henri iv of France took the pomegranate *Punica granatum* as his device with the motto 'Sour but Sweet' (which is just what the juice tastes like).

Punica Granatum.

At about the same time as this pair of exotic trees was beginning to make a mark, a shrub which exudes the warm delight of its Mediterranean homeland was also arriving here. It was *Rosmarinus officinalis*, the universally popular rosemary, whose earlier expressive names were Rose of the Sea and Sea Dew. It is a member of the *Labiatae* family, and its Latin name aptly illustrates where it is found growing wild, along the cliffs and within sight of the sea. John Evelyn reported that the perfume was so strong it could be detected 'thirty leagues off, at sea, upon the coast of Spain'.

The plant is believed to have been sent in about 1340 to Queen Philippa of Hainault, wife of Edward III, by her mother, the Countess Joan of Valois. There is evidence that the Queen was an enthusiastic gardener, and certainly a number of gardens at the royal residences and palaces were being modernised and altered at this time. Perhaps she wanted the rosemary for the walled garden at Eltham Palace in Kent, where she and the King sometimes dined al fresco, or possibly for the elaborate pleasure garden which was being created out of the park at Odiham Castle in Hampshire during 1332. She may also have needed to escape from her thirteen children, the eldest of whom was the Black Prince, the youngest Thomas 1st Earl of Buckingham, with John of Gaunt sandwiched somewhere in between.

A further queenly connection is attached to rosemary later in the century when Richard II (grandson of Queen Philippa) married Anne of Bohemia whose badge depicts the herb and is shown as such on the Wilton Diptych (displayed in the National Gallery). Two hundred years later, in 1543, Henry VIII had banks of rosemary and lavender planted in his garden at Chelsea. The herb obviously felt at home here, and is so much part of our garden now in its Englishness that it is surprising it has not gone native.

Bringing a souvenir home from the Crusades must have seemed quite daring, and to return with plants or seeds to see if they would grow showed confidence in the established order, and a curiosity and interest in the domestic surroundings at home. Since the lumbering carts or packhorses took nine months or more to complete the adventure, one wonders how many bulbs, seedlings or shrubs must have perished or seeds been lost on the long journey home.

As we have seen, however, introductions *were* made, whether planned or not, and two plants from the *Caryophyllaceae* family were certainly made welcome in our gardens from the fourteenth century. One is the aptly named Jerusalem Cross or Maltese Cross, *Lychnis chalcedonica*,

which, although collected from the Middle East, turned out rather surprisingly to come originally from European Russia. The other newcomer is its relative, *L. coronaria*, the Rose Campion, a native of southern Europe and one of the flowers used for making garlands and wreaths in ancient Greece; the flower is said to have sprung from the bath of Aphrodite. Crusaders would certainly have seen them both growing wild in the countryside, and *L. chalcedonica* is one of the plants that, since its introduction to western Europe, has always been linked with the returning pilgrims. The first name comes from *lychnos*, the Greek word for 'lamp' and is thought to refer to its very woolly leaves, which were used as wicks for oil lamps. Once more, Sir Thomas Hanmer (1612–78) offers advice about cultivating *L. chalcedonica*, or *Constantinopolitana* as he also called it: '. . . all are usually

Left: The Wilton Diptych was painted during the reign of Richard II (1377–99) by an unknown artist who may have been English, French, Flemish or Bohemian. The White Hart, which is shown lying on rosemary and appears on the back of the Diptych, is the emblem of Richard II.

Right: The name Chalcedon in *Lychnis chalcedonica* is the classical name for Kadekoy, the district of Turkey opposite Istanbul (Constantinople).

Lychnis Chalcedonica.
Flower of Conftantinople.

encreast from their rootes, which are easily divided, and planted betimes in the spring in good rich mold'.

Lychnis can be thought of as a mainly decorative flower for the fourteenth-century garden, but another plant which arrived about the same time was not only pretty but useful as well. *Lupinus albus* was grown by the Romans for animal fodder and the subsequent green manure, with its especially high nitrogen value, was then ploughed back into the ground. The European lupin is a native of the eastern Mediterranean. Another lupin, *Lupinus luteus*, was also used as a fodder crop, but it came from the opposite end of the Mediterranean, from south-western Europe; it had lovely sweet-smelling yellow flowers and was sometimes called the Spanish Violet. Understandably, it was also a favourite in gardens until it was superseded by the American perennial imports in the seventeenth, eighteenth and nineteenth centuries.

These American plants are the ancestors of our herbaceous lupins and it is a matter of conjecture whether the first to arrive, *L. perenne* (meaning 'perennial'), may have been part of the botanical loot which John Tradescant the Younger (1608–62) brought home from his voyage to Virginia in 1637. It was growing in the Oxford Botanic Garden by 1658, and a year later Sir Thomas Hanmer lists it in his *Garden Book*, calling it the 'Everlasting Lupine' and noting that 'it came not long since first hither from Virginia'. He thought the colour a 'good blew' but others found it rather washed out. It was eventually replaced by the yellow-flowered and sturdier Tree Lupin from California, *L. arboreus*.

The Tree Lupin was discovered in 1792 by Archibald Menzies (1754–1842) when he travelled as surgeon-botanist on Captain Vancouver's expedition to search for a sea passage between the North Pacific and the Atlantic. Botanists and plant collectors have to be opportunistic if nothing else when on expeditions. It too has a lovely honey scent and now grows wild in some areas of Britain, particularly around our coastline on sandy dunes and cliffs and the china clay tips in Cornwall. A second American perennial, this time the blue-flowered species *L. nootkatensis*, was also discovered by Menzies, in Nootka Sound. (Later, in 1831 and 1853, the Sitka Spruce and the Nootka Cypress were both to be collected from the same area.) It is now known as the wild lupin, and like its cousin, this plant has also escaped from northern gardens and may be seen growing wild along the banks of some Scottish rivers.

It was about fifty years later that another plant collector – a Scot like Menzies – discovered *L. polyphyllus* in British Columbia. His name was David Douglas (1799–1834), and we shall come across him later (see p.195) collecting in America and much involved with introducing fine firs and pines to this country. The lupin he found was one of twenty-one distinct forms he discovered.

The very English Russell Lupins were launched on the world by Bakers Nursery of Wolverhampton at the Royal Horticultural Society Show in 1937. The display caused a floral sensation and earned the nursery a rare Gold Medal. It was all the result of a twenty-three-year-long love affair (or obsession) with the development of the perfect lupin from the species *L. polyphyllus* by the Yorkshireman George Russell (1857–1951). Sometime in the early part of Russell's life, apparently, a display of blue and white lupins had caught his eye; years later, in the home of one of his gardening clients, he saw a vase of the same flowers. They must have taken his fancy again because he began in a quiet but determined way to try to improve them, enlarging the shape and the size of the flower and widening the colour range. This he eventually achieved in spectacular fashion, and his patience and skill were rewarded when, in his eightieth year, the Society awarded him the prestigious Veitch Memorial Medal.

The next flower introduced in the fourteenth century is one of the 900 species of salvia (a member of the *Labiatae* family) and amongst the earliest to reach Britain. It was *Salvia sclarea*, the Biennial Clary, which grows wild across Europe and Central Asia and, as its name suggests, blooms so vigorously that it exhausts itself, particularly on limy soils. Its metre-high (3 ft) spikes come in a variety of colours, cream, blue, pink or violet, and must have been an encouraging and cheerful sight in a fourteenth-century flower border. It was also what Hanmer would have called an 'Aromaticke'. Even today, if you smell the sage-like leaves on a hot summer's day, you will be reminded of Christmas, for they have a slight satsuma or mandarin fragrance.

The business part of *Salvia sclarea* is the sticky juice that comes from soaking the seeds in water, which was then used for bathing and soothing the eyes, hence its other vernacular names of Clear Eye, Eyeseed, Christ's Eye and Oculus-Christi. These last two, according to Nicholas Culpeper (1616–54), were the 'most blasphemous names', and were also used to describe our native Wild Clary, *S. verbenaca*, as well. He helpfully points out that the native clary could be found growing 'plentifully if you look in the fields near Gray's Inn and the field near Chelsea'. Londoners will surely be pleased to know this today.

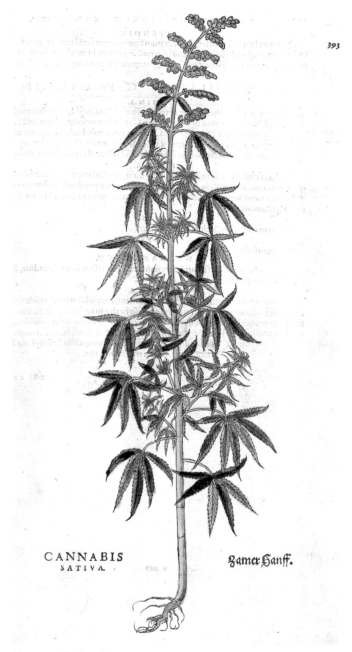

393

CANNABIS
SATIVA.

Zamer Hanff.

Hemp seed taken from *Cannabis sativa* was widely used along with bay, cowslip, goosegrass and ivy in love divinations until the early twentieth century.

The beautiful-smelling sage which we now use as part of the bouquet garni is *Salvia officinalis*, which was in all probability a Roman import since it was a native of Italy and was named in Abbot Aelfric's vocabulary of 995. To the Romans,

it was a sacred plant. It was collected with some ceremony and was never cut with an iron blade, since the iron could react with the chemicals in the leaf. It is a soothing herb which would surely have been grown for both medicinal and culinary purposes. Sage tea and ale were recommended to help the digestion, and the leaves were chewed to keep the teeth white. An Arabic proverb claims that: 'He who has sage in his garden will not die.'

Arriving during this century was a plant which has had a chequered history. Originating in India, it is one which has always caused both social and botanical difficulties. The plant is cannabis, the name given it by Dioscorides. The generic name covers two sub-species: the first is *Cannabis sativa* subsp. *indica*, which produces the now banned substance. The second is *Cannabis sativa* subsp. *sativa*, a plant apparently without narcotic implications and which is now widely grown commercially. The two sub-species have both been growing for centuries but have only recently, it seems, been botanically separated.

Cannabis was used by the Romans and for at least a thousand years before that by the Scythians, a tribe who inhabited the northern shores of the Black Sea. As it was in cultivation throughout the Roman Empire, it could well have been grown in England during the occupation, although there is no evidence for this. When it entered Britain from India during the fourteenth century, it was known as Canapus, Hanff Hemp or Devil's Flower, and was recommended for dropsy and distended stomachs. The seed itself could be distilled into a varnish, and, until comparatively recently, was also added to birdseed mixture. The plant was grown as a conventional field crop until the nineteenth century, being used in the manufacture of rope which was always in great demand, in particular by the Royal Navy. A tradition dictated that the seed should be sown on Good Friday, which in some places is known as Hangman's Day – the ultimate use of rope? Cannabis has given us the word 'canvas', which is made from hemp.

The constant need for cordage made hemp an important crop, and during the reign of Henry VIII, a royal edict made it compulsory for the plant to be grown, even though men working in the fields often complained of suffering from headaches; there was an understanding that women should not be allowed near the fields lest they be made barren. (Conversely, in Roman times the seeds were eaten as a contraceptive.)

A second Roman narcotic, and a rather sinister plant, also

made a reappearance during the middle years of the century. This was mandrake, or *Mandragora officinarum*. This old Greek name was a corruption of an Assyrian word for 'male drug of the plague god Namtar'. The plant is highly poisonous and was used to deaden pain, but was also considered an aphrodisiac. It is part of the *Solanaceae* family and one of 6 perennial species growing wild between the Mediterranean and the Himalayas. The root was believed to turn into a miniature human and scream in pain when being dug up (which had to be done in darkness at midnight), but Culpeper, in his eminently practical way, put it thus: 'The root formerly was supposed to have the human form, but it really resembles a carrot or a parsnip.'

Both cannabis and mandrake are plants which have been cultivated mainly in the herbal and medicinal gardens and as a curiosity. Somehow they seem the very essence of medieval life, with their associated half-myths and half-truths, their darker side suiting the timing of their entrance to Britain, which coincided with the raging of the Black Death.

The plague – its sinister Black Death name only came into being in the nineteenth century – first made itself known in Britain in 1348. The fleas of the black rat spread the disease from the Far East across Europe, arriving on the south coast of Britain in May of that year. It was so virulent that within six months almost the whole kingdom was at a standstill as the deadly virus cut a swath across the countryside. It has been estimated that between a third and a half of the population was decimated within that time. It had the most profound effect on rural society; there were just not enough people left to run the place. Travel in Britain would have come to an almost complete halt, too – both important business and merchant journeys would have been cancelled and no one who was as yet unafflicted would have wanted to visit relatives for fear of what they might find and bring home with them.

The year 1348 altered everything. In the space of just a few months, Britain went from being a confident, fast-growing, aggressive nation (the great victory of Crécy had only taken place two years earlier, in August 1346) to almost a non-country: bewildered, floundering, inward-looking and, eventually, rebellious. These circumstances were to lead to a 'black hole' of plant collecting, the effects of which would not be felt until the next century.

It is a relief to turn to a plant which is the very essence of the Mediterranean sun, and which, as a herb, is more popular today than ever. Sweet Basil or *Ocimum basilicum* (both

words derived from Greek and mean 'aromatic' and 'princely' or 'royal') is thought to have come to Britain about the same time as the plague arrived in the mid-1340s. It is a genus of some 35 species in the *Labiatae* family spread across tropical Africa and Asia. In India, where it is called Tulasi, the plant is sacred to the deities Vishnu and Krishna.

Sweet Basil had not long been in this country when two things were learned about it: first, that the young plants did not like being disturbed; and second, that a pot of the herb growing on the windowsill would deter flies. Perhaps some travellers from the south of Europe recounted how pots were displayed in bedrooms to deter fleas. Thomas Tusser (c. 1524–80) wrote an apt verse for basil:

Fine basil desireth it may be her lot
To grow as the gilliflower, trim in a pot
That ladies and gentles, to whom ye do serve
May help her, as needeth, poor life to preserve.

Later, in the seventeenth century, Nicholas Culpeper believed that basil was 'the herb which all authors are together by the ears about, and rail at one another, like lawyers. Galen and Dioscorides hold it not fitting to be taken inwardly, and Chrysippus rails at it with downright Billingsgate rhetoric.' Perhaps they should all have waited for the tomato, *Lycopersicon esculentum*, to arrive from South America in 1597, for, as the writer Jane Grigson remarks, basil and tomato are soul mates, complementing each other perfectly. Although there is a seventeenth-century recipe from a sausage-maker based in London's Fetter Lane for a sausage flavoured with basil, that is rare: there are few early recipes using the herb and it is only in the latter part of the twentieth century that we became uninhibited in our use of it and made the royal herb one of the delights of the kitchen.

Like basil and rosemary, the next plant also has the dual attributes of beauty and flavour, and has a long history of cultivation. *Crocus sativus*, the Saffron Crocus, a bulbous autumn-flowering plant, originated somewhere in the high mountains of Iran, Kashmir and Asia Minor. The flower is a rich lilac colour with purple veins, followed later by dull and insignificant leaves. The spice itself is obtained from the bright red pollen taken from the three-branch stigma within its goblet-like interior. It is one of the most ancient of named plants and betrays with its two names not only where it came from but a little of the history of ancient Asia Minor.

The word 'crocus' comes originally from the Greek *krokos*,

By 1670 such was the reputation of the golden spice saffron (taken from the stamens of *Crocus sativus*) as a cure for all the ills of the world that 300 pages of a work devoted to the whole of the Crocus genus, *Crocologia*, listed every known cure.

which in itself can be traced back to the old Semitic word *karkom* with associated words in both Sanskrit and Chaldean. This familiar name for such a familiar plant actually means 'saffron' in Greek, so 'Saffron Crocus' is something of a tautology. However, 'saffron' is a word which comes to us via a different linguistic route, the Arabic *zafaran*. So in talking of the Saffron Crocus, two different languages (and two different cultures) are joined together, with both words having the same meaning. Both they and the plant come from the same geographical area and the words reflect the linguistic maelstrom of the Greek and Arabic worlds, which met and melded throughout that area. There is still a reminder today of that ancient world, for in remote eastern Turkey, near to the

border with Syria, lies the thousand-year-old Syrian Orthodox monastery of Deir el-Zaferan, the Saffron Monastery.

The saffron spice had probably been here before (with the Phoenicians, when tin was being mined in Cornwall, and during the Roman period), but the cultivation of the flower as a saffron crop began somewhere in the 1340s at Chipping Walden in Essex. The spice is mentioned in a document of 1359 and the name Saffron became attached to Walden once cultivation had spread rapidly through the region. In 1375 at Peterhouse College, Cambridge (not far away), saffron was one of a long list of plants being grown there; others were leek, garlic, cress and parsley. There is a good story that a single crocus corm had been smuggled into Britain in the hollowed-out head of the staff of a pilgrim who had stolen it, presumably on the way home, 'proposing to do good to his country'. As this was related over two hundred years later, by the geographer Richard Hakluyt, not too much credence should be attached to it.

A second story credits Sir Thomas Smith (1514–77), statesman and Greek scholar, with the introduction of the crocus; he at least had the advantage of having been born at Saffron Walden, but long after the plant's arrival in this country. He could, however, have been responsible for organising the town's charter, granted by Edward VI in 1549, when, because of the commercial importance of the crocus to the town, three flowers were represented on its coat of arms.

Grown commercially throughout the Mediterranean and Arabic world for cooking, saffron was also used as a disinfectant, a dye and as a medicine – people were thought to laugh a great deal and 'be sprightly' when they had taken a dose. The streets of Rome are supposed to have been strewn with it for Nero to strut upon (perhaps in the hope that he might be more 'sprightly'?). Royal robes were dyed 'saffron yellow' and Hindu monasteries used it to dye the monks' robes, something that is occasionally still practised today. Saffron was sometimes substituted for gold in illuminated manuscripts.

It is said that it takes 4,300 flowers to yield 28.35 g (1 oz) of saffron; therefore approximately 152,000 flowers would have to be picked to produce 1 kg. Harvesting in the fourteenth century must have been a form of refined torture; nowadays, the plant is picked mechanically in September and October. However, since it is used in very tiny quantities in cooking, even a pinch goes a very long way. The most valuable spice on earth, saffron is so versatile that it was frequently adulterated; in Germany during the fourteenth century saffron inspectors were appointed, and punishment of those found guilty for tampering with the spice was to be buried alive.

Crocus sativus is a member of the *Iridaceae* family, and the 80 or so different dwarf species are indigenous across the eastern Mediterranean to China. None grows wild in England, but the spring-flowering blooms of **C. vernus** which arrived here at much the same time as *C. sativus* are one of the most welcome sights in the garden. They display themselves with a determined confidence, getting on with the job of blooming and banishing winter, but they are not show-offs, nor do they flaunt themselves to gain attention. The autumn-flowering corms behave similarly, but it is always a surprise to see them flowering during the latter part of the year, as if the seasons have got slightly out of balance.

All the plants for which there is evidence to justify their inclusion in our fourteenth-century garden were either the vernacular plants of Europe, or introduced first into southern Europe via the Arabic world and brought here by crusaders and travellers. Many other plants may well have been brought here in this way but did not thrive and have long vanished from our repertoire. The plants which did survive all have a long history in this country, having been proved worthy, by a process of trial and error, of being cultivated. Quite a number of these, once established domestically, have felt so comfortable that they have spread into our own countryside.

We must never forget, however, that other category which made an early and significant contribution to the glory of the garden, and that is our own domesticated native plants. The familiar plants of one's own patch of countryside would have been where every gardener began, and it was only later, with the growing confidence of travel and exploration, that imports gradually began to fill our gardens. During the previous three hundred years, from the Norman Conquest onwards, most of the population of Britain would have used the indigenous flowers growing around them to cheer up a cultivated piece of ground.

There would have been our native honeysuckle or woodbine, *Lonicera periclymenum*, draped over a wall or hedge (always twining from left to right); the pink Water Avens, *Geum rivale*; the lovely Yellow Corydalis, *Corydalis lutea*; Wild Angelica, *Angelica sylvestris* (some three hundred years later, in the early seventeenth century, John Tradescant senior would collect a better version of this from Russia; see p.131); the Guelder Rose, *Vibernum opulus*, *Juniperus communis*; and the delicate blue-flowered Granny's Bonnet, *Aquilegia vulgaris*. All these, and many more, would have played their part in making life that little bit more civilised.

The gardens of the monasteries and abbeys were still making the herbal and medicinal running, although Cambridge and Oxford Universities were also beginning to take gardening seriously. In spite of the terrible havoc caused by the plague, some of the country's wealthy and powerful families were busy building large new houses for themselves, and creating fine estates. Above all, it was the royal family that was in the forefront of the development of gardens and experimenting with newly acquired plants. They led the way in trying to tame the surrounding wilderness and creating civilised surroundings for their castles.

Plant Introductions in the period 1300–1399

nsd = no set date

nsd 1300   *Cannabis sativa* Hemp, Devil's Flower. Asia possibly India. Reintroduction; (see p.35).

Lupinus albus Lupin. E. Mediterranean.

Lupinus luteus Yellow Lupin. S. W. Europe.

Lychnis chalcedonica Jerusalem Cross, Maltese Cross. European Russia.

Lychnis coronaria Rose Campion. S. E. Europe.

Picea abies Common or Norway Spruce. S. Scandinavia to C. and S. Europe.

Populus alba White Poplar, Abele. N. Africa, Turkey, C. and S. USSR, including S. W. Siberia.

c. 1340   *Ocimum basilicum* Sweet Basil. Asia.

Rosmarinus officinalis Rosemary. Mediterranean.

c. 1350   *Crocus sativus* (syn. *C. sativus* var. *cashmirianus)* Saffron Crocus. W. Asia to Kashmir.

Crocus vernus Dutch Crocus. W. Russia, E. Europe to Italy.

Mandragora officinarum Mandrake. Devil's Apple, Love Apple. W. Balkans, N. Italy, W. Turkey, Greece. Reintroduction; (see p.35).

Platanus orientalis Oriental Plane. S. E. Europe, W. Asia.

Prunus dulcis Almond. W. Asia, N. Africa.

Punica granatum Pomegranate. S.W. Asia.

Salvia sclarea Biennial Clary. Europe and C. Asia.

c. 1375   *Cupressus sempervirens* Italian Cypress, Mediterranean Cypress. E. Mediterranean to Iran, *c.* 1375.

The Black Hole of Horticulture

1400–1499

The Portuguese used caravels to explore the sea routes to the Cape of Good Hope, India, the Spice Islands and South America.

Significant dates

1404 Owen Glendower (*c.* 1355–*c.* 1417), last independent Prince of Wales, holds his own Parliament

1407 Merchants Adventurers' Company founded

1413 Henry v (1413–22)

Portuguese voyages of discovery begin under Henry the Navigator (1394–1460)

1415 Henry v victorious at Battle of Agincourt, 25 October

1422 Henry vi (1422–61); murdered 1471

1429 Joan of Arc relieves Orleans; dies at stake 1431

1440 *The Feate of Gardening* published by Master Jon Gardener

1448 First European printing press

1453 Constantinople falls to the Ottoman Turks

End of Hundred Years War

1455 Wars of the Roses begin

1461 Edward of York seizes throne: crowned Edward iv (1461–83); deposed 1470; restored 1471

1474 William Caxton (*c.* 1422–91), the first English printer, prints his first English text at Westminster

1483 Edward v (uncrowned); Richard iii (1483–5)

1485 Battle of Bosworth Field; Richard iii killed; Henry vii (1485–1509)

1492 Christopher Columbus (1451–1506) sets sail and discovers the New World

1497 John Cabot (*c.* 1450–98), with his son Sebastian (*c.* 1474–1557), sails from Bristol, landing in Newfoundland

1499 Vasco da Gama (*c.* 1469–1524) rounds Cape of Good Hope, reaching Calicut in India; beginning of the Portuguese empire

COMPARED BOTANICALLY with the previous hundred years, the fifteenth century could not be more different, for the simple reason that it was almost devoid of new plant arrivals. No really exciting flowers were given hospitality, no great trees laid their shadow across the grass, and no beautifully scented shrubs perfumed our gardens. It seemed as if the route from the Mediterranean to our English gardens had all but dried up. Which of course, in a way, was exactly what had happened.

Two events of European magnitude which had taken place during the previous century had the knock-on effect during this one of shutting off almost completely the source of new plants to Britain. One was that the great Christian upsurge of the Crusades which had lasted on and off for more than two hundred and fifty years, and which had contributed so handsomely to English gardens during that time, was over. The war-weary Christians were defeated by the Muslims, who established themselves firmly at the crossroads of the known

world at the eastern end of the Mediterranean. No longer were the byways which criss-crossed the Continent busy with Crusaders and pilgrims coming and going to the Holy Land; 'travelling for Christ' was not an option. 'Travelling for trade' did not begin for the British in earnest for almost another hundred years.

The final death knell of western Christendom's adventure in the east came in 1453, the year that the Ottoman Turks captured the last remaining piece of the Byzantine Empire, the fabled city of Constantinople. Western Europe turned in on itself and meekly abandoned its questing routes into Asia Minor and beyond. Only Portugal, living as she did on the edge of the known world, was the exception. In 1418, the Portuguese discovered and colonised Madeira and the smaller Porto Santo. The latter had to be abandoned within a few years because rabbits which had been introduced by the colonists were steadily eating their way through the island's vegetable patches. Plump rabbits for the cooking pot were no good without the greens to accompany them on the plate.

Under the guidance of Henry the Navigator (1394–1460), Portugal's high-sided caravels lay, like so many spaceships, waiting to explore the unknown worlds beyond the horizon. Voyaging across oceans and along strange coastlines was just the start of an adventure which would eventually lead to a cornucopia of plants streaming their way to continental Europe and Britain, but that was all in the future.

The second event, and the far more serious one, was the decimation of the English and European population as a result of the plague. This catastrophe had struck three times in the previous century – in the 1340s, 1350s and 1370s. Travel anywhere, especially the undertaking of long continental journeys, was a terrifying experience, and for a time must have virtually ceased. The last thing on anyone's mind was the bringing home of seeds or bulbs to grow in the garden. The intention would have been to return as fast as possible – praying no doubt that the dread disease had not struck the family at home.

The lack of any real horticultural evidence in this century highlights the problems of finding out about early plant history. As the garden historian Richard Gorer so succinctly put it, plant collecting can be divided into three distinct portions, just as Caesar divided All Gaul into three. First and usually easiest, the plant must delight and enthuse you; second, live material or seed must be brought home; and third, it must like its environment and grow and reproduce. There is also a fourth criterion which applies particularly to the early periods and that is the recording either in writing or painting of the plant, and then the survival of that material. This last brings in one of the imponderables – how long was it before the horticultural immigrant made such an impact or was so profuse as to be worthy of note? In earlier times, a plant may have become so familiar in gardens or perhaps so locally profuse that no one thought it worth while to record it, whereas the nearer our own time the discovery, the more likely it was that a swift record was made.

Some life, of course, did keep going, but, as the medievalist John Harvey writes, following the abdication of Richard II in 1399 and the ensuing political turmoil, as well as the economic chaos of the Black Death, England died a little. There were wars still being fought: the Hundred Years War with France, which had begun in 1338, did not end until 1453, with the Battle of Castillion on 17 July in south-west France. Two years later, in 1455, and for the following three decades, there was the internal Wars of the Roses, a surreal and long-

The Tudor Rose Award is the premier award of the Hampton Court Flower Show.

drawn-out bitter fight: its name – so inappropriate – was popularised by Sir Walter Scott.

When Eleanor of Provence married the English Henry III in 1236, she brought with her as her emblem the *Rosa x alba*, the white rose. Later, her son Edward I continued to use this flower, although he upped the stakes and made it into a gold rose. Eleanor's second son Edmund became 1st Earl of Lancaster in 1275 and he brought from Provence his red rose emblem, based on *Rosa gallica* var. *officinalis*, see illustration p.44. Much later these two roses, with their opposing dynasties, became the warrior symbols of the Wars of the Roses, which ended in 1485 when Elizabeth of York (white rose) married Henry VII (red rose), and the two were combined to create the emblematic 'Tudor Rose'.

This artificial creation which represented the House of

The White Rose of York *Rosa alba semi-plena* came to be the
badge of the Jacobites and in the Language of Flowers, white roses
(with two buds) were a symbol of secrecy.

Tudor had all the hallmarks of a modern-day logo. First, its
design was clearly recognisable; second, it carried a clear
message (we are the Tudors); and third, it could be reproduced in many different mediums – carved in wood, embroidered on uniforms, created in jewels, woven into material or
used in paintings, all essential imprints when the majority of
the population used images and pictures to understand a
message.

Much like two great warlords with their followers (the
medieval equivalent of a favoured football team and its fans,
perhaps), there were several dynastic skirmishes between the
House of York – the white rose – and the House of
Lancaster – the red rose. Serious and bloody fighting took
place at a number of battles, including one in a snowstorm at
Towton in Yorkshire in 1461, when the home side proved too

much for the 'red roses'; later in the year, however, near St
Albans, the Lancastrians took the upper hand. That was the
second Battle of St Albans; the first had been won six years
earlier by the 'white roses'. Ten years later, in 1471, at the
Battle of Tewkesbury, so many men on both sides were
killed – although it was judged a Yorkist victory – that the
site became known as 'the Bloody Meadow'.

The fearful clash which took place on 22 August 1485 at
Bosworth Field brought Henry Tudor to the throne of
England as Henry VII when he defeated the Yorkist Richard
III, and this more or less brought to a close the civil war and
the batterings of two great dynasties.

By the very nature of life, events have to be seen in retro-
spect and set in context, and historians usually consider this
century as one of transition: from the old order to the faint
stirrings of the modern age. Certainly this can be seen in
horticulture and the development of the garden (if not in the
acquisition of new plants). It has been calculated that at the
beginning of the fifteenth century there were probably about
a hundred plants or so, both native and introductions, that
had been 'domesticated' for growing in gardens, and it was
not until well into the reign of Henry VIII, *c.* 1530, that the
figure reached some two hundred plants.

It was, however, during this century that saintly intervention was invoked to help gardeners, with not one but two
patron saints being venerated for their helpful horticultural
skills. Although both St Phocas and St Fiacre had been
called upon by gardeners on the Continent for centuries, it
was only now that they seemed to be needed in England.
Indeed, St Fiacre, despite having a French name, was
believed to be a Scots or Irish prince who lived sometime
during the seventh century. He left home to preach to the
Gauls and his journey took him to Meaux (near Paris) where
he settled as an anchorite, making a beautiful garden deep in
the forest. Legend has it that a magical hedge grew up
around him and that he lived unmolested by wolves and wild
boar. The devout Fiacre built an oratory and a hospice for
travellers where now the town of Saint-Fiacre-en-Brie
stands.

Phocas, the second gardening saint, lived earlier (probably
in the fourth century) at Sinope on the Black Sea near
Pontus – from where the much maligned *Rhododendron ponticum* originates (see p.122). He was by all accounts a good
market gardener but combined that with being a hermit. At
a time when Christians were being persecuted, Phocas had
the misfortune to have to dig his own grave during which

This painting by Henry Payne entitled 'Choosing the Red and White Roses' in the Pre-Raphaelite style takes its inspiration from Shakespeare's play *Henry VI Part II* where the nobles of the Houses of York and Lancaster meet in the gardens of the Inner Temple in London to pluck white and red roses.

task his fortitude so impressed the soldiers overseeing him that afterwards they started talking about this holy but now unfortunately dead man. Almost immediately pilgrims began flocking to Phocas's grave and his prowess as a gardener began to be celebrated.

The seeming paucity of new plants, however, did not stop the development of gardens. Gradually it was realised that, as with the contemporary house interiors, space could be allotted within gardens for specific uses. Thus items like 'kechengardyn', 'the great garden' and the 'pleasaunce' began to appear on plans and in leases of the time, confidently showing the beginnings of divergence between the working garden and the pleasure grounds.

Likewise, symbolism in the late medieval garden became as complicated as working a modern-day website, with both gardens and plants being interpreted in a highly pietistic way. The most mystical and idealistic of these symbols and the one which inspired painters of the Middle Ages, was the idea of the 'Mary Garden', where the glory and image of the Virgin's holiness was seen through the beauty of flowers. Even from the pre-Christian era flora (and fauna) had been used to interpret and inspire the burgeoning thoughts of an emerging nation. As we have seen, some of the images, like the pomegranate, were easily assimilated into the early Christian world. By the fifteenth century, the coded use by painters of the Madonna lily and its image of the celestial joy of the Virgin Mary was both supreme and enduring, but the

iris, the violet, the hollyhock, the marigold, the daisy (which Chaucer called 'the floure of floures'), the lily of the valley and the wild strawberry were, among many others, all being used as well almost as a matter of course to exemplify the Christian message. The rose was the other great inspirational flower. In its early days it signified the spilt blood of the martyrs, and St Bernard of Clairvaux (1090–1153), in his treatise on the Love of God, declared it to be the symbol of Our Lord's Passion. Gradually this beautiful bloom took on the association of Divine Love, but through the centuries, as religious ideals gave way to more humanistic interpretations, it became a symbol of that most powerful of man's own declarations of love. This century was to prove the apogee of the floral device as the message of the Church, since such images would, from now on, become progressively more secular.

The enclosed pleasure gardens became places in which to pray, to linger and play out the enticing games of courtly love. They were set within what often looks in paintings of the time like an outdoor room – something very familiar to us today. Thus ornamental buildings and structures began to play an increasing part in the design of gardens. An 'herber' or arbour in the corner of the gardens at Windsor Castle was described by the future King James I of Scotland when he was held prisoner there from 1413 (when he was nineteen) to 1424. Sometime during those eleven years he saw Lady Joan Beaufort (granddaughter of John of Gaunt) walking in the garden and, like all good romances, composed a poem about

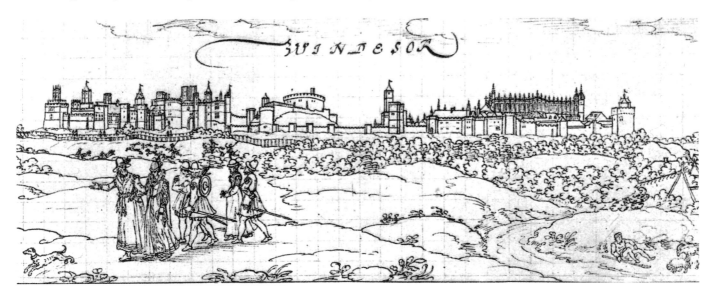

Windsor Castle has been a royal residence since it began life as a stockaded earthwork in the eleventh century following the Norman invasion. By the thirteenth century the gardens and the orchards of the Castle were well developed as existing documents show.

Calendar illustration of gathering flowers from *The Playfair Hours*, a fifteenth-century manuscript from Rouen.

her great beauty and his love for her. The following two verses describe the scene:

> Now was there made fast by the tower's wall
> A garden fair, and in the corner set
> A herber green with withies long and small
> Railed about; and so with trees set
> Was all the place, and hawthorn hedges knit,
> That not a one was walking forby
> That might within scarce any wight espy.
> So thick the boughs and the leaves green
> Beshaded all the alleys that there were,

> And midst every herber might be seen
> The sharp, green, sweet juniper,
> Growing so fair with branches here and there,
> That, as it seemed to anyone without,
> The boughs spread the herber all about.

It certainly sounds a very secluded arbour for a tryst. Perhaps James managed to send her his tender love poem, for on his release from Windsor in 1424, the couple married at the Priory Church of St Mary Overy in Southwark, London.

Despite the traumas of the plague and the wars, progress was being made; gardens were taking on the gloss of modernity and adapting the old to the new. Paths were gravelled (as at Kenilworth in 1440) or sanded; seats were set in place, often encased in wattle hurdling, either turf benches or surrounds to trees. There were more solid buildings too: a pavilion or 'gloriet' was sometimes set in a favourable place from which to view a feature – the innovative knot garden (sometimes called a maze), or a bed of fine flowers. Often the gardens were decorated even further with very complicated three- or four-tiered topiary trees, intricate trellises or pergolas, or fountains. All these noble gardens and their ornamentation were a far cry from gardening in the rest of the country, where the growing of vegetables was the pre-eminent garden task. If the new complicated patterns and the sophisticated crispness of these creations had ever been glimpsed by the vast majority of the population, they would have seemed outlandish, much as a Barbara Hepworth sculpture appears avant-garde to us today.

So it was that, despite death and destruction stalking the land, the beautifying of the outside space around the home – or castle – was being developed and the seeds were being sown for the great onslaught of the Tudor garden in the next century.

Just prior to the invention of the movable metal-type printing press (in Germany in 1448 by Johannes Gutenberg), the first English garden book, *The Feate of Gardening* by Master Jon Gardener (or Gardyner), is believed to have been written. It survives as five manuscript pages written in doggerel verse. It can be dated to *c.* 1440, although it has been argued that it was actually written about a hundred years previously, and that it was an earlier 'Master Jon' who was the author. This earlier Jon (whose real name was John Penalowne) was in charge of Edward III's gardens during the 1350s at Windsor and then in London (for which he earned 2¼d a day). We know he was pensioned off in 1365 to the monastic house of St Germans in Cornwall where, in 1400, he apparently accepted a cash payment in lieu of living there. The second Master Jon worked at the King's great Eltham Palace some forty years later, in the reign of Henry VI.

Whoever the real author was, there are two surviving but slightly different manuscripts, and there is no doubt that the writings are based on the practicalities of working with plants in a real garden. One hundred herbs are described, well over a quarter of which are native to Britain. There are also sections on trees, grafting, viticulture, onions, parsley and coleworts (greens), but, disappointingly, there is no mention of the royal rosemary which had enjoyed such a

spectacular entrance in the previous century. Perhaps this omission points to the manuscript having been written earlier rather than later, since we know rosemary only arrived during the 1340s (see p.82), and would thus have been too new on the scene to be mentioned in a manuscript dating from the same decade. On the other hand, *Crocus sativus,* the Saffron Crocus, arrived sometime after 1350 (see p.82) and there are detailed instructions in the manuscript regarding 'Saferowne', indicating that the manuscript might have been written at the later date.

There is another crocus which, against all the odds, seems to have made its entrance about this time. It is ***Crocus nudiflorus****,* which flowers in the autumn and is neatly described by its second name, 'flower blooming before the leaves' – just like the Saffron Crocus. There is a little mystery about this corm, which originated in south-western France and Spain; it relates to the worthy Yorkshire town of Halifax, where, from the Middle Ages on, it was being grown as a field crop and called the Halifax Autumn Crocus. The dried stigma apparently has the same properties as *C. sativus,* so it was presumably grown because it was sturdier and fared better in that area. The Saffron Crocus flourishes in a dry climate – hence it did well in East Anglia, but is generally much more difficult to grow. One wonders if the Halifax Autumn Crocus was used to bulk up the true saffron (a serious and punishable offence; see p.84) or used simply as a straight sub-

Crocus nudiflorus was introduced into England from the Iberian peninsula. It is autumn flowering and, once established, has great tenacity and will spread to make bold splashes of colour.

One species of the Star of
Bethlehem *Ornithogalum
umbellatum* that can be
cultivated in England is
O. longibracteatum False Sea
Onion which grows in the
Northern and Eastern Cape
of South Africa.
O. pyrenaicum which is
known to have been
growing here since at least
1548, is a native of Turkey.

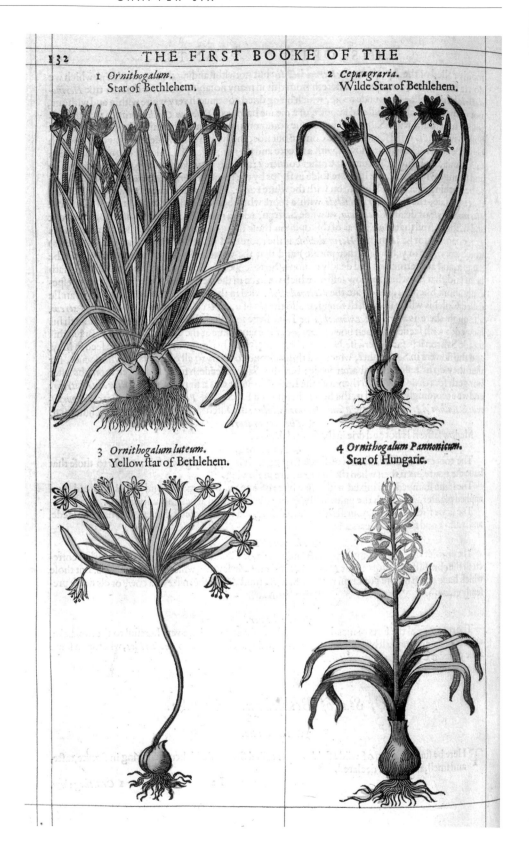

1 *Ornithogalum.*
Star of Bethlehem.

2 *Cepaagraria.*
Wilde Star of Bethlehem.

3 *Ornithogalum luteum.*
Yellow ſtar of Bethlehem.

4 *Ornithogalum Pannonicum.*
Star of Hungarie.

stitute. It is still to be found naturalised around the north Midlands, in isolated groups and usually in association with buildings which once belonged to the Knights Templar or the Hospitallers, the interpretation of which seems to point to an association with their crusading activities. Sir Thomas Hanmer calls the 'Autumnall Crocuses' the true crocus and warns that 'field mice doe often eat the rootes in the ground, and therefore must bee taken in traps or otherwise destroyed, with poyson'd kernells or apples'.

The only other probable introduction for this period also has a tradition of having caught the eye of the Crusaders, and that is the aptly named Star of Bethlehem, *Ornithogalum umbellatum.* Translated from the Greek this means a milk-white flower which rises from a single stalk and looks like a bird. Just so. Although it does grow wild in Britain, along with three other Ornithogalums, it has never been considered a true native, its natural home being the Near East. It is in a genus of about 80 species in the *Liliaceae* family which nearly all come from southern Africa. One we all know today and which has been used as a cut flower for some years on account of its long vase life is the Chincherinchee, the common name for *O. thyrsoides*, which arrived from the Western Cape in 1757. Apparently, the flower spikes could stand the long sea journey from South Africa to England and still be in good decorative order for some weeks thereafter.

The Star of Bethlehem is not an energetic flowerer like some plants, but awakes at about eleven o'clock in the morning and by three p.m. is exhausted and retires for the night. It was known throughout Europe as the Eleven O'Clock Lady, or Wake-at-Noon . . . what a reputation! Like many bulbs, it has been considered edible and was once part of the diet in Palestine, where it was dried and stored then used for food on journeys across the desert. However, the writer Alice Coats issues a stiff word of warning: 'It is now considered "intensely poisonous" unless well cooked, and the leaves are said to be avoided by grazing animals, so dietetic experiments should be conducted with caution.' You have been warned.

It does seem extraordinary that there appears to be no other firm evidence for any further plant introductions during the whole of this hundred years. On the other hand, if one remembers the events which were preoccupying the country – from the endless Wars of the Roses to the ongoing effect of the Black Death (the population apparently did not regain or pass the five million figure until the seventeenth century) – it is not that surprising.

There was one traveller, however, who came to England in 1446 and left a record of his travels, in the southern half of the country at least. Leo von Rozmital was a wealthy Bohemian nobleman who had already visited a number of countries in Europe on a fifteenth-century version of the eighteenth-century Grand Tour. His impression of England is quite extraordinarily modern: he found it '. . . like a small garden, surrounded by the sea on all sides'. When he got to London he remarked on 'the elegant gardens, planted with various trees and flowers, which are not found in other countries'; again a most interesting statement, indicating that already Britain was showing signs of the precocious horticultural knowledge it was to display in the coming centuries.

If the English did but know it, towards the end of this troublesome century the very seeds of our new botanical life were being laid thousands of miles away by the Spanish and Portuguese explorers of Africa, India, the West Indies and South America. The discoveries of these lands were to lead to the spirit of adventure gripping the whole of Europe, and for Britain, waiting in the arbour was not an option. The Tudor dynasty was about to take centre stage.

Plant Introductions in the period 1400–1499

c. 1400 *Crocus nudiflorus* Autumn Crocus. N. and E. Spain, S.W. France.

Ornithogalum umbellatum Star of Bethlehem. Europe, Asia Minor, N. Africa.

Over the Horizon
1500–1599

4 Flos Aphricanus maior simplici flore.
The great single French Marigold.

5 Flos Aphricanus minor simplici flore.
The small French Marigolde.

John Parkinson grew *Tagetes erecta*, the African Marigold, and thought it a '…goodly double flower which is the grace and glory of the garden…'.

Significant dates

1509	Henry VIII (1509–47)
1514	Building of Hampton Court begun; completed 1540
1516	Publication in Latin of *Utopia* by Thomas More (1478–1535); English translation 1556
1534	English Act of Supremacy; break with papacy as Henry VIII declares himself Head of the English Church
1536–40	Suppression of English and Welsh monasteries
1538	Publication in English of *Libellus de Re Herbaria Novus* by William Turner (*c.* 1508–68)
1543	First botanical garden created, at Pisa
1545	Botanical garden started at Padua
1547	Edward VI (1547–53)
	Portuguese settle on coast of Brazil
1550	Formation of the Muscovy Company
1551	First edition of *A New Herball*, published by William Turner
1553	Mary I (1553–8)
1557	*Hundredth Pointes of Good Husbandrie* published by Thomas Tusser (*c.* 1524–80)
1558	Elizabeth I (1558–1603)
1563	*A most Briefe and Pleasaunt treatyse: The proffitable Arte of Gardening* published by Thomas Hyll (*fl.* 1540s–70s)
1564	William Shakespeare born (d.1616)
1569	Gerardus Mercator (1512–94) produces world map
c. 1570	John Tradescant the Elder born (*c.* 1570–1638), gardener and plant collector
1573	*Five Hundredth Pointes of Good Husbandrie* published by Thomas Tusser
1577–81	Sir Francis Drake (*c.* 1540–96), circumnavigates the world
1578	*A Niewe Herball or Historie of Plantes* published by Henry Lyte (*c.* 1529–1607)
1579	Formation of the Eastland Company
1581	Formation of the Dutch East India Company; received its charter 1602
1587	Execution of Mary Queen of Scots
1588	Defeat of the Spanish Armada
1594	*The Jewell House of Art and Nature* published by Sir Hugh Platt; contains first mention of a window box
1596	*Catalogue of Plants Cultivated in the Garden of John Gerard* published by John Gerard (1545–1612)
1597	*Herball or General Historie of Plantes* published by John Gerard

THE CLAMOUR OF EXPLORATION and discovery of the New and Old Worlds by the Spanish and the Portuguese towards the end of the previous century is reflected in this new century, the sixteenth, by a delicious bouquet of exotic plants being delivered to grow and flower in the gardens of Europe. But at the beginning of the century all eyes in Britain were focused on the evolving power struggle taking place between France and the octopus-like Habsburg Empire.

Foreign exploration for the English at this time was of little interest. John Cabot's voyage to Newfoundland in 1497 had hardly caused a ripple; the place was simply 'a new fishing-station for cod off the Canadian coast'. (In fact the mapping of Newfoundland would be one of James Cook's most extraordinary surveys. The island is larger than Ireland and has some 9,600 km (6,000 miles) of coastline. His hydrographic work, which took place before he took command of his own ship, the *Endeavour*, lasted from 1763 to 1767 and was much admired.) It took England another sixty years to break out of its self-imposed isolation and join the exploration race for overseas power, wealth and possessions. By then, trading possibilities beyond the boundaries of Europe were the overwhelming lure for the English.

The cloth and woollen industries were of the greatest economic value to Britain, most of the produce being sold via Antwerp (which had taken over the trade from Bruges) by the Merchants Adventurers, an English trading company set up in 1407. They came to rival the Hanseatic League, a much older organisation of North German trading towns set up in the thirteenth century. Both trading companies were acrimoniously dissolved during the mid-seventeenth century. Later in this century, trade with the Russian market was put on a more formal basis with the founding of the joint-stock Muscovy Company in 1550. Their headquarters were based on Rose Island in the White Sea and the trade was mainly in commodities to be used by our navy – train oil (fish, seal and whale oil), tallow, tar, timber for masts, and hemp.

The Russian trade monopoly was lost in 1698 when, after several hair-raising adventures, it was realised that Siberia offered little in the way of commerce, and that the north-east passage to Asia and the Orient was nonexistent. By then anyway the Muscovy Company, which was only finally dissolved in 1917, had transferred its trading prowess to Persia. That trade, along the Silk Route and beyond, was firmly in the hands of the Venetians who plied their wares eastwards across the central Asian lands with the co-operation of Turkey. Yet a further trading conglomerate was set up in 1579 to cover Scandinavia, the Baltic and Poland and called the Eastland Company. Then in 1600 the greatest and most powerful trading company of all, the East India Company was incorporated. This and its fellow rival, the Dutch East India Company, which began its life two years later, had the most profound influence not only on trade but on political power in India and the Far East. It almost became a government in its own right in the eighteenth and nineteenth centuries, and gardeners in particular during that period have much to be thankful for in the Company's diligent searching and transportation of the wonderful plants from the Far East.

These different trading organisations finally pushed the British into seizing not only the opportunities for making money through trade but for exploring new and unseen lands. The expansion outwards from Europe both westwards and eastwards led inevitably to the horticultural expansion of our gardens. After five hundred years of sampling mainly the

THE CITRUS SPECIES

The sixteen Citrus species in the *Rutaceae* family are probably the most important genus of subtropical and tropical fruits. Originating from Asia, particularly China, a few are native to some of the islands in the eastern Pacific, but without exception they are all evergreen and make a small tree or shrub.

One of the tallest, at 6–9 m (20–30 ft), and the first to be recorded as growing in England, was the Seville Orange, *Citrus aurantium*, which makes such tasty marmalade; the rind also provides the perfume industry with Oil of Neroli (an essential oil said to have been discovered about 1670 by Anne Maria de la Tremoille, the Italian Princess of Neroli). Another species which also produces an essential oil, in particular for eau-de-Cologne, is *Citrus bergamia*, the Bergamot Tree, a close relative of the orange. This was grown particularly in Calabria and Sicily, and the small tree was brought to England sometime in the latter part of the seventeenth century.

The next three species to arrive – the citron, lemon and lime – were all recorded as growing in the Oxford Physic Garden in 1648. *C. medica*, the citron, was described by Theophrastus as being cultivated during his lifetime (the fourth century bc), especially around the ancient region south-west of the Caspian Sea where the Medes had once lived, hence the *medica* part of the name. The tree also has the reputation for being the first of the Citrus species to have arrived in Europe. It is certainly one of the thorniest, and the edible fruits are very acid; they are long rather than round, have a rough yellow rind and were used in the Jewish ceremonies of the Feast of the Tabernacles.

The lemon (*C. limon*) has been found growing wild in northern India (although there is some dispute as to where its true home is) but by the thirteenth century it was growing in the Middle East. The Crusaders are believed to have introduced the tree into Italy, where it flourished and spread throughout the Mediterranean region, and by the nineteenth century it was naturalised in both the West Indies and Florida. The lime (*C. aurantifolia*) was the third fruit to have arrived by 1648 and, just like the lemon, was found to be an essential part of ships' stores, and was used the world over to ward off the ghastly effects of scurvy. The lime had originally been brought to south-west Europe probably by the Arabs, and by the fifteenth century it was well enough established for Columbus to take to America – presumably so that eventually fresh limes could be taken on board for the journeys home.

Because of the climate, limes really only have a toehold in Europe, and are only able to survive with any degree of aplomb in the hottest and most southerly parts of the Mediterranean. They are much more sensitive to temperature change than lemons (which have tender skin and which thrive in southern Europe).

C. sinensis, the Sweet Orange, seems to have been a relative latecomer to Europe, with a reputation for having been originally brought to southern Europe in 1498 by Vasco da Gama from the Cape of Good Hope – presumably having started life in Asia, then been traded and grown down the east coast of Africa. (The Dutch East India Company was the first to make a European garden at the Cape; it was planted on the north slope of Table Mountain in 1652, and was used to supply their ships with fresh vegetables and fruit, including all the Citrus tribe, on the long voyages to and from the East Indies.) In England the orange was eaten, rather apprehensively, by Samuel Pepys, just as he tried sipping tea, that other new import of the seventeenth century. Like the lime and lemon, the Sweet Orange tree quickly spread to the West Indies and the warmer regions of America, both South and North. It was from Brazil during the nineteenth century that one of the best sweet varieties was introduced into the USA, the Navel Orange, *C. sinensis* 'Washington', so-called that because that was the state where it was first grown.

Citrus reticulata is the species which has produced the loose-skinned varieties like the mandarin and satsuma, while the shaddock, *C. grandis*, a native of Java and Malaysia, was brought to England by a Captain Shaddock in 1722. It is however better known by its native name of pomelo and is probably the forerunner of the grapefruit, *C. paradisi*. The first mention of this fruit in the eighteenth century referred to it as 'the forbidden fruit of Barbados'. The grapefruit's antecedents are not at all clear; the fruit may be a hybrid between the pomelo and the orange, or a sport of the pomelo. However, by 1830 it was recognised as a species, and by 1880 the fruit was being grown commercially in Florida and had begun its seduction of the world's breakfast tables.

Like all Citrus trees, neither the pomelo nor the grapefruit is winter-hardy in northern Europe but must be pampered and needs considerable warmth. It is just as well that most Citrus species are tough and 'good doers' otherwise they would never have survived the early attempts to keep them frost-free through the winter: dark cellars, braziers, coal fires, enclosed sheds, caves and who knows what else were all experimented with.

The Citron *Citrus Medica* is reputed to have been one of the trees chosen for the Hanging Gardens which Nebuchadnezzar II created at Babylon.

The very first use of glass for the growing of Citrus fruit was at Heidelberg in Germany in 1619, which housed about 400 trees. The idea spread, and in England the specially built structures were first called 'oringe gardins' and then later 'orangeries'. The idea was helped by the writings of John Evelyn, who charged that 'light is half their nourishment philosophically speaking'. Buildings or structures using glass in England often came to be called conservatories or greenhouses – the first 'greenhouse' so named was built at Chatsworth in 1697. Orangeries in the succeeding centuries gradually became more and more elaborate, and by 1804 at least one, that at Nuneham Courtenay in Oxfordshire, was quite celebrated, having a removable roof, sides and front, designed thus so that when the structure was dismantled and the ground turfed over, the oranges appeared as though growing in the lawn.

floral delights of Europe, suddenly other possibilities were made available, and the exposure of both different lands and of plants found growing in the Near East and the outer fringes of Europe, excited the experimental gardeners of the day. This was the start of the second great period of plant introductions, which took place between 1560 and 1620.

The future Henry VIII was nine years old at the beginning of the century, and by the time he was eighteen he was the most highly regarded Renaissance prince in Europe, a sort of Prince Charming with brains, who, seven weeks after his coronation in 1509, married his brother Arthur's widow, Catherine of Aragon.

For most of the century there was little time to look around or, indeed, forward. With so much kingly muscle being flexed in the royal drive to exert total control over the Church and to seize the wealth of the monasteries, time was best spent protecting one's back – or one's head from execution – thus leaving little energy for the entrepreneurial spirit to flourish. It was not until the century was almost over that the rage with Spain trumpeted into the Spanish Armada and a more aggressive attitude was taken towards seaborne adventure. This in itself was to have enormous influence on the arrival of new plants in this country.

Horticulturally, the century began quietly enough, but gradually wound itself up into a rush of newly acquired plants, ending with Queen Elizabeth I being offered, to considerable royal amazement, an out-of-season freshly ripened cherry (then a much admired fruit) when she visited the home of Sir Francis Carew of Beddington in Surrey in 1599. The fruit's ripening had been delayed by about two months by keeping the tree covered and cool between June and August in order that the Virgin Queen could be offered the succulent morsel.

Sir Francis could also have shown her another singular attraction, his orange trees; he was reputedly the first person in England to have planted them. The orange trees growing at Beddington would have been the Seville Orange (*Citrus aurantium*), the bitter fruit we know today as the marmalade orange (seemingly they are not used in Spanish cooking at all). When John Evelyn visited Beddington in September 1658 – about the same time as the Sweet or 'Portugal' Orange (*Citrus sinensis*) arrived in Britain – he found the original trees 'being now over grown . . . and planted in the ground and secured in winter with a wooden tabernacle and stoves'. The orange trees at Beddington eventually succumbed to the vagaries of our English weather but not until 1739, when they

must have been well over a hundred and fifty years old.

It was during this century that four men, by their writings, came to influence the gardening world of the sixteenth and seventeenth centuries and beyond. Three of them were practical, and obviously keenly observed their gardens and the natural world about them; two of them were acquainted and exchanged ideas; and one of them, the Lord Chancellor of England, wrote a book on the philosophy of life which, 500 years on, still has as much relevance as when it was first written.

Sir Thomas More (1478–1535), Henry VIII's unfortunate Treasurer and Chancellor, wrote *Utopia* in 1516. It was first published abroad, in Louvain. In it he presents his thoughts as one of the most eminent humanists of the Renaissance. He seems to have had a most serenely simple private existence, living as a Carthusian (but without taking the vows) at the Charterhouse in London. His book recommends, in

In his book, *Utopia*, Sir Thomas More envisaged a town in which all the residents had the space for a garden around their home.

effect, the creation of garden cities to help with the stress of living in large conurbations. *Utopia* was not translated from the original Latin into English, nor published here, until 1556, twenty-one years after More's death, but his philosophy was echoed in the twentieth century with the idea of a 'Green Belt' being retained between towns, and the creation of country parks to help city dwellers retain (or regain) their equilibrium. Unfortunately, Thomas More did not escape stress in his own life, and was executed in 1535 for refusing to acknowledge Henry VIII as Head of the Church.

The second writer of influence was the great botanist William Turner (*c.*1508–68), who was born at Morpeth in Northumberland. In 1526 he went up as an undergraduate to Cambridge, where he soon became so frustrated at the standard of botanical tuition he received that he wrote he 'could learn never one Greek, neither Latin nor English name even amongst the Physicians, of any herb or tree such was the ignorance in simples at that time'. He felt so impatient about the lack of available knowledge that in 1538 he wrote his own plant list, which has become a unique record of 238 of our native plants. The huge advantage of this list over all earlier ones was that it was recorded for the first time in the vernacular instead of in Latin, although it was published under the title of *Libellus de Re Herbaria Novus*.

Turner, a Protestant, was an ardent supporter of the Reformation, and because of his strongly held beliefs he spent two long periods of exile on the Continent, the first during the reign of Henry VIII from 1541 to 1548 and then again during Mary's reign (1553–8). However, having studied at Cambridge, he was able to take his degree in medicine at Bologna during the 1540s. Here he met or corresponded with all the leading botanists and naturalists of the day.

Turner greatly admired the long-time Professor of Medicine at Tubingen University, Leonard Fuchs (see p.152), and it was at Tubingen that he began to embark on his most serious work of plant identification and classification. The first part of *A New Herball* was published in 1551; eleven years later, in 1562, delayed partly by his enforced absence abroad, the second part was published, while the third and final part was completed in 1568, the year of his death. After all the religious problems he had encountered in his life, he perhaps wisely dedicated *A New Herball* to Queen Elizabeth I.

It is these two books which have caused Turner to be given the title of the 'Father of British Botany'. He was made a Fellow of Pembroke Hall (now Pembroke College),

Cambridge, and was probably the first in a long and honourable line of ecclesiastical botanists and horticulturists who have enlivened the floral world by their studies and erudition. Eventually he became Dean of Wells Cathedral.

Nearly two hundred years later, Linnaeus honoured Turner by naming a genus for him: Turnera is a group of 90 species, mainly shrubs and mostly from South America, but with a few found in Madagascar, Mauritius and South Africa. None was brought to this country until the eighteenth century and the only one available now appears to be *Turnera ulmifolia* (meaning 'with leaves like an elm') – the West Indian Holly or Sage Rose – which arrived here in 1733. Coming from South America, it needs hot conditions; it is a sub-shrub with yellow flowers which can be grown from seed.

The third botanist of note is someone we would now call a polymath and who, like Turner, was also at Cambridge. His name was Thomas Tusser (*c.*1520–80) and he came from Essex. He was educated at Eton and Trinity Hall, and wrote on agricultural matters. He was successively a musician, schoolmaster, servingman, husbandman (or cultivator of the soil), grazier, poet and writer, and was 'more skilled in all than thriving in any'. Being very keen to impart his husbandmanly knowledge, he published in 1557 his *Hundredth Pointes of Good Husbandrie*, and later, in 1573, an enlarged version entitled *Five Hundredth Pointes of Good Husbandrie*. To make the 'pointes' easier to assimilate, both books were written in verse – a form which, in her book *Flowers and their Histories*, Alice Coats considers 'lumbers along like a trotting carthorse'. However, one of the nicest points is:

> Good huswifes in Sommer will save their owne seedes
> Against the next yere, as occasion nedes
> One sede for another, to make exchange
> With fellowlie neighbourhood, seemeth not strange.

Sensible and generous behaviour which is followed by all good gardeners today.

John Gerard (1545–1612) compiled what was probably the first compendium of garden plants. Like William Turner's list, it was first written in English. Gerard was a Cheshire-born herbalist and gardener, who moved to London and was an enthusiastic collector of what were then called 'exotic' plants, which he grew in his garden in the fashionable suburb of Holborn (probably on the corner of Fetter Lane or Furnival Street). He seems to have been acquainted with

Thomas Tusser, and undertook the long training to become a barber-surgeon; he became Master of the Honourable Company of Barber-Surgeons during the last few years of his life. Earlier he had been gardener to Lord Burghley (1520–98), the loyal and close adviser to Queen Elizabeth. There were two sumptuous gardens for Gerard to look after: one was in London, Cecil House in the Strand, and the other was in the country, at Theobalds Park in Hertfordshire. The garden here, created over a period of ten years to 1585, was more important than the one in the Strand, and was certainly the grander of the two.

Gerard published both his *Catalogue* – which contained a list of over one thousand plants growing in his Holborn garden, including the first printed reference to the potato – and his *Herball or General Historie of Plantes* right at the end of the century, in 1596 and 1597 respectively. Thus, for the first time, one learns of the flowers and herbs that were being grown domestically, and sometimes even about their cultivation.

As a reward for all this diligence, and like William Turner before him, Linnaeus named a genus in his honour, the Gerardia. There are about 30 species in the genus, and they belong in the *Scrophulariaceae* family. The flowers are quite showy and look rather like penstemons, and they range in colour from pale violet through rose pink to yellow. They are all native to North and South America. The only one seemingly available in Britain now is **Gerardia tenuifolia** (meaning 'with slender leaves') which was brought over from Mexico in the nineteenth century; see p.219.

The plants from Africa, Asia, India and the New World which began to contribute so much to the look of our gardens in the sixteenth century were also intermingled with a vast number of European flora, in particular from around the Mediterranean area. Italy was leading the way with the creation of the first botanical garden which was made at Pisa in 1543; the botanical garden, at Padua followed in 1545. At this time, too, the continental universities began to study and catalogue the plant life and natural history of their surroundings. So before we venture too far from these shores, we should perhaps look in some detail at the relatively local flora which travelled a short way across the Continent and over the Channel to fill the herbaceous borders and later the rockeries which are still found in our British gardens today.

The 'bulbous violet' does not have much of a ring about it, nor does it sound very inviting, but it appeared in our gardens sometime during this century and was the earliest name given to the snowdrop, the Galanthus. John Gerard called them by their Latin name in his *Herball*, but it was in a revised edition of this book, published later in 1633, that we read that 'some call them also Snowdrops', surely a more descriptive name. A Christian legend revolves around an angel, who was helping Eve after her Fall from Paradise when it snowed; catching a snowflake, the angel breathed life into it, and the snowflake fell to earth as the first snowdrop.

The word *Galanthus* is derived from the Greek, meaning 'milk flower', and the first of the 15 or so species from the eastern Mediterranean to reach us was given a further apt name, this time the Latin *nivalis* which means 'growing near snow', and so **Galanthus nivalis**. The seeming frailty of snowdrops defies their toughness, as the bulbs push up through the soil at what can only be described as the most inhospitable time of the year, the very act of blooming appearing to crush the winter and welcome the spring. And yet they have a most unfortunate reputation. One would think that the joy of seeing flowers blooming at the sterile turn of the year would enhance their character, but not a bit of it. Although the bulb has properties which aid the healing of wounds and bruising, the cutting of the flower for the vase makes many people even today feel uncomfortable, for by bringing them indoors it is feared a death will occur in the family. They do indeed look far more lovely living *au naturel* on banks and tucked into the bottom of hedges, anywhere really so long as they are outside the home.

Nowadays the interest in growing the species and different varieties means that there are several National Collections that can be viewed, and usually a 'Galanthus Gala' is held to celebrate their flowering . . . which takes us back to the original Greek word to describe them – *gala* meaning 'milk', *anthos*, 'flower'.

A flower that everyone can recognise by its Latin name so aptly given is the Campanula, or Bellflower. The whole family (a very large one of some 300 species) comes mainly from the Mediterranean area and spreads over the whole range of the landscape, from sea level to the highest Alps. Our native bellflower, *Campanula trachelium*, was one of a number of Campanulas with which the Romans were acquainted, and was accorded the privilege of being called the Canterbury Bell, because the shape of the flower was supposed to resemble the small bells with which the Canterbury pilgrims used to decorate their horses. Another of its English names refers to its use as a gargle: Throatwort, from the Latin word for 'throat', *trachelium*. By the sixteenth and seventeenth centuries however they were being called Coventry Bells.

Both the native and the foreign bells were growing in gardens by the time Henry Lyte (*c.*1529–1607) published his *Niewe Herball or Historie of Plantes* in 1578. Lyte lived at Lytes Cary in Somerset and was an antiquarian. The book he published was a translation from the French of an original Flemish book written by Rembert Dodoens. (One of Henry Lyte's nineteenth-century descendants was the Scottish pastor, also Henry, who wrote the hymns 'Abide with me' and 'Praise my soul, the King of Heaven'.) Another native Campanula, *C. specularium,* flaunted itself at Greenhythe near Gravesend 'in a field among the corne', according to Gerard, who discovered it one day and brought the seed home to his garden in Holborn. It was usually grown as an annual. It is now called *Legousia speculum veneris,* nowhere near as pretty a name as the direct English translation of Venus's Looking Glass. Sir Thomas Hanmer (1612–78) tells the story of Venus, who, silly girl, lost her magical mirror, in which everything appeared so beautiful. A shepherd found it – perhaps in the field at Greenhythe – and when Cupid came to claim it on behalf of Venus, so entranced was the shepherd with the vision of himself in the acclaimed mirror that he could not let go. In the struggle for ownership, Cupid, rather clumsily one feels, allowed it to smash to the ground, whereupon the splinters of glass all turned into the 'very pretty violet-coloured flowers'.

Another native bellflower had a reputation for making children quarrelsome (as if they needed any help): the Rampion or *C. rapunculus.* The name means 'like a little turnip' and gives a clue to what it must taste like. The leaves and grated root could be used in a salad or 'boyled and then eaten with oyle and vinegar, a little salt and pepper'. Mrs Beeton, in the nineteenth century, refers to it forming 'a valuable addition to the materials in season for salad making' and in winter it turned into a root vegetable. Most Campanula roots are edible but one gets the impression that they were only used when there was little else around; they were not eaten as a great delicacy.

The tintinnabulation of bellflowers continues with *C. persicifolia,* the Peach-Leafed Bellflower, which is native to a huge area anywhere between southern Europe, Asia and Russia and which was cultivated in both blue and white, and eventually in double forms as well. Another, *C. pyramidalis,* was originally named Steeple-bells but changed its name to the Chimney Bellflower to reflect its use. Following its arrival sometime before 1597 from northern Italy, it later became the fashion to grow it in pots set in the fireless grate during the summer, with an arrangement of Sweet Basil, Origanum and, from the 1630s, the very fragrant tuberose *Polianthes tuberosa,* a practical but pretty display which remained popular into the nineteenth century. The biennial bellflower still does best grown in a pot.

A hedgehog with a round head is the descriptive name for the Globe Thistle, *Echinops sphaerocephalus,* which arrived here from Siberia in 1542. The writer Christopher Lloyd (1921–) has called it 'an undistinguished member of a large family with ugly foliage'. A second species, *E. ritro,* was growing here by 1570; this one is more compact (*E. sphaerocephalus* is over 2 m/6 ft tall), with a clear blue head. A number of other Echinops arrived from Mongolia, the Caucuses and Asia as well during the eighteenth and nineteenth centuries (there are at least 120 species). They were mainly thought of as being suitable seaside plants (and look somewhat similar to our native Sea Holly, *Eryngium maritimun*) and were popular in maritime gardens. Now that they have the adjective 'architectural' attached to them, and the hybridisers have realised their potential, we have Globe Thistles the size of footballs and of the deepest azure, and a range of sizes and colours to suit the most discerning of tastes.

The lovely but very poisonous Monkshood or 'Old woman in bed, with her shoes on', *Aconitum napellus,* brings us back down to European earth. It arrived from southern Europe shortly after *Cistus salviifolius,* the tissue-paper Rock Rose with its salvia-like leaf; then there was *Galega officinalis,* Goat's Rue, which reached Britain in 1548. Again we come across the Greek word for milk (see Galanthus above), and, true to its name, nanny goats were fed on the plant to improve their flow of milk. Goat's Rue is a member of the Pea family, and among modern hybrids one to note is a delicious soft lilac-blue called 'His Majesty' (syn. *G.* 'Her Majesty'); another is the rather stronger-coloured 'Lady Wilson'.

Monkshood, Goat's Rue and the Cistus among many others appear to have been growing here by the 1550s, but one plant that has a set date for its entrance, 1562, is *Glycyrrhiza glabra,* better known as Liquorice. Its horticultural name is taken from the Greek words *glykys,* meaning 'sweet', and *rhiza,* 'a root'; our word 'liquorice' seems to be a corruption of the Greek word and comes via Anglo-Saxon and Old French. It is the root that is used in the commercial manufacture of liquorice and also in the tobacco industry. Market

Workmen in the liquorice fields around Pontefract, Yorkshire, digging up the tap-rooted perennial. When the juice of liquorice
Glycyrrhiza glabra is boiled it thickens and produces liquorice sticks, a now more or less forgotten confection.

gardens around Pontefract used until quite recently to grow
acres of Liquorice to make the famous Pontefract Cakes.

One flower that our gardens would be the poorer without
and which arrived here in 1570, again from the
Mediterranean, is the wonderfully named Love-in-a-Mist,
Nigella damascena – *Nigella* from its black seeds and
damascena because it was thought to have first arrived from
Damascus. Although no one seems to know where or when
that English name first occurred, the plant has a string of
other names including Devil-in-the-Bush, Love-in-a-Puzzle
and Jack-in-Prison, all presumably because the delicate
fern-like leaves encase the pretty pale blue flower. Its official
name, according to Gerard, was 'gith', an Anglo-Saxon word
for the similar-looking corn-cockle (now *Agrostemma
githago*).

The seeds of *N. damascena* are aromatic and are apparently slightly narcotic, whereas the seeds of *N. sativa*, which
arrived prior to Love-in-a-Mist (by 1548), can be used as a
spice. It was first grown in the gardens of Syon House on the
banks of the Thames in west London and is sometimes
called the Nutmeg Flower or Black Cumin. It can be sprinkled on bread and cakes (much favoured in Egypt). In
France the seeds were crushed with cinnamon, coffee and
chocolate and used to flavour cream. In England a recipe of
1561 put them to a different use: one was encouraged to burn
the seeds to 'ashes, put swynes grese thereto, and strake or
kemme the heyres therewyth, that dryveth away lyse and
nittes'. In other words, comb bacon fat mixed with the
nigella seed ash through one's hair to get rid of unwelcome
visitors.

The Greek word *klema* means a 'twig' or a 'shoot of the vine' and eventually was used to describe clematis. It was during the latter part of this century that the European members of what in the end turned out to be a very large family indeed began to scramble through our gardens. *Clematis viticella*, a bell-shaped late flowerer, was introduced into England towards the end of the 1560s by Hugh Morgan (*c*.1540–1613), who had not one but two 'botanic' gardens in London. He was Apothecary to Queen Elizabeth and was later rewarded with the naming of a genus for him, that of *Morgania*, although Gerard called him 'a curious conserver of simples'. 'Simples' were just what the name implied, a simple or single herb used for medicinal purposes, while a 'compound' was when two or more herbs were used – all part of the very familiar Doctrine of Signatures (see p.14).

Another clematis, *C. flammula* (meaning 'small flame') arrived from southern Europe before 1597 and is a small and dainty scrambler with heavily scented white flowers. John Gerard grew it and found that if you crush the leaves on a hot summer day and smell them 'it causes a smell and pain like a flame'. His name for it was Purging Periwinkle. Gerard also grew *C. cirrhosa*, which sounds like a disease of the liver but simply means 'with tendrils'. This one is evergreen, flowers in the winter and has silky seed heads, a double bonus. It comes from southern Europe and Asia and so is slightly tender. Some 400 species of clematis have been discovered all told, distributed throughout the world, and it is now one of the most highly developed of all plants. The Clematis family is similar to the Rhododendron family in the twentieth century: the climbing ability of the plants, combined with the fervent nature of the enthusiast for growing and breeding them ensures that there will be no clematis-free zone anywhere.

Arriving in the sixteenth century were two species of lavender which quickly found a place in our gardens. These were *Lavandula dentata* (meaning 'toothed', with reference to the leaves) and *L. stoechas*. This latter plant is easily recognisable since its flower is so distinctive, with its 'hare's ear' blooms. Unsure how to pronounce the word *stoechas*, it soon became 'Stickadove', a name which remained with it for a long time. Now it is simply known as French Lavender. Its date of arrival here is usually given as 1550, and its name refers to where it comes from, the Islands of Stoechades, off the coast of France in the Mediterranean; they too have changed their name and are now called the Iles de Hyeres. With *L. angustifolia* having been available to gardeners since

One of the Viticella group of Clematis which is a native of Central and Southern Europe, here painted by Gillian Barlow.

the thirteenth century, as was another, then called *L. vera* but now known to be a type of *L. angustifolia* (see p.66), it is not surprising that by 1568 William Turner could write that they 'are so well knowen al redy'. A third plant, called *L. multifida* ('many-flowered'), was being grown by Thomas Hanmer in 1659 in his garden in Wales, and this is considered to be the same plant as the earlier arrival *L. dentata*.

The plant we know as English Lavender is a cross between *L. angustifolia* and *L. latifolia*, the latter a lavender originally from Dalmatia which is not quite hardy enough for our climate. The Latin name for 'our lavender' is now, most unoriginally but accurately, **L. x *intermedia***: it has had a variety of other names as well – lavender has always challenged the botanist. The main growing area in England for the manu-

Lavandula angustifolia is seen growing here in the Parc Naturel at Mongfrague in Spain. This is the species the Romans would have known, and which they may well have brought to England with them.

facture of lavender water and dried flowers for sachets and potpourri is Norfolk. A hundred years ago Hertfordshire and Surrey were also lavender centres, but are no more. On the Continent it is the south of France which produces the essential oil for the perfume industry.

As a garden plant, lavender helped fill the early knot and parterre designs of the Tudor garden, as well as being culti-

vated for its culinary and medicinal qualities. Over the centuries it has probably been grown in more gardens than almost any other plant. In the seventeenth century it was sometimes planted as a lawn, and later, in the nineteenth century, lavender contributed to the idyll of the cottage garden. Even today it helps give the illusion of permanent warmth and sunshine to our sometimes soggy and cold climate.

Towards the end of the seventeenth century at Moira Castle in County Down, an Irish botanist Sir Arthur Rawdon (*c*.1660-95), experimented with using lavender plants to make a lawn. He laid down just under half a hectare (1 acre), which was kept neat by scything. Following his innovative idea, word spread and lavender lawns were quite the vogue for a while in Ireland. Earlier, in the 1680s, Sir Arthur had visited John Watts, the apothecary and gardener then in charge of the Chelsea Physic Garden, and between them they agreed to send a young gardener, James Harlow, across the Atlantic to collect plants for Watts from Virginia, and for Sir Arthur from Jamaica. Harlow returned with a large number of exotic plants including ferns which were all placed in the newly constructed 'hot-house' at Chelsea.

From southern Europe and Asia came a tree we can all recognise, the laburnum. It is a member of the Pea family, *Leguminosae*. There are only 2 species in the genus, both of which entered our gardens *c*.1560. *Laburnum anagyroides* is the Common Laburnum or Golden Rain, a most descriptive name, for that is just what the long drooping yellow racemes look like. In some early books it was sometimes referred to as Golden Chain. The genus name itself comes from the Latin, and its second name refers to it looking similar to another small tree which arrived from the Mediterrenean (in 1750), *Anagyris foetida*. Golden Rain suits it much better. All parts of the tree are poisonous, and it is said that if another tree is cut down near it, the laburnum will go into mourning for a whole season and will not flower. The second species to arrive, somewhat later, was *L. alpinum*, which means 'coming from high mountains', above the tree line, which might be how it acquired its common name of Scotch Laburnum, although its ancestry lies in central and southern Europe.

An oddity on the edge of the Laburnum family is a plant which occurred in 1825 at the Paris nursery of Monsieur Adam; it was a cross-breed – a graft hybrid between *Laburnum anagyroiides* and *Chamaecytisus purpureus*, the Purple Broom (the latter arriving from south-east Europe in 1792) – and was called, logically enough, *Laburnocytisus adamii*. The flowers reflect the three colours of its parents, borne on separate racemes of wholly purple, wholly yellow, and purple-pink with yellow. Ugh!

Many of the plants we enjoy in our gardens today were first listed in John Gerard's *Catalogue* of 1596 and his *Herball* of 1597, most of them of course still arriving here from Europe. One example is Cupid's Dart, the Greek-named *Catananche caerulea*. This was used traditionally as the basis

John Gerard, who travelled widely in Europe and met and corresponded with eminent continental botanists such as the Frenchman Jean Robin, the Royal Arborist and Gardener to three Kings Henri III, Henri IV and Louis XIII.

of all love potions – which is probably what made it so popular. Gerard tells us he got his seed from Padua; we can get ours via the *Plant Finder* and *The Seed Search*. Then came the lovely bright yellow Winter Aconite, which is a native of southern Europe extending from France through to Bulgaria. Its Greek name, *Eranthis hyemalis*, explains that it is the 'earliest flower that blooms in winter'.

From western France came the pine, *Pinus pinaster*, a source of turpentine and resin and known in England as the Maritime Pine or the Bournemouth Pine, for the obvious reason that they were planted in profusion by the seaside and especially at that resort.

Astrantia major, a native of central and eastern Europe, was one of a small group of about 10 woodland-happy species acquired by Gerard from Austria. He called it Black Masterwort, but it sometimes goes under the name of

Hattie's Pincushion or Melancholy Gentleman; alas, one does not know who the Hattie in question was or whether her pins had anything to do with the gentleman's melancholia. A rather acerbic comment made about the plant by the Rev William Hanbury in the eighteenth century seems somewhat uncalled-for: 'being flowers of no great beauty, the very worst part of the garden should be assigned them'. Thankfully, today we can appreciate their dignified and tranquil charm.

Just like the *Tagetes patula* we read of later, the next plant, **Primula auricula**, is believed to have been brought by the Huguenots to England from France at the end of the century. The old name for it was *Auricula Ursi*, Bear's Ears, reflecting the shape of the leaves; John Gerard grew it and called it the Mountain Cowslip. This was a plant whose native home was the upper pastures of the Alps and the Dolomites and whose 'threddy' roots, if eaten, would cure vertigo – a real plus for such a high-altitude flower, one supposes; not that it needed any help with its personality, for the Auricula in all its finery has the ability to make grown men tremble when they behold its serenity, beauty and charm.

Both sides of the Channel succumbed: in eighteenth-century France it was recommended that the cultivation of Auriculas could be a source of spiritual satisfaction and might encourage growers to 'lead a saintly life, free from all reproach'. Here in England, Gerard listed thirty different sorts, John Parkinson later grew twenty or so; then Sir Thomas Hanmer, in his *Garden Book*, recorded the forty he grew, and by 1665 John Rea, the nurseryman and author of *Flora, Ceres et Pomona*, catalogued four different classes of Auricula. By then, 'these lovely ornaments' were being prinked and prized as a 'florist flower' and were given splendid names such as 'Black Imperial' and 'Mistress Buggs Her Fine Purple'. Neither of these species, alas, has survived, nor have any of the most highly-prized blooms, the striped flowers; an admirer at the time once paid £20 for one of them.

As if that was not enough excitement, a most sensational horticultural event took place, being first reported in 1757 by James Thompson, a florist from Newcastle upon Tyne. In his book *Distinguishing Properties of a Fine Auricula*, he described the 'break' that had occurred in a flower as being almost unnatural. In fact, there were two changes: one was the colour – a clear green which had never before been seen in a flower – and then there was the almost unreal centre ring with its meal, or paste as it was called. Eventually the new hybrid Auriculas which were all developed from this one

'break' showed colours completely new in flowers – slate blue, cinnamon, and perhaps most characteristic of all, the greenish-grey colour so beloved by Auricula fanciers. The names grew equally grand – 'Honour and Glory', 'Glory of the East' and even 'Marvel of the World'. In the early nineteenth century, they were popularised by those living in the East End of London. Later, between 1840 and 1850, the connoisseurs of the show Auricula were apparently the silkweavers of Lancashire and Cheshire. Considering there was so much enthusiasm for this plant, it is surprising that the National Auricula Society was not founded until 1872.

Although plants continued to arrive here from Europe, we must turn our attention further afield now to the wonderful world of the 'exotick' – although this word, which comes from the Greek, was not recorded as being in use in Britain until 1599. Exploring, trading and the annexing of lands which, as we have seen, began during the previous century continued in earnest in the 1500s, leading Portuguese and Spanish explorers into the world's then most outlandish places. Among these explorers was Christopher Columbus, who, between 1502 and 1504, made his final voyage down the coast of Central America.

A few years earlier, in the spring of 1500 and almost by accident, landfall had been made on the coast of South America by the Portuguese Pedro Cabral (1460–1526). He was actually on his way to India leading thirteen heavily laden ships on the first major trading mission to the East. In the Atlantic he set a course well to westward and found land, subsequently named Brazil. Such a stupendous and rich find resulted in one of the ships being dispatched back to Portugal with the news, while two men were left behind to explore the new land. Cabral himself sailed on, around the Cape of Good Hope, up the eastern coast of Africa to Mozambique, and then across the Indian Ocean where he reached Calicut, a port lying on the south-west coast of India. Here he made treaties with the rulers of Cochin to the south and Cannanore to the north. He returned to Portugal well laden: spices, porcelain, incense, aromatic woods, pearls, diamonds and rubies were all part of the treasure which was brought back to Lisbon in June 1501. Thus began the Orient's seduction of western Europe.

Cabral's success was fully exploited by the brilliant Portuguese explorer Alfonso d'Albuquerque (1453–1515) who, after attacking Hormuz, at the entrance to the Persian Gulf, set up Portuguese fortresses all along the coast from East

The annuals *Tropaeolum majus* and *T. minus*, the nasturtiums, are easy to grow and the leaves make a peppery addition to salads.

Africa to Malaya. In 1510 he captured Goa and a year later Malacca in Malaya, thus establishing Portuguese hegemony around the Indian Ocean.

Twenty years later the Pacific was being mapped by both the Spanish explorer Vasco Nuñez de Balboa (*c.*1475–1519) and by Ferdinand Magellan (*c.*1480–1521). Magellan, a Portuguese veteran of the capture of Malacca, set out in 1519 in the service of Spain with five vessels to find a westward route to India, and thus began his epic circumnavigation of the world. The only ship to complete the voyage was the *Victoria*, a vessel of less than 130 tons, which returned to the Spanish harbour of Sanlucar de Barrameda on 6 September 1522. Although Magellan is often given the accolade of being the first person to circumnavigate the globe, he in fact was killed on 27 April 1521 in a skirmish with the men of the island of Mactan in the Philippines. It is therefore more likely that it is his Asian interpreter who had been with him on previous voyages, who can claim the honour.

In 1532 Peru felt the conquering might of the Spanish conquistador Francisco Pizarro (1478–1541); eight years later the Spaniard Francesco Vásquez de Coronado (*c.*1510–54) began the systematic exploration of North America and incidentally discovered the Grand Canyon in Arizona. Almost at the same time, in 1542, on the other side of the Pacific, the Portuguese reached Japan. China was the next country on their itinerary; the port of Macao became a permanent European base, first for the Portuguese in 1557, and then later for all European commerce with China, although the Chinese population themselves were allowed little trade and no contact with the foreigners.

These land and sea discoveries eventually led to much of the world's natural flora being introduced into Europe. Understandably, the ports on the Iberian peninsula were often the first recipients of the seeds, corms and bulbs which the returning mariners brought home with them. In England, Elizabeth I ascended the throne in 1558, and nearly twenty years later, in 1577, Sir Francis Drake began the first of his voyages around the world. Metaphorically, we are beginning to use the next century's major invention, the telescope, to see further than our own island's horizon.

Following these mighty explorations, a small step in the globalisation of botany took place in the 1530s when probably the first new seeds were being brought back to Europe from the New World of South America. They were recorded as then having been brought from Spain to Britain as early as 1535. The records show that it was *Tropaeolum minus* that had the honour of arriving first. The plant's native home is Peru, which had been conquered by the Spanish a few years previously.

We know the Tropaeolum genus as the nasturtium which comprises a group of about 80 species. There is, not unusually with new plant arrivals, some confusion over the botanical and common names. *Tropaeolum* is the word coined by Linnaeus from the Greek, and is the Latin name for a 'trophy' or a sign of victory; it refers to the shape of the leaf, which resembles a shield. Linnaeus noted in the eighteenth century how gardeners put up a pyramid of poles or grew the plant up a tree trunk, and how the leaves hung down to look just like shields. The flowers too are supposed to ape the golden helmet on the statue of the Roman goddess Victory. The leaves when eaten have a very pungent taste which gives rise in Latin to its second name of *nasturtium*, meaning a 'twisted nose' That word is actually part of the Latin name for watercress, *Rorippa nasturtium-aquaticum,* of which the first word is a Latinised version of an ancient German word originating in Saxony – all very confusing.

But to return to *T. minus,* it was a sprawler not a climber, and had deep yellow flowers. It seems to have disappeared not long after it was introduced, but was collected and growing again from the eighteenth century to the early part of the twentieth, when it disappeared again. It seems not to be listed in any of the current catalogues. It was first described in 1576 by a friend of John Tradescant, Mathias de l'Obel (1538–1616) of later lobelia fame, who called it *Nasturtii indici genuina effgies.* There exists in the herbarium of the Botanic Garden at Genoa in Italy a dried specimen dating from about 1585.

The climbing plant we generally know as the nasturtium is *T. majus,* which, again, is from Peru and arrived here in the second half of the seventeenth century. This plant is a pretty tough cookie, gorgeous to look at, easy to grow and not bad to eat; the seeds can be used in place of capers or, as John Evelyn tells us, they may be ground down to make a tasty mustard. The flowers and the leaves make a fine addition to 'a sallad', and as if that were not enough, they are the most caterpillar friendly plants too, having apparently more than ten times the amount of vitamin C than lettuce. Other species, all of which originated in South or Central America, were collected by plant hunters during the nineteenth century.

The shrill trumpets of the African and French Marigolds announced their arrival in Britain in 1535 and 1572 respective-ly, and it is not surprising that ever since then there has been confusion about these two bright yellow and bronze beauties. Their Latin name, *Tagetes,* is derived from the name of a demi-god called Tages who apparently had Etruscan leanings. The African Marigold, *Tagetes erecta*, as it eventually came to be called, certainly had an identity crisis. It was first thought to have originated in India, so was called Rose of the Indies, then was believed to be a native of either Peru or even China. However, the flower initially caught the attention of the battle-weary corsairs who went to Tunis in 1535 with Emperor Charles V to help free the 22,000 Christian slaves held captive by the Moors. When the plant arrived in England it was assumed to be a native of north Africa and given the self-explanatory name of Flos Africanus. This it was called until well into the eighteenth century when, to everyone's amazement, it was found to be indigenous to the New World, in fact to Mexico. Obviously it had travelled to Spain, incognito, with the conquistadors, and from there had taken the short step to invade the north African littoral. In Mexico its name is bound up with the blood spilt during the Spanish conquest; there the plants are called Deathflowers. The botanical name of *Tagetes erecta* reflects neither its colour nor its origins, but its everyday name, African Marigold, certainly reflects its history.

There is much the same confusion over the French Marigold, *Tagetes patula* (meaning 'spreading'). It too was found on the northern edges of Africa and originates from Mexico, and just as we saw earlier with *Primula auricula,* the tradition has long been held that it was brought to England from France by the Huguenot refugees following the Massacre of St Bartholomew in 1572; it is supposed to have first flowered here in 1573. The wholly embarrassing smell – some say stench – eventually appears to have been ignored; perhaps the witty and pert demeanour of its flower has something to do with it. The Pot Marigold or *Calendula officinalis* is, as we have seen in a previous chapter (see p.59), a quite different plant, having descended from European rather than New World stock.

One of the first trees to arrive in Europe, and then Britain, from the New World was introduced by French explorers from Canada in 1536. It was the Northern White Cedar, or the American Arbor-vitae, *Thuja occidentalis, thuja* being the old Greek word for the 'juniper' and *occidentalis* meaning 'western'. It is an extremely hardy conifer and has proved a very important timber tree in America. There are only 6 species in total, but they are spread out over Japan and

Yucca gloriosa is a native of the warmer reaches of the Americas, thriving in places like Florida where it is sometimes called the Palmetto Royal (Fan Palm – which it is not). The plant John Gerard was given in 1593 came as a gift from a '…skilful Apothecary of Exeter, named Master Thomas Edwards'. It had not bloomed by the time his *Herball* was published in 1597.

Yucca Gloriosa   Yucca à fuilles entières

China, as well as North America. At least 5 of the species have the Latin words Arbor-vitae, meaning 'Tree of Life', attached to them.

A plant with personality, a real Yankee, arrived in the 1550s, spiking its way into our gardens and creating such a sensation that even today heads turn when it blooms. *Yucca* were in the *Liliaceae* family; over the years, the species has also been put with the spiky Cordylines, the Dracaenas, and the Phormiums. Even its name is not its own and like other horticultural and botanical confusions the yucca must surely suffer from an identity crisis.

It is native to the southern United States, Central America and Mexico, so only a few species are truly hardy in England. They thrive of course around the Mediterranean, and during the early part of the twentieth century the Italian nurseryman Carl Sprenger raised over 100 different varieties. But to return to its entry into Britain in the 1550s: one name it was given was Adam's Needle, but others, like the Spanish Bayonet or Dagger, reflect the history that was being played out in its homeland when it made its debut.

John Gerard was given one of the plants in 1593 and it was he who promptly misnamed it *Yucca*. He thought it was 'Iucca', which is the Carib word for the manihot (the root from which cassava and tapioca are made) and a member of the *Euphorbiaceae* family. *Y. gloriosa* is believed to have first bloomed in the Essex garden of William Coys (*c.*1560–1627) in 1604 – over fifty years after it first arrived in this country – and when the 1–3 m (3 ft 3 in–6 ft 6 in) creamy-white spike flowered, it caused such a sensation it was truly named '*gloriosa*'. Nearly two centuries later, Henry Andrews of Knightsbridge in London (fl. 1790s–1830s), who was a botanical artist and engraver responsible for producing the ten massive volumes of the *Botanists' Repository* between 1797 and 1811, had some sharp words to say about its name. He claimed in Volume 7 that the word '*gloriosa*' was a 'metaphysical hyperbole, very inapplicable to any plant, however beautiful'. Oh dear, what a sheltered life he must have led – and if anything is '*gloriosa*' it surely is this plant.

There was a long gap before another yucca appeared in Britain – over a hundred years, in fact – and this was a hardier variety. It was one of the many delights brought into England by John Tradescant the Younger (1608–62), and was named *Y. filamentosa*, the Silk Grass. Like *Y. gloriosa*, it originated from the south-eastern United States, and was named for the curly white threads which come from the leaf margin. This variety, too, has a wonderful pyramid of creamy-white flowers. Since

The luscious tropical look of *Canna indica* certainly came into its own during the late nineteenth century when the plant with its bold and handsome foliage came to be used as a centre point of a flower bed.

it blooms at an earlier age, it has consequently been used in breeding a number of hybrids. The development of all the yuccas took on a new lease of life in the latter part of the twentieth century with the development, or rather the redevelopment, of the conservatory and greenhouse.

William Coys was a skilled gardener, an acknowledged plant collector and, like John Gerard, a friend of Mathias de l'Obel. He grew several of the rare and difficult plants arriving from the Americas, including the sweet potato *Ipomoea batatas*: batatas, its vernacular Carib name, gives us the word 'potato'. This form of Morning Glory comes from tropical America. Gerard also grew it, and found that 'in summer it flourished, but rotted in winter, this climate being too cold for its cultivation as a food plant'. After its initial introduction it must have died out, as it was reintroduced in 1797 and is still (just) available as seed today.

Coys also grew the 'common potato', as *Solanum tuberosum* was called. Although there are many legends about its introduction, there is no evidence for either Sir John Hawkins or Sir Walter Raleigh to have been involved. There is, however, just a possibility that it could have been brought from Chile in 1586 by Thomas Heriot, a botanist travelling with Sir Francis Drake.

Another rarity which William Coys apparently grew in his Essex garden was the persimmon, *Diospyros virginiana*. Records show it was not supposed to have arrived until 1699, but it appears Coys had access to the newest of new plants. Collected from Virginia, it is a native of the eastern USA, and is a member of *Ebenaceae*, the ebony-producing family. The edible fruit is the Chinese Persimmon, *Diospyros kaki*, which arrived in the eighteenth century. Coys also grew, via a connection in Spain, a number of new plants from the Spanish peninsula, such as the Ivy-Leaved Toadflax, *Cymbalaria muralis*, which is now naturalised over most of Britain.

Rather like the African Marigold, which was seen growing wild along the coast of north Africa, so Indian Shot eventually naturalised itself in Spain and Portugal. This was found to be *Canna indica* and was an early introduction from the West Indies towards the end of the 1560s. It may have attracted attention when Spain was earlier castanetting itself through Central America between 1511 and about 1530. The genus has about 50 species in the family of *Cannaceae* and is spread over tropical South and Central America and also Asia. When it was introduced into Britain it was considered a great rarity and it was some time before anyone understood how to cultivate it. The plant feels most comfortable growing on forest margins in moist open forest areas, but seems to be domestically quite adaptable. The word *Canna* comes from the Greek for 'reed', and *indica* is because of its connection with the West Indies. All the species, and now the hybrids, carry the most spectacularly coloured flowers and give a zing to any garden; as with the yucca, they seem sophisticated and foreign.

A number of the Canna species were introduced during the eighteenth and nineteenth centuries, but they were always considered specialist plants, until the idea was developed in France of using subtropical and tropical plants for summer splendour. This was pounced upon by gardeners in Britain and is a style of summer bedding still used in much municipal planting today. A mass of them blooming together looks rather like a flock of exotic parrots, and their exuberance at least cheers up what maybe an otherwise dull summer.

Another plant which must have added to the fun, but which didn't actually land on our shores until the beginning of the following century, is ginger. Similar to the Canna in its flaunting quality, and making a spectacular statement during the summer months, *Zingiber officinale* came from the East Indies, and is, of course, the source of our root ginger, and used in cooking in all sorts of ways.

A 'wonderful purge' is the translation of *Mirabilis jalapa*, a half-hardy annual which found its way here from Peru via Spain in 1568. We know it as the Marvel of Peru or the Four O'Clock Flower. Its botanical name again is a cause of confusion – *mirabilis* indeed means 'wonderful', but its second name we now know to be a mistake, as the plant was first thought to be the source of the drug jalap, a strong purgative which is actually made from the powdered root of *Ipomoea jalapa* (a relation of the sweet potato which arrived here in 1733) and got its name from the town of Jalapa (once Xalapan) Veracruz in Mexico. Gerard wrote glowingly of the Marvel of Peru, proclaiming it to be the marvel of the entire world, not just of Peru. Its Four O'Clock name needs a little explanation too: the blooms remain closed until mid-afternoon, during the sunniest part of the day, and then open at about four o'clock and remain open all night, closing the following morning before the sun gets going again.

This is similar to the 67 species of *Nicotiana*, the tobacco plant, some of which come from the same continent. Tobacco itself (the word actually refers to the pipe the Carib Indians used) is believed first to have been seen by Columbus in 1492, but was not introduced here until early in the seventeenth century. In our not so hot climate and particularly if it is overcast, *M. jalapa* can remain half open all day. They flower naturally in a most remarkable range of colours and are best treated in England like dahlias, which also came from the same part of the New World but not until much later, in the nineteenth century.

One plant that must be included in the exotics, and which settled here in the same year as the Marvel of Peru, is the Passion Flower. It was first called *Flos Passionis*, or, as John Tradescant named it in the following century, *Amaracock sive Clematis Virginiana*; Linnaeus in the eighteenth century gave it the name with which we are familiar, *Passiflora incarnata*. The plant, which comes from a very large group of over 400 species, nearly all of them climbers, has never lost its foreign

As Captain Winter discovered in 1579, the healing properties of the shrub *Drimys winteri* were efficacious in combating scurvy, although it was almost another 250 years before the plant itself was introduced into Britain.

good looks. It is sometimes known as 'Maypops' or 'May Apple'. Discovered by the Spanish missionaries, it comes from the southern United States and was illustrated fifty or so years later by John Parkinson in his book *Paradisi in Sole Paradisus Terrestris.* There it is quite clearly shown how the different parts of the flower could be made to represent the Passion of Christ. It has three stigmas, which are the Nails; the five stamens become the five Wounds; the Crown of Thorns is the distinctive corona, and the sepals and the petals of the corolla are ten of the Apostles, without counting either Peter or Judas. If ever a plant represented the history of its discovery and the time of its finding it is the *Passiflora,* a most glamorous addition to our gardens.

Scurvy was the ever-present problem on long sea voyages,

and was experienced on one of the ships which set sail with Sir Francis Drake on his round-the-world voyage in 1577. The crew of the *Elizabeth,* with Captain Winter in command, was badly afflicted by the disease whilst sailing along the coast of South America. Because of illness on board, the ship appears to have lagged behind and eventually left the fleet in 1579, setting sail to return home through the Straits of Magellan. Here, on this bleak and desolate coastline, Captain Winter went ashore to search for medicinal herbs. He discovered a tree, the bark of which contained an antiscorbutic which, when given to his crew, alleviated their scurvy. He returned to England with some of the bitter aromatic bark, which came to be known as 'Winter's Bark'. Its botanical name is now ***Drimys winteri***, which is derived from the Greek and means 'acrid or pungent'. It was a sure-fire cure for scurvy; even Captain Cook used it during his voyages two hundred years later. Cook also used sauerkraut to deter the disease, and – if they could be obtained – limes and lemons too.

The whole genus of about 30 species is now in the family of *Winteraceae*, named in honour of Captain Winter, but they were first placed in the Magnolia genus. The shrubs and trees come from the forests and mountains of Malaysia and Australasia, as well as from South and Central America, where the first one was discovered. Even though it proved so valuable medicinally, no one thought to experiment to see if the plant would grow in England until 1827. Even at that time, not everyone seemed to recognise the benefits of fresh vegetables and fruit, let alone using *Drimys winteri* to combat scurvy, for as late as 1846, as reported in the *Royal Cornwall Gazette* for January of that year, a schooner, the *Mary Pope,* entered Fowey harbour in Cornwall with her crew 'in a most distressed state . . . and reduced to a skeleton by scurvy'.

A second species, *D. aromatica* (now *D. lanceolata),* which came from Tasmania in 1843, is the only other species of the genus which at least manages to thrive in our climate, with *D. winteri* proving the more hardy of the two.

A plant from the other side of the semi-tropical world which has been ingratiating itself into our gardens since it first arrived from India very late in the sixteenth century is *Impatiens balsamina*, the first of a huge tribe of annuals whose name aptly describes their habit of firing their seeds as if from a cannon. This has led to at least two streetwise (or rather garden-wise) names, Touch-me-not and Jumping Betty. They are, of course, known to us as Busy Lizzies. There are around 850 species in the genus, and they have never given botanists an easy time; three hundred years after their arrival, even the great Sir Joseph Hooker (1817–1911) thought them 'a terror to botanists'.

Another tropical plant which found its way here from Africa is *Hibiscus trionum*, the Venice Mallow, Goodnight at Noon, or Flower of an Hour, all names which in one way or another describe its habits. It belongs in the *Malvaceae* family and is a fast-growing, short-lived perennial. This one used to be called *Hibiscus Africanus*, which at least denoted its origin. There are about 200 species and they are quite the most accommodating group imaginable. Their provenance ranges from the tropics to warm-temperate areas; some are happy in moist conditions, others favour dry rocky sites;

some are deciduous and others are evergreen. They can be either annuals, shrubs or trees, and some of the flowers can be eaten. One that we do eat is the vegetable okra, or Ladies' Fingers, which was known until recently as *H. esculentus*, now it is *Abelmoschus esculentus*, the former apparently a word derived from the Arabic and meaning 'father of musk', of which the seeds smell. Another species yields face powder. They are obviously a family which will try anything.

A second species arrived at about the same time as *H. trionum* and was on John Gerard's long list of plants made at the end of the century. This was called Rose of China, perhaps an echo of where it originally came from. Its provenance is repeated in its botanical name, *Hibiscus rosa-sinensis*, although it is indigenous right through the tropics of India and Asia as well. This species is a large, vigorous, shiny-leaved shrub, and a number of modern cultivars have been raised from it. The large five-petalled flowers nearly all do a 'Benny Goodman' with their stamens – as if the clarinet is being played from the flower centre. Two more modern names for it are the Hawaiian Hibiscus and, in Jamaica, the Shoe Flower or Shoe Black; this is because the petals are mucilaginous: when squeezed they extrude a viscous gum which makes a very good shoe polish.

The flowers of an Hawaiian species have an altogether different connotation for the young ladies there, where it was and still is the custom for the blooms to be placed behind their ears or used to decorate their hair. What is perhaps not quite so widely known is the different messages this adornment sends. For instance, if you tuck a red hibiscus flower behind your left ear, it is clear to everyone that you desire a lover; worn over your right ear, it signifies that you have a lover already. However, if both ears were bedecked with these gorgeous red flowers, the message is that you already have one lover, but would like to find another. For those of us not skilled in remembering our right from our left, it is perhaps advisable if we just enjoy these truly versatile exotics in the garden.

What a contrast to the previous century these one hundred years of plant introductions make, and how much more sophisticated the growing and cultivation of plants was by the end of the 1500s compared with the beginning. The monastery herbers and gardens, as well as suffering the upheavals of Henry VIII's reign, were being superseded by the new botanic gardens – at least, they were on the Continent; the first one was not established in this country until 1621.

Ladies' Fingers, *Abelmoschus esculentus*, is a native of Africa and arrived in Europe, via Egypt, probably in the thirteenth century. In the Hibiscus genus, Okra (its West African name), has yellow flowers and the vegetable is an essential ingredient in making Creole gumbo.

Hibiscus. Abelmoschus.

Sydney Parkinson pinxt 1769.

Members of the new nobility were beginning to create their own particular knot gardens and allées. Plants and fashion were becoming entwined in the design of the garden. The demands of the moment, therefore, encouraged the growing and propagation of, in particular, the newly arrived plants. To be a gardener not only became an acceptable occupation, but also brought scientific knowledge and creativity into the equation. The naming and cataloguing of our own natural history was underway, and the influx of seeds and plants from around the world acted not only as the beginnings of the horticultural powerhouse here, but as a catalyst on the awakening ideas about world power and empire. It took about four hundred years for this idea to reach its apogee, but this is the century when metaphorically the seeds were being sown. For the first time in our history, there were recognisable gardeners, people who loved growing plants and observing them for their own sake; not just for medicinal or culinary purposes, but because they looked beautiful or were perfumed or had endearing habits.

John Gerard is the first person we know of who catalogued what he grew in his garden: actually he made at least two lists, one in 1596 and the other in 1597. What gardener among us has not made a New Year's resolution to list the plants growing in our own small patch or at least record our new acquisitions. One usually ends up with a few scraps of paper, some labels, a marked nurseryman's catalogue and a promise that 'When I have the time I must make a list or, to be really up to date, put them onto the computer.' John Gerard is to be admired for sticking to his self-imposed task and producing his two lists. In a way he can be called the pioneer of the National Plant Collection movement, where designated holders of a particular genus collect as many of those species, with their varieties and cultivars, as is feasible. The fundamental difference is that Gerard's enthusiasm was for what was modern and new, while National Plant Collection holders search out, save and propagate as many of the old species and varieties as they can, as well as staying abreast of new hybrids and varieties.

As we shall see, the idea of recording and listing gardened and introduced plants reached new heights in the following century, as a father and son set off on their travels around the northern hemisphere in search of new delights.

Plant Introductions in the period 1500–1599

nsd = no set date   pr = prior to

nsd   *Galanthus nivalis* Snowdrop. Pyrenees to the Ukraine.

c. 1500   *Anemone pavonina* Peacock Anemone. Mediterranean.

Gentiana lutea Bitterwort, Yellow Gentian. Pyrenees, Alps, Apennines, Carpathians; see p.230.

1510   *Polemonium caeruleum* Jacob's Ladder, Greek Valerian. N. and C. Europe, W. North America.

1530   *Sorbus domestica* Service Tree. S. and C. Europe, N. Africa, Turkey, Caucasus, Ukraine, Moldavia. One was planted at the request of Henry VIII at Hampton Court.

c. 1530   *Lilium chalcedonicum* (syn *L. heldreichii*) Scarlet Turkscap Lily. Greece, Albania; see p.50.

1535   *Tagetes erecta* African Marigold. Mexico.

Tropaeolum minus Nasturtium. Peru.

1536   *Thuja occidentalis* Tree of Life, Northern White Cedar. E. North America.

1542   *Echinops sphaerocephalus* Globe Thistle. C. and S. Europe, W. Asia, Caucasus, Siberia.

Prunus armeniaca Apricot. N China; see p.76.

pr 1548   *Artemisia abrotanum* Southernwood, Lad's Love, Old Man. S. Europe. Reintroduction (see p.45).

Nigella sativa Black Cumin. N. Africa, Asia Minor.

1548   *Acanthus mollis* Bear's Breeches. S. W. Europe. Reintroduction (see p.34).

Cistus salviifolius Rock Rose. S. Europe

Cynara scolymus Globe Artichoke. Reintroduction (see p.35).

Galega officinalis Goat's Rue. C. to S. Europe, Turkey to Pakistan.

Ornithogalum pyrenaicum Bath asparagus. Europe, Turkey, Caucasus.

Physalis alkekengi Winter Cherry, Chinese Lantern. C. and S. Europe, W. Asia to Japan. Reintroduction (see p.67).

Pinus pinea Stone Pine, Umbrella Pine. Mediterranean; see p.72.

1550 *Lavandula stoechas* French Lavender, Stickadove. S.W. France.

c. 1550 *Aconitum napellus* Monkshood. N. and C. Europe.

Yucca gloriosa Spanish Dagger, Spanish Bayonet. N. Carolina to Florida, Mexico, Central America.

1551 *Cornus mas* Cornelian Cherry. Europe, W. Asia.

c. 1560 *Dianthus plumarius* Common Pink, Sops in Wine. C. and E. Europe. Reintroduction (see p.39).

Laburnum anagyroides (syn. *L. vulgare*) Common Laburnum, Golden Rain. E. France to Italy, S. C. Europe, Slovenia, Croatia.

1561 *Anthemis tinctoria* Dyer's Chamomile, Golden Marguerite. Europe, Caucasus, Turkey, Iran. A yellow dye is extracted from the flower.

1562 *Glycyrrhiza glabra* (syn. *G. glandulifera*) Liquorice. Mediterranean to S.W. Asia.

Laurus nobilis Sweet Bay Tree, Bay Laurel. Mediterranean. See p.168.

Ruta graveolens Rue, Herb-of-Grace. S Europe. Reintroduction (see p.35).

Satureja hortensis Summer Savory. S. Europe. Reintroduction (see p.35).

Satureja montana Winter Savory. S. Europe. N. Africa. Re-introduction (see p.35).

Teucrium polium Mediterranean to W.Asia; see p.52.

1568 *Mirabilis jalapa* Marvel of Peru, Four O'Clock Flower. Peru, tropical America.

Passiflora incarnata Passion Flower, May Apple, Maypops. S. USA.

c. 1568 *Canna indica* (syn. *C. edulis*) Indian Shot. Tropical America.

c. 1569 *Clematis viticella* Vine Clematis. S. E. Europe.

1570 *Celosia argentea* Common Coxcomb, Prince's Feather. Tropical Asia, Africa, N., C. and S. America. Now used in breeding the Cristata or Cockscomb annuals for growing on in containers

Echinops ritro Globe Thistle. S. C. and S. E. Europe to C. Asia.

Euphorbia palustris Grows naturally in marshland of Europe, S. Scandinavia to Spain to W. Caucasus, W. Asia, W. Siberia.

Nigella damascena Love-in-a-Mist. S. Europe. N. Africa.

c. 1570 *Atriplex halimus* Tree Purslane. S. Europe.

Biarum tenuifolium Mediterranean.

Hemerocallis fulva Orange Day-lily. China, Japan.

Hemerocallis lilioasphodelus (syn. *H. flava*) Yellow Day-lily. China, where the flowers are cultivated to eat and the foliage is supposed to be excellent fodder for cows, especially when in milk.

Lavatera olbia Tree Mallow. S. France. Now naturalised over W. Europe. Named for eighteenth-century Zurich naturalist brothers.

1572 *Tagetes patula* French Marigold. Mexico and C. America.

1573 *Consolida ajacis* (syn. *C. ambigua/Delphinium ajacis*) Larkspur. Mediterranean.

Dianthus barbatus Sweet William. S. and E. Europe; see p.204.

c. 1573 *Iris susiana* Mourning Iris. W. Asia. Named after Susa, a town in Iran.

Rosa damascena Damask Rose. Middle East; see p.43.

1575 *Geranium macrorrhizum* Cranesbill. Europe. Effective ground cover in shade.

c. 1576 *Prunus laurocerasus* Cherry Laurel. E. Europe. S.W. Asia; see p.140.

c. 1576 *Eranthis hyemalis* Winter Aconite. S. France to Bulgaria, Greece, Italy.

1577 *Tulipa gesneriana* E. Europe and Asia; see p.139.

1578 *Delphinium elatum* Europe and Siberia. First perennial delphinium to be grown here and one of the parents-in-chief of the modern border delphinium.

c. 1579 *Crocus flavus* Yellow Crocus. S.E. Europe and W. Asia. Used for growing in knot gardens and symmetrical raised beds.

c. 1580 *Fritillaria imperialis* Crown Imperial. S. Turkey to Kashmir; see p.127.

Quercus ilex Holm Oak. S.W. Europe, Mediterranean region.

1582 *Corylus colurna* Turkish Hazel. S.E. and W. Asia.

c. 1586 *Solanum tuberosum* Potato. Chile, Peru.

1588 *Genista germanica* Dwarf Spiny Broom. C. and W. Europe to C. Russia.

c. 1590 *Clematis cirrhosa* (syn. *C. calycina*) Fern-Leaved Clematis. Mediterranean, Asia.

Rosa moschata Musk Rose. W. Asia; see p.43.

1594 *Corydalis ochroleuca* Italy, S.E. Europe.

1595 *Citrus aurantium* Seville Orange. S.E. Asia.

1596 *Anemone ranunculoides* Native of woodlands of N. and C. Europe.

Aster amellus C. and E. Europe to W. Russia and Turkey; see p.155.

Cercis siliquastrum Judas Tree. S.E. Europe, S.W. Asia.

Digitalis grandiflora (syn. *D. ambigua/D. ochroleuca*) Yellow Foxglove. C. and S. Europe to Siberia, Turkey.

Impatiens balsamina Busy Lizzie. India, China, Malaya.

Helianthus annuus Sunflower. USA to C. America; see p.160.

Laburnum alpinum Scotch Laburnum. S. C. Europe, Italy, W. Balkans.

Lilium martagon Common Turkscap Lily. Europe, Mongolia; see p.50.

Morus alba (syn. *M. bombycis*) White Mulberry. China; see p.136.

Muscari comosum (syn. *Leopoldia comosa*) Tassel Grape Hyacinth. S. Europe, Turkey, Iran.

Pinus pinaster Maritime Pine, Bournemouth Pine. S.W. Europe, Mediterranean.

Primula auricula Alps, Apennines, Carpathians; see p.108.

pr 1597 *Alcea ficifolia* (originally *Althaea ficifolia*) Antwerp Hollyhock. Siberia; see p.64.

Althaea cannabina C. S. and E. Europe; see p.64.

Astrantia major Black Masterwort, Hattie's Pincushion, Melancholy Gentleman. C. and E. Europe.

Campanula pyramidalis Steeple-bells, Chimney Bellflower. N. Italy, Balkans.

Clematis flammula S. Europe, N. Africa, W. Syria, Turkey.

Lagenaria vulgaris Bottle Gourd, Calabash Cucumber. Reintroduction (see p.35).

1597 *Catananche caerulea* Cupid's Dart, Blue Cupidone, S.W. Europe, Italy.

Cyclamen hederifolium (syn. *C. neapolitanum*) Sowbread. Italy to Turkey. The root resembled a loaf and pigs were believed to enjoy eating it.

Ipomoea batatas (syn. *Batatas edulis*) Sweet Potato. Tropical America.

Lycopersicon esculentum Tomato. Peru, Ecuador; see p.82.

c. 1597 *Diospyros virginiana* Persimmon. E. USA.

Hibisicus rosa-sinensis Rose of China, Chinese Hibiscus, Hawaiian Hibiscus. Tropical Asia.

Hibiscus trionum Flower-of-an-Hour, Bladder Ketmia. Old World tropics but origin uncertain.

Lavandula dentata French Lavender. W. Mediterranean, Arabian peninsula, Atlantic islands.

Plants across the Pond
1600–1699

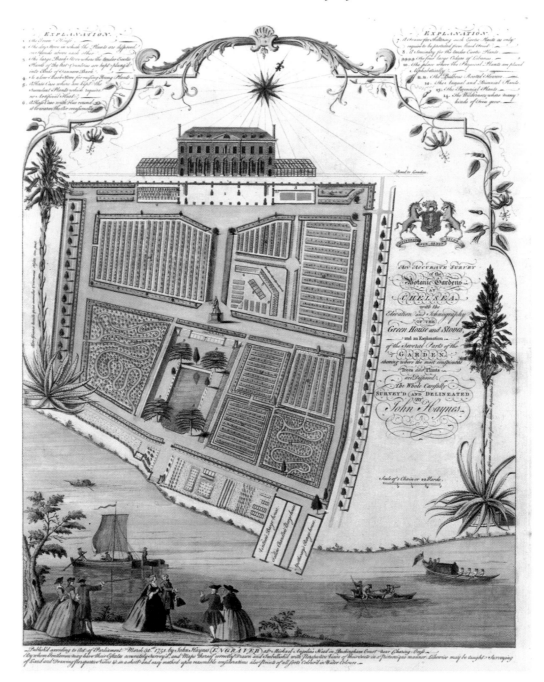

Four Cedars of Lebanon *Cedrus libani* were planted at the Chelsea Physic Garden in 1683.

Significant dates

1603	James VI of Scotland becomes James I of England (1603–25)
1605	The Gunpowder Plot
1607	First British settlement in Virginia
1608	John Tradescant the Younger born (1608–62), gardener and plant collector
1618	Beginning of the Thirty Years War in Europe
1619	First greenhouse erected by the Cistercians at Heidelberg
1620	Pilgrim Fathers arrive in America on the *Mayflower*
1621	Oxford Physic Garden, the first English botanic garden, founded by 1st Earl of Danby
1625	Charles I (1625–49)
	Francis Bacon (1561–1626) publishes his essay *On Gardens*
1629	Publication of *Paradisi in Sole Paradisus Terrestris* by John Parkinson (1567–1650)
1630	John Tradescant the Elder appointed Keeper of His Majesty's Gardens, Vines and Silkworms, at Oatlands Palace
1634	*Catalogue of Plants* published by John Tradescant the Elder
1639	Publication of Parkinson's *Theatrum Botanicum: The Theatre of Plants*
1642	First English Civil War begins; Battle of Edgehill
1649	Trial and execution of Charles I; Charles II proclaimed King in Scotland and Ireland
1653	Oliver Cromwell (1599–1658) made Lord Protector
	Publication of *The Complete Herbal* (under the original title of *The English Physician*) by Nicholas Culpeper (1616–54)
1656	*Catalogue of Plants* (2nd edition) published by John Tradescant the Younger

1659	Sir Thomas Hanmer (1612–78) wrote his *Garden Book*, which remained in manuscript form until 1932, when it was first published
1660	Charles II regains throne of England (1660–85)
1662	Foundation of the Royal Society
1664	*Sylva: A Discourse of Forest Trees* published by John Evelyn (1620–76)
1665	Great Plague of London
	Flora, Ceres et Pomona published by John Rea (fl.1620–77)
1666	Great Fire of London
1670	Edinburgh Botanical Garden founded by Sir Andrew Balfour
	Foundation of Hudson's Bay Company
1673	Foundation of Chelsea Physic Garden by the Society of Apothecaries
1681	Greenhouse built at Chelsea Physic Garden; first to be built in England
1682	*The Anatomy of Plants* published by Nehemiah Grew (1641–1712)
1684	William Kent born (1684–1748), landscape gardener and Palladian architect
1685	James II (1685–88)
	Celia Fiennes (1662–1741) begins her journey through England on horseback; described in her *Tour Through England*
1686	First volume of *Historia Plantarum Generalis* published by John Ray (1627–1705); three volumes in all published
1688	Richard Bradley born (1688–1732), first Professor of Botany at Cambridge University, 1724–32
1688–9	'Glorious Revolution'; crown offered to William III (1689–1702) and Mary II (1689–94)
1690	*Synopsis Methodica Stirpium Britannicarum* published by John Ray

WITH HINDSIGHT, the seventeenth century has turned out to be the floristic fulcrum of the last thousand years in Britain, mainly because the newly expanding sciences of botany and horticulture were, for the first time, subjects to be studied in their own right, following the pattern across the rest of Europe. This brought the cultivation of plants and flowers for their own sake into the art of gardening, leaving herbal remedies and medicine to become the prerogative of the apothecary. It was a time of high intellectual activity, despite political and social upheavals, the apogee of which was the founding of the Royal Society, the first and most important scientific society in the country. It met originally during the 1640s to discuss 'the new or experimental philosophy' of Francis Bacon, and received its Royal Charter from Charles II in 1662.

The first of many books on garden flowers was published during this century and the hocus-pocus jumble of fantasy and pseudo-facts was finally swept away on to the compost heap of what we are now pleased to call 'plant lore'. Nehemiah Grew (1641–1712), describing himself as a 'plant anatomist', wrote three books: the first two discussed the structure of seeds and the structure of roots; the third, and most important, published in 1682, was *The Anatomy of Plants*, in which Grew accurately described for the first time the sexual relationship of plants. His theory caused a furore in European scientific circles. The leading French botanist, Joseph Pitton de Tournefort (1656–1708), who was Professor of Botany at the Jardin du Roi in Paris and under whom Sir Hans Sloane studied, just could not believe the facts and thought pollen existed for 'merely excrementitious' purposes.

It would be unfair, however, to de Tournefort's reputation not to mention that he devised an important system of plant classification using petals as the special structural arbiter. This classification, which lasted for nearly a hundred years, finally fell out of favour, to be replaced by the system devised in 1736 by Carl Linné (1707–78), known as Linnaeus. On the orders of the French King Louis XIV, de Tournefort led a two year botanical expedition to explore the eastern Mediterranean. Among his team was the botanical artist Claude Aubriet (1668–1743). Between them they discovered over 1,300 plants; one of them was a small evergreen trailing plant with pretty violet or purple flowers, native to mountain regions from Sicily to Iran. It received the name of M. Aubriet and thus became *Aubrieta deltoidea* (the petals are triangular, like the Greek letter delta). The genus consists of about 12 species and is allied to both Alyssum and Arabis, all

three genera being in the same family of *Brassicaceae/Cruciferae*.

Amongst the other discoveries was *Rhododendron ponticum*, which was described for the first time and named in honour of the Turkish town Pontus, near its first sighting. This rampageous species took its time to arrive in Britain, not being recorded here until 1763; see also p.170. De Tournefort's book *Institutiones Rei Herbariae*, originally published in 1700 (and later translated into English), described and illustrated the new plants collected by the professor and his team. Like so many pioneers in the early botanical world, he was eventually honoured by having a genus named in his honour: the Tournefortia genus, belonging to the *Boraginaceae* family. These plants comprise a group of about 100 species of trees and shrubs; later they were swallowed up and placed in the genus containing the Heliotropiums, although still remaining in the same family.

The memory of Nehemiah Grew has fared better botanically, since the group of subtropical shrubs named after him, Grewia, in the *Tiliaceae* family, has retained its name. *Grewia occidentalis* was an early arrival from southern Africa, being recorded here by 1690, and *G. biloba* (*G. parviflora*) was collected and introduced from Korea in 1889; both can still be found growing in England.

A second botanical writer, a little older than Grew and who, like him, had studied at Cambridge, was the Reverend John Ray (1627–1705). His scientific knowledge of the structure and classification of plants was published in a refreshingly straightforward way in his 1690 *Synopsis Methodica Stirpium Britannicarum*, and in his major work, which was published in three volumes between 1686 and 1704, *Historia Plantarum Generalis*. He corresponded, always in Latin, with other botanists from all over the world. One was the Jesuit priest Father George Joseph Kamel, a missionary in the Philippines, after whom Linnaeus later named the Camellia (see p.164). Although neither Grew nor Ray, as far as it is known, introduced any new plants into England, they should be acknowledged as having made a significant contribution to scientific and botanical learning in the century. Indeed, John Ray through his writings was regarded as the 'Father of English natural history'.

For such an important horticultural century, the weather for the first few years was apparently rather cold, particularly the summer of 1601. Despite that, it was the year that the Autumn Crocus, *Colchicum byzantinum*, arrived from Asia Minor. Although not a crocus at all, it is sometimes put in its

The Colchicum *Colchicum byzantinum* came to Britain from the alpine and sub-alpine meadows and stony hillsides of Europe, Asia and Western China.

tion or not, but we do know that Sir Theodore remained the King's medico. (Gout could, apparently, be eased by a number of other plants as well, including horseradish, tansy and alder, none of which was swallowed but which were warmed and placed against the offending ache.) The French equivalent of the Meadow Saffron's poisonous nature is known as Tue-chien or Mort-au-chien; in Alsace the drug was combed through the hair to kill off 'vermin'. Now, it is used as a tool for hybridisation and plant breeding, having the unusual property of being able to duplicate chromosomes and thus speeding up the development of new plants.

The home of a number of different Colchicum species is around the area of Colchis in the Black Sea region of Georgia in the Caucasus, hence the reason for their name. As with nearly all plants indigenous to that part of the world, there was a godly connection, this time with Medea, the sorceress daughter of the King of Colchis, who, in encouraging Jason to steal the Golden Fleece, concocted a brew to help make him look younger. It is supposedly from some drops of this liquid that the plant sprang.

Research does not always show the exact date of a plant's introduction, so all sorts of circumstantial evidence and devices have to be employed in tracking down the most likely period of the chosen plant. Usually, the larger and more important-looking the plant, shrub or tree, the easier it is to pinpoint, and the more likely it is to have been recorded in some way – by diary, or letter, or in a painting perhaps; the smaller and more insignificant the plant, the harder it often is to find out its previous history. It is therefore all the more surprising that two of the greatest trees in our landscape both crept into England during this century with no record of the specific date of their introduction or indeed of who was perspicacious enough to return to England with the seed or seedlings.

The nearest we can get to the arrival of the 'conker' tree, the Horse Chestnut, *Aesculus hippocastanum*, is around 1616. It is native to Macedonia and Albania, and with its majestic white-flowered 'candles' it quite quickly established itself here. Four hundred years later it has almost naturalised itself in our landscape. When planted as a specimen tree in parkland, it has 'spherical appeal' and contrasts well with our own native trees, so many of which are round-headed.

The name 'chestnut' came from the resemblance of the conker to the fruit of the Sweet Chestnut, even though they come from different families – the Sweet Chestnut, *Castanea sativa*, as we have seen (p.22), probably arrived in England

own *Colchicaceae* family, or in the *Liliaceae* family: botanically, it is still not quite sure where it should be. Its common names, however, usually cause some comment: Naked Ladies, The Son Before the Father, Naked Nannies or Star-Naked Boys, all highly suggestive and referring to the startling theory that the seeds were thought to be produced before the flowers. The flowers are luscious to look at and do indeed appear naked, but the plants are highly poisonous.

It is the widespread European (and native) *C. autumnale*, the Meadow Saffron, that secrets an alkaloid called colchicine in the roots. From around 1800 BC, the Egyptians knew all about the efficacy of taking it for gout. It was still in use nearly three thousand years later, but with the added refinement of ground-up unburied human skull bones. The drug was taken, or at least offered to, James I by his physician Sir Theodore Mayerne, precisely to cure his gout. There is no report whether the added calcium improved his condi-

with the Romans. The 'horse' part of this chestnut is believed to record the horseshoe-shaped mark left at the end of the stalk when a leaf falls or is pulled off – one of the few facts that some of us can perhaps remember from our earliest nature study lessons at school. However, long before the tree arrived in Britain, the Turks were treating bruising to their horses with the natural chemical aescin, derived from the seed. Perhaps this titbit of plant lore arrived here with the tree, as ever since then the extract has been used to relieve aches and pains, not just in animals but in humans too. Its efficacy now seems to be such that there were reports – in the 1990s – that Horse Chestnuts were to be grown commercially so that the extracted aescin could be made available to remedy the aches, bruises and strains sustained by sportsmen and women (and the rest of us). Bearing in mind that one of the tree's original roots was in northern Greece, one can envisage the first Olympian athletes rubbing the ointment into their tired limbs.

The second tree of great importance to come into Britain almost unrecorded during this century was from the Near East and is equally as commanding and majestic as the Horse Chestnut, if not more so. It was the Cedar of Lebanon, *Cedrus libani*, and the nearest we can get to the date of entry is sometime during the ten years between 1630 and 1640. It takes two years for the cone to release its seeds and it is known that there were seedlings in evidence during the 1650s, and that they were scarce. Even then we do not know whether it was gathered as seed, seedlings or cones, but since cones are easiest to transport, that was probably the way the seed arrived here, with a returning traveller.

The Cedar of Lebanon is reported to have reached France

CONKERS

One cannot talk about the Horse Chestnut without mentioning its inedible fruit, the conker. The word itself is supposed to be derived from 'oblionkers', a word which may itself be a reflection of either conquerors (a Cheshire name for conkers) or Constantinople. In the 1660s, the Horse Chestnut growing in the Lambeth garden of the great plant collector Bishop Compton was known as the 'Constantinople Nut'. On the other hand, it may just be that the word is a corruption of 'conch' – the snail-shell. The game of conkers, or 'oblionkers', seems peculiarly British and did not appear until the nineteenth century. A rhyme which was used, at least up until the Second World War, in the Worcestershire area, makes use of the original name, and was chanted before the battle of the conkers began:

Obly, obly-onker
My best conker,
Obly, obly O,
My best go!

Following success in a game of conkers, one could perhaps show off a bit more to one's vanquished opponent by showing him or her how the tree apparently possesses magical powers to turn water blue. This is achieved by stripping the bark off some twigs and soaking it in a glass of water. Then, by shining a light through the liquid, the water appears sky-blue; it is not magic, but an effect caused by a fluorescent substance from the bark altering the wavelength of the light.

A bizarre chain of events during the First World War involving the use of conkers led indirectly to the creation of the state of Israel. Vast numbers of artillery shells were being made, and cordite was required in their manufacture; in turn, the making of cordite needed acetone, which had to be extracted from starch, an organic compound found in plant cells. Most of the huge amount of starch required was shipped over from America, but that was a costly operation, and the supply was unpredictable due to U-boat activity. A more accessible supply had to be found. It was Chaim Weizmann, then the Director of Chemical Research at the Admiralty in London (and later Professor of Chemistry at the University of Manchester), who suggested that conkers could supply the necessary starch. Autumnal collections for the war effort were put in hand at once. However, the starch provided by the conkers was not enough to supply the required amount of acetone, and research was begun to find an alternative. After numerous experiments, a successful formula was found using coal-tar.

Weizmann made such a significant contribution in safeguarding the manufacture of cordite during the war that the Prime Minister, Lloyd George, wished to honour the professor for his contribution. He, in response, spoke only of his 'aspirations as to the repatriation of the Jews to the sacred land they had made famous'. This statement eventually led to the Balfour Declaration of 1917, which promised a home in Palestine for the Zionists. Chaim Weizmann was eventually rewarded by being made the first President of Israel, in 1948; he died in 1952. Meanwhile, conkers slipped back to their 'oblionker' role.

This Cedar of Lebanon was planted by the Revd Robert Uvedale, Headmaster of the Grammar School, in the grounds of Enfield 'Palace', which was the name given to the Manor House and which he leased in order to accommodate boarders at the school.

hundreds of years earlier, in the middle of the twelfth century, and to have been planted by Louis VII, the first husband of Eleanor of Aquitaine, on his return from the Second Crusade to Jerusalem. He had taken part in the Crusade as an act of reparation for the terrible massacre that had occurred at Vitry-sur-Marne in Champagne in January 1143. To commemorate the awful event, he had arranged for some small cedar trees to be transported especially from the Holy Land and planted on the hillside overlooking the town. Their descendants still grow in the area today.

John Evelyn, quick to encourage the planting of these great trees, wrote in 1664 of their thriving 'in the moist Barbados, the hot Bermudas, the cold New-England; even where the snow lies as (I am assured) almost half the year. Why then it should not thrive in Old England? I conceive is from our want of industry.'

He need not have worried, for they were adopted by all the newest of grand houses and parks, just as *Aesculus hippocastanum* was. Wilton Park, the home of the Earls of Pembroke, was soon planted with cedars (and still is) to commemorate significant family events. They were also fashionably planted in Middlesex at Chiswick House; and in Sussex at Petworth House and Goodwood House – where over a thousand are reported to have been planted across the downland of the estate. At Childrey in Oxfordshire, a tree that was thought to have been planted in 1646 bore its first cones eighty-six years later, in 1732.

In Derbyshire, at Bretby Park, the planting of a cedar was first recorded in February 1676. Bretby was the home of the Earls of Chesterfield, and the gardens and parkland were being extensively remodelled at about this time, including the installation of the most elaborate water features – said to be more complex than nearby Chatsworth. Celia Fiennes (1662–1741), who made several journeys throughout England between 1685 and c. 1712, described in her journals a number of the newly created gardens and their houses, recording that at Bretby 'The front garden, which has the largest fountaine has also a fine greenhouse and very fine flowers, and the beds and borders and the hedges cut in severall forms.' She also mentioned that one of the fountains, when playing, chimed the tune 'Lilibolaro'. Of the cedar she noted that: 'here was one tree not much unlike the cypress green, but the branches were more spread and of a little yellower green, the barke of the limbs yellow, it was the Cedar of Lebonus'. If the date for its planting is correct (1676), then the tree would have been about twenty years old by the time she saw it in 1698.

It was not only on the large estates that cedars were planted. In 1683, under the auspices of John Watts (fl. 1670s–1690s), the curator of the Chelsea Physic Garden (then on the rural outskirts of London), four seedling trees were planted. These flourished and grew so large that two of them had to be felled in 1771, before they were even ninety years old. Cedars seem to look their best growing on a large lawn near (but not too near) the house. Curiously for such a tree that looks so weighty in the landscape and so self-confident, its life expectancy is comparatively short, at least in this country, for the last of the 'Chelsea four' had to be cut down in 1903, a mere 220 years after their planting (see illustration on p.120).

The whole character of the tree is of an ancient biding his time. They tend to shed branches in gales and high winds from what is quite a brittle wood, and as a result they can take on a somewhat battered appearance. The limbs and trunks can be chained and braced to stabilise them and help prolong their life. A legend has grown up around the cedar in Bretby Park that the breaking off of a limb foretells a death in the

IOHN PARKINSON.

Publish'd 1810. by W. Richardson Strand London.

John Parkinson had his garden in Long Acre close to Covent Garden, an area that had long been owned and cultivated as the market garden of the Abbey of Westminster.

family. The last time this was noted was in 1923 when the 5th Earl of Caernarvon, grandson of the sixth Earl of Chesterfield, and Howard Carter had just excavated the tomb of Tutankhamun. The Earl's sudden death after the tomb was opened was said to have been caused by the 'curse of the Pharaoh'.

The growing of flowers for adornment in gardens in the seventeenth century was hastened by four very remarkable people, all with the Christian name John; all were friends, and they exchanged plants and corresponded with one another.

The oldest of the four was John Parkinson (1567–1650); the next two were the Tradescants, a father-and-son team (c. 1570–1638 and 1608–1662 respectively); and the last was John Evelyn (1620–1706), who became the great 'tree man' of the century. These four – and many others too – were keen to

THE WORSHIPFUL SOCIETY OF APOTHECARIES

The Worshipful Society of Apothecaries of London was formed from a breakaway group of the Grocers Company. The reasons for its declaration of independence are given in its charter of 1617:

James, by the grace of God, King, Defender of the Faith …To all whom these present shall come greetings. Whereas … very many Empiricks [quacks or charlatans] and unskilful and ignorant Men … do abide in our City of London … which are not well instructed in the Art or Mystery of Apothecaries, but … do make and compound many unwholesome, hurtful, deceitful, corrupt, and unwholesome, dangerous medicines and the same do sell … and daily transmit … to the great peril and daily hazard of the lives of our subjects. … We therefore … weighing with ourselves how to prevent wicked persons … thought necessary to disunite and dissociate the Apothecaries of our City of London from the Freemen of the Mystery of Grocers…

King James stood by the Society and informed the Speaker of the House of Commons that the break was for 'the general good'. Despite that, it was a number of years before the Apothecaries Society was invited by the Lord Mayor of London to participate in the river procession on Lord Mayor's Day.

The members of the Society would organise 'herbarizing' parties, and records report that one such party of ten companions went off to ramble around the fields of Kent in 1629, while another group two years later visited Hampstead Heath. It was also noted that Wild Bugloss (*Lycopsis arvensis*) grew in 'Piccadilla'.

Cobham House, on the banks of the Thames at Blackfriars, became the home of the Apothecaries Society, but in 1666 almost everything, including the library, was lost in the Great Fire of London. The Company rebuilt their Hall on its old foundations, but soon after this it was decided that the study of botany could be better advanced by investigating living plants rather than dried ones as previously, and a search was begun to find a garden or a plot of land on which to cultivate rare plants and sow seeds coming from 'foreign lands'. Three and a half acres were found by the River Thames in rural Chelsea, and in 1673 the Society obtained a lease from Charles Cheyne for an annual rent of £5. Thus began the long cultivation of the Chelsea Physic Garden. In its early years there were financial and personality problems, but in 1722, Sir Hans Sloane, their landlord of the time, most generously conveyed the freehold of the Physic Garden itself, together with its greenhouses, stoves, and barge-house, to the Apothecaries Society, thus enabling them 'to hold the same for ever'.

impart their meticulous observation of the plant world to an expanding public enthusiastic about taking up this new form of floral gardening.

John Parkinson was an apothecary – a cross between a chemist and a herbalist. He is believed to have been born in Nottinghamshire and apprenticed there, but later he moved to London. By 1617 he was instrumental in founding the Society of Apothecaries.

Parkinson was appointed Apothecary to James I and later to Charles I. His greatest love, however, was his beloved garden in Long Acre in London, and he eventually resigned as Warden of the Society of Apothecaries in order to devote more time to its development. He grew so many wonderful and rare plants that his garden was visited by all those who were caught up with the new fascination of all things floral.

His fame today rests on his authorship of the first great British gardening book, published in 1629, in which he describes some 1,000 plants, embellished with nearly 800 woodcuts. With prose that is simple but persuasive, he writes of the cultivation of plants for their beauty and pleasure: The Crown Imperial for his stately beautifulnesse, deserveth the first place in this our garden of delight.'

Fritillaria imperialis, the Crown Imperial, had arrived from Turkey *c*. 1580. Its name implies a noble bearing, and indeed it first grew in the Imperial Gardens of Vienna, the Palace of the Holy Roman Emperor. But not everyone agreed with Parkinson about its beauty; Sir Thomas Hanmer, referring to the colour, calls it 'of an ill dead orenge colour'. It also aquired the name of Stink Lily, as the corms smelt strongly of fox, 'a very nauseous smell', yet were often added to stews (by the Persians). Parkinson could find no good medicinal use for them at all – which perhaps is just as well, since the corm uncooked is poisonous. The legend attached to its flowers and the way it appears to hang its head in shame is said to have derived from the bold way the flower originally stared at Christ on His way to the Crucifixion. It is a native of the Holy Land.

Parkinson is clear concerning how to plant up flower beds: 'The vernal crocus or saffron flowers of the spring, white, purple, yellow and striped, with some vernal colchicum or meadow saffron among them … all planted in some proportion as near one another as is fit for them, will give such grace to the garden, that the place will seem like a piece of tapestry of many glorious colours, to increase every one's delight.' One could not follow more down-to-earth instructions; you can almost smell the soil being dug ready for planting.

The Crown Imperial *Fritillaria imperialis* was originally known as 'Corona imperialis' and was often used to decorate May Day garlands.

Not all gardeners are blessed with such a delightful sense of humour as John Parkinson. The Latin title of his wondrous book, *Paradisi in Sole Paradisus Terrestris*, is a schoolboyish pun on his name, 'Park-in-Sun, Park-on-Earth'. The title continues: *A Garden of all Sorts of Pleasant Flowers, which our English Air will permit to be noursed up*. The book, of 600 pages, appeared like an early version of the admirable *RHS Encyclopaedia of Garden Plants and Flowers* of modern times, and was dedicated to Queen Henrietta Maria. This was perhaps a clever move, since Charles I was so impressed with it that he gave Parkinson the title of the Botanicus Regius Primarius, the King's First Botanist. His second book, written ten years later and entitled *Theatrum Botanicum – The Theatre of Plants; or an Universal and Complete Herbal*, described nearly 4,000 plants, many of them herbs for use by apothecaries. This demonstrated how garden horticulture and medicinal horticulture were drawing apart and becoming specialist subjects. Parkinson's name is now firmly entwined with both books, and the list of plants he describes in them has become increasingly important as the story of plant introductions and the history of botany have established themselves as worthy subjects for study.

Some plants included in his compendium had only just arrived on these shores but are still in favour today; for instance, *Lobelia cardinalis* arrived from the eastern United States in 1626, only three years before his first book was published. The whole Lobelia family was named in honour of Mathias de l'Obel, the French physician and gardener who settled in London and was acquainted with both Parkinson and the Tradescant family. The *cardinalis* part of the name is supposed to have come from an observation of the Queen who, when shown the reddish-stemmed and bright red flowers, thought they looked just like the red stockings worn by cardinals. A second *Lobelia*, which arrived too late for Parkinson to list, landed from America in 1665. American Indians believed it to be a cure for syphilis, hence its Latin name, *L. siphilitica*.

Ipomoea purpurea, Morning Glory, or, as Parkinson called it, the 'Great Blew Bindeweed', twined its way here from Mexico, via Italy, in 1621. Like all great beauties it has led an unpredictable life, changing its name and genus a number of times. The annual began domesticity as *Pharbitis purpurea incarnata*; it was then named *Convolvus major*, which was quickly followed by *C. purpurea*, and it was not until the twentieth century that it received its present name. Whatever it might change to in the future, it is related to our native Hedge Bindweed *(Calystegia sepium)*. There are about 500 species of Ipomoea in the *Convolvulaceae* family, spread all over the world, mostly in the warmer climates. The word *Convolvus* is self-explanatory – 'to twine around' – and *Ipomoea* is from the Greek word for 'worms' – *ips* – which aptly describes the plant's bindweedy ways too. In 1963 horticultural suppliers had to suspend the sale of the seeds, which were being chewed by American drug addicts for their alleged hallucinogenic effects. After investigation the seeds proved to have no such properties, and sale of the 'Great Blew Bindeweed' was resumed.

Of *Amberboa moschata*, Parkinson wrote that it was 'a stranger of much beautie, and but lately come from Constantinople'. In fact, it arrived in 1630 and, like the Morning Glory, has several relations native to Britain. These are the knapweeds and star thistles, most of which have a rather rank smell but a stately air about their bloom. The 'stranger of much beautie' also had a change of name, to *Centaurea moschata* (by which name it is probably much better known, though, confusingly, it has recently returned to its original name). This is the musk-smelling daisy from western Asia and the Caucasus known in the vernacular as Sweet Sultan, and earlier as Bachelor's Button (perhaps that is what the flowers resemble); another of its early names was Blackamoor's Beauty. The Greek derivation of *Centaurea* recalls the mythical beast, the centaur, *Kentauros*. Thirty years after Parkinson, Sir Thomas Hanmer described the flowers as 'very sweet and double'. It is a good cutting flower and is still a favourite in many gardens.

While some plants listed by Parkinson are lost to modern cultivation, others did not settle immediately and were reintroduced much later: for instance, the beautiful but rather difficult Pyrenean Alpine, *Auricula ursi flore et folio boraginis* (or, as Linnaeus later named it and as we still know it, *Ramonda myconi*). It was recorded in his book in 1629 but did not survive; it reappeared in 1731 and is still with us. It is a rosette-forming stemless evergreen perennial with hairy leaves; the flower is a lovely violet colour and resembles a primula. There are only 3 species of Ramonda, which are in the *Gesneriaceae* family (the same family as the African Violet), and they can all, with care, be grown in England in the rock garden or greenhouse. The name commemorates two people: Louis Ramona, Baron de Carbonnière (1753–1827), who was a French botanist and traveller in the Pyrenees; and a Spanish botanist, Francisco Mico (b. 1528). Also in the Parkinson list of 1629 was *Rhododendron fer-*

rugineum, from Switzerland, but this plant did not remain for long either, and it was not until 1739 that it became established here permanently. On its return to England from the Alps, it was botanically described by Linnaeus and became the type species for the whole genus in 1753. It is now in the subgenus Rhododendron, which contains the mainly hardier species (another subgenus is Hymenanthes), and within that it has been placed in the small subsection *Rhododendron*, as has the closely related *R. hirsutum*, which was introduced from the same area in 1656. Both plants bloom in the high mountains during July and August and both came to be called the Alpine Rose.

Shortly after this, there arrived from North America a shrub known by three different names – Labrador Tea, Marsh Ledum and Wild Rosemary. It is *Ledum palustre*, and is common over most of the northern hemisphere; it was not a native but has made itself comfortable in boggy places in the countryside. Strangely for such a widely distributed shrub, it too did not stay long, but reappeared on a list of reintroductions in 1762, along with a new Ledum. This second plant was collected a year later in 1763 from Greenland and was thus called *L. groenlandicum*; it is this that is now known as Labrador Tea.

John Parkinson died in London in 1650 aged eighty-three and was buried at St Martin-in-the-Fields. Although, as far as we know, he never travelled abroad to discover new plants himself, he was an enthusiastic receiver and grower of all plants, loving his garden, and above all writing down his observations, in which task his training as an apothecary no doubt stood him in good stead. He was a gardener and recorder par excellence.

Parkinson's influence lives on in the naming of the Jerusalem Thorn, *Parkinsonia aculeata*, in his honour, but, as is so often the case in the naming of plants, he did not have the pleasure of seeing it during his lifetime (its date of entry was 1739). The plant's indigenous home is tropical America, not the Middle East as its name might suggest. It is a prickly evergreen shrub (hence *aculeata*) with sweet-smelling yellow racemes, and is frost-tender. Like some of the plants in Parkinson's list of 1629, it appears today to be just about clinging on and is seldom seen in cultivation in Britain.

Parkinson's two books remain of immense value in helping to trace the history of our garden plants. This was recognised in 1884, when gardener and children's author Juliana Ewing formed the Parkinson Society. Its aims were simple: to encourage the cultivation of old garden flowers, so as to prevent their extinction; and to plant flowers on uncultivated ground, or, as Parkinson himself wrote, 'noursed up to furnish waste places'. John Parkinson would surely have been an enthusiastic founder member of such a society. The Parkinson Society flourished for only a few years and was then merged with the Selborne Society, whose aims were not dissimilar. Almost a hundred years later the National Council for the Conservation of Plants and Gardens came into existence with very similar aims to those of the Parkinson Society. The NCCPG is the plant heritage organisation which looks after nearly seven hundred National Plant Collections and encourages people to search out and, in the old phrase, 'nourse up' and grow old and rare species and varieties of plants.

The legacy of co-operation and the exchange of plants and cuttings which began during the Middle Ages when monastic life was spreading through Europe, and which is still practised today, was really established among the enthusiastic coterie of the seventeenth century and in particular by the two Tradescants. Both men were, in succession, Gardeners by Royal Appointment, both knew and corresponded with Parkinson and John Evelyn, and were keen to share the 'curiosities and rarities' they had collected, and both were travellers. John Tradescant the Elder visited the Continent twice, once in 1609 and again in 1611, for his employer Robert Cecil, 1st Earl of Salisbury. His mission was to buy fruit trees for the gardens of the newly built house at Hatfield. The quality, we are assured, was of the best, and the quantities were quite staggering: hundreds of each species were purchased, including at least 20,000 vines.

Parkinson wrote of the elder Tradescant: 'He hath wonderfully laboured to obtain all the rarest fruits he can heare off in any place in Christendome, Turky, yea or the whole world.' Also for the Earl of Salisbury, he travelled throughout the Low Countries and to France, buying an eclectic range of plants including 800 tulip bulbs. In 1612 Robert Cecil died, but William, his eldest son, who succeeded to the earldom, was just as absorbed with the Hatfield gardens as his father had been. However, he also wanted the London garden of Salisbury House in the Strand remodelled (mostly with roses, it would seem), and he sent Tradescant off again on a further buying spree to Europe. Despite his obvious expertise and enthusiasm for the sumptuousness of the Cecils' gardens, by 1615 John Tradescant had moved his family to Kent, to work in the equally beautiful and, by then, very

John Tradescant the Elder grew a large number of introduced plants, recording them in his *Plantarum in Horto*, published in 1634.

old Canterbury garden of St Augustine's Palace belonging to Edward, Lord Wotton.

Three years later, on 3 June 1618, Tradescant had his first real taste of adventure when he set sail from Gravesend for Russia with the Ambassador, Sir Dudley Digges. On the instructions of the King, James I, Sir Dudley was carrying £20,000 in cash to help fund the Russians, who were at war with Poland. The new Romanov dynasty in Russia had appealed to the West for help in its fight with Poland, and Britain, looking for trading possibilities, took the pragmatic view that helping them might assist this objective, but all ended in failure as the 'perpetuall and invyolable league of Brotherly love' dissolved into enmity. No doubt to the relief of the Muscovy Company and the East India Company (who had jointly raised the loan in the first place) the money was brought safely back to England.

John Tradescant obviously took the opportunity to botanise whenever the chance arose, and it seems that the expedition must have spent some time within the Arctic Circle, around the town of Archangel in northern Russia, because he brought back several plants which thrived there. One or two of them were natives of Britain that he just could not resist gathering, but the majority were new. He collected *Rosa moscovita* – or, as Parkinson called it, '*Rosa sylvestris Russica,* the wild bryer of Muscovia'. Under its modern name, *Rosa acicularis*, the Arctic Rose is found growing in all the northern parts of Asia, America and Europe. However, it stayed only briefly in Britain until apparently being reintroduced in 1808. The plant, with its sharply pointed leaves (the epithet *acicularis*), is now seen rarely in cultivation.

Another of Tradescant's finds, *Angelica archangelica*, is very similar to our own native *A. sylvestris*. The latter has pinkish-white flowers and purple-flushed stems, while the cultivated angelica he brought home has a ribbed green stem, with creamy yellow flowers. It belongs in the same family as dill (*Apiaceae/Umbelliferae*), and all parts of the plant can be taken to relieve stomach ache or flatulence, including the root, which was considered to have the strongest flavour, and was especially recommended during 1665 when the plague was raging in London. The hollow stems were and still are candied and used for the familiar cake decoration, angelica. Denmark was the first country to candy them. The leaves may be used to make a non-alcoholic drink which tastes rather like China tea, but, conversely, the seeds are sometimes put into gin – the whole plant has a gin or juniper aroma and taste. It is one of the many introductions which are now recorded as thriving in Britain's countryside, especially by the sea. Authorities disagree over the 'archangel' name: some say its name commemorates the town from where it was collected; others claim it comes from the 'heavenly' use of the plant, and yet others that its use was revealed by an archangel. Regardless of how it came by its name, it is known that gypsies used it 'to ward off dark spirits' and as a specific against the plague; it was also thought to have aphrodisiacal qualities. You may take your choice, but birthday cakes decorated with angelica should always taste just that bit more exotic from now on.

Two years later, Tradescant set out on a much bolder and more dangerous adventure, joining the expedition under Sir Robert Mansell which set sail from Plymouth Sound on the morning of 12 October 'against the Pyrates in the yeare 1620' to Algiers and the north African coast. In fact, Tradescant sailed in December in the brand new 240-ton pinnace *Mercury*, which was carrying twenty guns and sixty-five men.

Her captain was Phineas Pett, who, as the Commissioner of the Navy, was responsible for every new ship that was brought into the service, so Tradescant was in safe hands. The object was to try and winkle out the Barbary pirates who were proving an aggressive nuisance to British trading interests. Once more, Tradescant took the opportunity to explore wherever they landed and successfully brought home a number of new plants, no doubt writing to John Parkinson as soon as he was home in the autumn of 1621 telling him of the delights he had seen.

It is thought that at this time Tradescant introduced a clover, *Trifolium barbaricum stellatum*, the Starre-Flowered Trefoil from Barbary, now known as *Trifolium stellatum*, telling Parkinson that he had collected it from Formentera, one of the Balearic islands; and *Pistacia terebinthus*, the Chian Turpentine Tree (the turpentine was collected mainly from the trees growing on the island of Scios; the word 'Chian' was the seventeenth-century corruption for the Aegean island), now known just as the Turpentine Tree or the Terebinth. The latter is a native of Asia Minor and all around the Mediterranean, and its two Greek names are words which can both be traced back to old Persian It is a small tree or large shrub with aromatic glossy leaves, small greenish flowers and purple-brown nuts; the fragrant sap is now used in some cancer treatments. Another member of the genus is *P. vera*, the Pistachio Tree, which produces the nuts and was introduced into the country in 1770. It comes from western Asia, and although it will grow here, it needs a hot, dry, sheltered position or a greenhouse. Both trees are in the same family as the cashew nut, the *Anacardiaceae* family.

John Tradescant also brought back from north Africa some Rock Roses (*Cistus* spp.) and the Green Briar, *Smilax aspera*. This last is an evergreen climber which requires a warm, sunny position and is not suitable for cold areas; consequently, it is rarely cultivated here. It would however make a good anti-personnel plant since it produces a dense tangle of very prickly angular stems.

Tradescant's next patron was George Villiers, 1st Duke of Buckingham and, as before, his gardening magic made a great impression at the Duke's home, New Hall in Essex, recently purchased (for £30,000) from the Earl of Sussex. When Tradescant arrived, probably in 1623, the house and garden were already well established – a hundred years previously it had been a royal palace. Henry VIII's second wife Anne Boleyn and her daughter Elizabeth had lived there, and Elizabeth, when Queen, gave it to the Earl of Sussex.

Tradescant set about modernising the grounds with arbours, walks, knots and avenues – the New Hall lime avenue, or, as John Evelyn later called it, 'the fair avenue planted with stately lime trees', was considered to be one of the earliest in England. Tradescant had visited Holland several times on buying trips for his employers, and he probably copied the idea from the Dutch who were very good at 'dressing a landscape'.

It was not long before adventure and travel again became part of Tradescant's life. On his next journey, however, in 1625, there was no opportunity for plant hunting, since he was accompanying the Duke of Buckingham, who had been sent in glorious array to escort the fifteen-year-old Princess Henrietta Maria from her wedding in Paris (where the Duke of Chevreuse had stood proxy) to her husband, King Charles. The new queen became known as the Rose and Lily Queen, joining the heraldic lily of France and the rose of England. Tradescant could not have guessed at the importance of this marriage to his gardening life. The memorial to both him and his son in Lambeth Church acknowledges this:

Those famous Antiquarians that had been
Both gardeners to the rose and lily Queen

It was undoubtedly Tradescant's connection with the Duke of Buckingham which brought him into contact with the court and, for the first time, to the notice of the King. Perhaps the King's First Botanist, John Parkinson, had also mentioned Tradescant and his collection of 'curiosities' and newly introduced plants growing at Lambeth, to where he had moved in 1626.

The publication of Parkinson's book three years later must have given the friends much to talk about. Indeed, Tradescant began to record in the back of his copy of *Paradisus* his own list of plants growing in his garden. This list grew and grew, and in 1634 it was eventually incorporated into his important *Catalogue of Plants*, with a further list produced by Tradescant the Younger in 1656.

From now on, John Tradescant the Elder really was his own man. His collections of plants and 'things' – as he called his curiosities and rarities – flourished in his Lambeth garden. Following Buckingham's assassination in 1628, Tradescant was appointed Keeper of His Majesty's Gardens, Vines and Silkworms, at Oatlands Palace in Surrey, another old estate and garden which by 1630 needed attention. Originally built as a hunting lodge for Henry VIII, Oatlands

Rhus copallina the Shining or Dwarf Sumach. It arrived from America in 1688 and was sent to the Hon. Revd Henry Compton who cultivated it in his garden at the Bishop's Palace in Fulham.

had been given by Charles I to Henrietta Maria as a jointure as part of her marriage settlement.

Tradescant the Younger (1608–62), also called John, was, by the time he was twenty, showing the same characteristics as his father: penchants for gardening and botanising, keeping lists of his plants, and for travel and adventure. He was admitted into the Worshipful Company of Gardeners on 19 December 1634, and by 1637 the lure of botanical discovery and plant collecting found him voyaging for the first time to the New World. He was so enthralled with what he saw there that he returned in 1642 and again in 1653. Virginia had begun to be settled some thirty years prior to his first visit (Jamestown was founded in 1607 and the Pilgrim Fathers sailed from Southampton in August 1620), and the American flora (or at least that along the eastern seaboard) was starting to make an impact on the British gardening scene.

One of the earliest of American invaders, which has never lost its garden appeal, is *Parthenocissus quinquefolia*, the Virginia Creeper, which was listed by John Parkinson in 1629. At that time it was thought to be a vine so went under the name of *Vitis quinquefolia*. Its botanical name today reflects, unusually, its English name, rather than the other way round, and is derived from the Greek words for 'virgin', *parthenos*, and 'ivy', *kissos*; *quinquefolia* describes the five-fingered leaves. Virginia itself was named sycophantically in honour of the 'Virgin Queen', Elizabeth I, and 'Virginia Creeper' accurately describes both the plant's nature and where it comes from, though giving no indication of the fiery nature of its autumnal colouring.

The Stag's Horn Sumach, *Rhus typhina*, belongs to the same family as the Pistachio Tree, *Anacardiaceae*, and was another early arrival from the eastern United States which Tradescant the Younger could well have seen before he left on his voyage, since Parkinson had described it in 1629. It was known in America as the Dyer's Sumach because a yellow dye could be extracted from the bark. The year 1688 saw the introduction of a second sumach, *Rhus copallina*, the Shining Sumach; 'sumach' is a derivative of an Arabic word. In all the species, the leaves and the bark have been used in tanning, leather work and dyeing; all parts of the tree are toxic. In gardening terms the sumachs are of a domestic size and can easily be accommodated in quite small gardens, but their look is decidedly foreign. There are about 200 species and they are spread across the temperate and subtropical zones of North America, South Africa and East Asia. Like Virginia Creeper, their pinnate feathery leaves suddenly produce a firework display of autumnal colour.

Both the Tradescants had an unerring eye for discovering good garden plants on their respective travels. Father and son were so important in the story of horticulture that it was inevitable that they should have been recorded for posterity, not only in the naming of one plant, but in the naming of a whole genus to honour them. It is particularly apt that the genus chosen should include a plant which Tradescant the Younger brought back from America, and which has stood the test of time. This is *Tradescantia virginiana*, now *T. x andersoniana*. Its common name is either Flower-for-a-Day or Tradescant's Spiderwort. The plant was believed to be the antidote to the supposedly poisonous bite of an American spider, the Phalangium spider, which turned out to be quite harmless. However, until at least 1718 the plant was botanically called *Phalangium Ephemerum Virginiana Joanna Tradescantium*. It is one of about 65 creeping and climbing species found all over the New World with names like Moses-in-the-Cradle (*T. spathacea*, which arrived here in about 1750) and Wandering Jew (*T. fluminensis*, which arrived in about 1834); the latter is a native of Argentina and Brazil, with *fluminensis* reflecting the Latin name for Rio de Janeiro, Flumen Januarii.

John Tradescant the Younger succeeded his father as Royal Gardener to Charles I.

It is a sure sign that if plants have a number of synonyms attached to them, they are a complicated genus which the nomenclature botanists have had difficulties in sorting out. Such is the case with the Tradescantia family, and it was to honour Edward Anderson from the Arnold Arboretum in America – who, in 1935, wrote a monograph on the North American Tradescantia – that the name was altered from *T. virginiana* to *T. x andersoniana*.

Tradescant the Younger took his first trip across the Atlantic in 1637, the year prior to his father's death. He visited first Barbados, from where he is thought to have gathered *Mimosa pudica*, literally the Bashful Mimic but known as the Humble or Sensitive Plant as the leaves curl together at the slightest touch. A true curiosity which would have delighted his father was also collected from Barbados, and that was one of the pitcher plants, *Sarracenia purpurea*, named for Michel Sarrasin, a French botanist and doctor from Quebec who sent several species to Europe. The jug-

like leaves hold a deadly but insect-enticing juice in them; the offering, when accepted, kills the insect and gives the plant the added nitrogen it needs.

Aster tradescantia, the first of the North American Michaelmas daisies, was brought home on this trip. It is a fairly late bloomer and did not seem to make much of an impression. Sir Thomas Hanmer, who can usually be relied on to pass a pertinent note on most new plant arrivals, does not mention it. It was *A. novi-belgiae*, collected by seed in 1710 from the area around New York, which proved the sturdier of the species for helping to make the numerous crosses we now grow in our gardens.

Rudbeckia laciniata, the Coneflower, is a Tradescant introduction but via France, as it was reported by John Parkinson to have come 'from the French colony about the river of Canada and was noursed up by Vespasian Robin, the French King's Herbarist at Paris, who gave Mr Tradescant some rootes that had encreased well with him'. There is an interesting parallel here between the Tradescant father-and-son team and the Robins who were also a father and son. Like the Tradescants in England, the Robins, Jean (1550–1629) and Vespasian (1579–1662), were successively royal gardeners to the French court. There was much exchange of plants between the two families, particularly as each of the sons undertook horticultural voyages (Vespasian to the Guinea coast and John to Virginia).

Rudbeckias liked the conditions here and they increased so well that they are now occasionally found outside the garden wall. All the Rudbeckia species are native to North America, and this time it is a Swedish father and son who are remembered in the name. The Rudbecks were successive Professors of Botany at the University of Uppsala during the latter part of the seventeenth and the early eighteenth centuries. According to John Claudius Loudon (who lived nearly a century later), Rudbeck senior was celebrated 'for having made the discovery that the Paradise of Scripture was situated somewhere in Sweden'.

John Tradescant the Younger was appointed to succeed his father as Keeper of His Majesty's Gardens, Vines and Silkworms on his father's death in 1638. The silkworm in the title gives a rather exotic touch, and was added early in the reign of James I when John Bonnell was Keeper. The King was enthused by the silks coming from China and, in an

Sarracenia purpurea is one of a genus of eight species whose natural habitat is the acid and nutrient-deficient bogs in North America.

effort to promote a silk industry in England, passed a decree that mulberry trees were to be established all over the country as food for silkworms. They were planted by him in what are now the gardens of Buckingham Palace, and later, in 1721, the Park Company of Chelsea Park planted numerous mulberry trees as well. It is probable that the Huguenot settlers who had market gardens in Chelsea and silk-looms in Spitalfields may have suggested the venture.

Of the three varieties of mulberry which were known about by the mid-seventeenth century, the leaves of only

Left: Mimosa pudica the Humble or Sensitive Plant must have been a cause of great excitement in 1637 when John Tradescant the Younger returned with it from his first voyage to America.

one, *Morus alba*, the White Mulberry – or, as it was first known, *M. bombycis*, the Silky Mulberry – were suitable for silkworms, but the species did not do particularly well in our damp climate. The tree originates from China and must have been brought to Europe very early, as it had long been growing around the eastern part of the Mediterranean (especially Beirut, where there was a thriving silk industry). The tree, which has a more spreading habit than the Black Mulberry, was first noted growing here in 1596. It has bright glossy leaves but the white ovoid fruit taste quite insipid when eaten raw; however, when they are dried, they can be used in both sweet and savoury sauces.

Morus nigra, the Black Mulberry, was by this time a well-established tree (see p.55), while *Morus rubra*, the Red Mulberry, was recorded by John Parkinson for the first time in 1629. Its natural home was the eastern seaboard of America, so it must have been one of the early entrants brought back on a returning boat, and it is entirely probable that Tradescant the Younger was familiar with it before he left on his first journey. It seems to have made a tentative entrance but was not a great success and is now rarely planted. Like the White Mulberry, its fruits are tasteless, and even in its homeland are fed either to pigs or poultry, though not alas to silkworms.

At Oatlands Palace in 1616, prior to John Tradescant's time there, Anne of Denmark, consort of James I, had commanded Inigo Jones to design a Silkworm House for her, but even that sumptuous abode did not ensure their survival, and gradually the silkworm enterprise petered out. What has remained is a legacy of aged mulberry trees in a few of the country's older gardens, and a belief, in the west of England at least, that once mulberry leaves are unfurled there will be no more frosts in that year. Silk production, according to John Evelyn, was, however, flourishing in America, since he reported that 'Virginia has already given Silk for the clothing of our King; and it may happen hereafter to give Cloathes to a great part of Europe, and a vast Treasure to our King.' (The Emperor Justinian, in about AD 550, was much taken with the beauty of the mulberry and the shape of its leaf, and it is said that he called part of the Peloponnese after it – the Morea area, from the Greek *moran*; see also p.73. He was so entranced with the bright sheen on the silk that he had silkworm eggs smuggled into Constantinople from China in a bamboo stick, thus making silk culture a royal monopoly.)

Tradescant the Younger's second and third visits to Virginia brought a rich haul of now very familiar plants.

Several trees of great significance were collected from the Virginian forests, among them the aptly named Tulip Tree, *Liriodendron tulipifera*, whose trunks always grow very straight and were thus much favoured by the early pioneers for use in building their canoes. John Evelyn considered it a difficult tree to grow, and certainly none in this country have reached the height or lushness of their American relations. In later years, according to Philip Miller (1691–1771), who looked after the Chelsea Physic Garden in the eighteenth century, the first of them to flower in this country did so during the early years of that century in the west London gardens of the Earl of Peterborough at Parsons Green (see also p.32).

A tree with a completely different character which was collected by Tradescant the Younger was the conifer *Taxodium distichum*. The first word is a mixture of Latin and Greek, 'looking like a yew', and *distichum* refers to the two habits of the branches, with some spreading outwards and other upwards. It is the dominant tree of the Florida Everglades, which might give a clue to its English name, the Swamp Cypress.

Another tree which has had an impact on our landscape and is believed to have had the Tradescant seal of approval is the False Acacia – *Robinia pseudoacacia*; this was named for Jean Robin, the French royal gardener mentioned earlier. The tree had been discovered and brought to France much earlier, and was reportedly growing in the royal gardens of the Louvre by 1606. It must have been one of the plant gifts given to the elder Tradescant by the Robins, as by 1630 it was making itself at home here. It was originally classed as a mimosa and called *Acacia americana Robinii*, although some botanists more cautiously called it *Pseudo-acacia*, and that is what it has remained ever since. The tree has pretty pea-like flowers which are very attractive to bees. In the autumn it quite glows with its golden leaves and is more robust than it looks. Being very tolerant of smoke and soot, it was often planted on the embankments in the early days of the railways.

Thank goodness the two John Tradescants were blessed with such a lively curiosity and were obsessive about keeping lists to record what they grew and what they had seen. Their findings have immeasurably enriched our gardens, and they thoroughly deserved the honour of the Tradescantia genus being named after them.

The fourth John on the list, John Evelyn (1620–1706), although passionate about trees (one might almost say obsessive), was also honoured by an orchid being named for him.

The Tulip Tree *Liriodendron tulipifera* was one of the earliest tree introductions from North America. In 1745 a specimen at Waltham Abbey was already 30 m (96 ft) high with a girth of 3 m (9 ft), indicating it was of a substantial age.

Although it is now called Sobralia (after a Spanish physician of the eighteenth century, Francisco Sobral), its original name was Evelyna, following which it became *Elleanthus caravata*. The change of name is a pity, since John Evelyn

deserves to be remembered. He was an early advocate of the idea of the garden city and was immersed and influential in the horticultural life of the country. He travelled on the Continent from 1641 until 1653 (he espoused the cause of

Charles I), and was a polymath, a prolific writer and diarist.

Evelyn's first book was *Sylva: A Discourse of Forest Trees* (with its appendant *Kalendarium Hortense*), which appeared in 1664; this was followed in 1679 by *Pomona*. He also translated two French garden books as a result of his travels, these becoming *The French Gardener* and *The Complete Gardener*. He wrote much else and his diary is full of his wide interests in the form of architectural, political and religious observations. His descriptions of both European and English gardens have proved intriguing and invaluable, since so many of them have changed or disappeared over the centuries.

Although Evelyn is not usually credited with the introduction of any new plants, it appears from his writings that he may after all have brought something interesting home with him from his Continental travels, for in his first book he writes of the *Alaternus*: 'I have had the honour to be the first who brought it into use and reputation in this Kingdom for the most beautiful, and useful of Hedges, and Verdure in the world (the swiftness of the growth considered) and propagated it from Cornwall even to Cumberland: The seed grows ripe with us in August; and the honey breathing blossoms, afford an early, and marvellous relief to the Bees.' He is describing the Italian Buckthorn, now **Rhamnus alaternus**, which the Tradescants had growing in their Lambeth garden as well. The shrub is in a genus consisting of over 120 species. Mostly they are considered as having little garden value, but they are, as Evelyn observed, very good material for hedging.

The writing of his great work, *Sylva*, came about because of the acute shortage of mature trees for shipbuilding. It was a constant problem, not only for the original construction of the ships but also for their repair. In the first year of the Royal Society (1662), the timber problem was put to a small committee, of which John Evelyn was a member. He delivered his conclusions a few months later, on 15 October, in a paper read to a meeting of the Society, which included such able scientific thinkers as Sir Christopher Wren, Samuel Pepys, Robert Boyle and Isaac Newton (later to become the president). Pepys (1633–1703) was a long-standing friend of Evelyn, and they must have discussed the problem together on many occasions, and probably with the Tradescants as well. It is known that Pepys visited their house in Lambeth.

It must have been a tremendous meeting, the result of which, even today, is instilled in our psyche. As a country, we have always desired to be self-sufficient in timber (although apparently near the bottom of the European league for the amount of wooded countryside we possess), and there are

John Evelyn published his *Kalendarium Hortense or the Gardener's Almanac* in 1664, which is full of modern and relevant advice.

today various government schemes and grants giving encouragement to the planting of trees. There was 'Plant a Tree in '73', then the annual National Tree Week was instigated which has been running now for some years, and at the end of the twentieth century the 'Year of the Tree' was celebrated. The idea of planting a 'Millennium Forest' to mark both the ending of the second millennium and the commencement of the third was an exciting and visionary development. Who knows in a thousand years what ancient woodland will welcome in the fourth millennium. John Evelyn would have thoroughly approved of the idea.

Trees are so important in the landscape but it is flowers which are the real stars of the seventeenth century, and the tulip must surely be one of the most important blooms we think of in association with the 1600s. The first bulbs came from the Continent about 1578, according to Richard Hakluyt, the Welsh traveller, who wrote: 'There have been brought into England from Vienna in Austria divers kinds of flowers called Tulips and these and others procured thither a little before from Constantinople.' The name reflects the plant's Near East ancestry but, like so many others, it became muddled in translation. The Turkish name for the tulip is *lalé* – actually a Persian word – but the word 'tulip' is the Turkish *duliband*, the turban, the shape of which is echoed in the flower. One can see where the muddle occurred, for the flower *lalé* was quite often worn as a decoration on the *duliband*.

Whatever the name, the tulip's popularity grew rapidly all over the Continent. They were known in France in 1608, and when John Tradescant the Elder visited Archangel in Russia in 1618 he was told 'that thear groweth in the land both tulipes and narsisus'. It was, of course, in Holland that the tulip took the strongest hold, resulting in the famous *Tulpenwoede* (literally 'tulip fury'). For three years, from 1634 to 1637, the Dutch acted in a most un-Dutchlike way, gambling away their guilders for possession of the bulbs with the newest blooms. (At the same time, they also collected and paid enormous sums for sea shells, particularly those brought back from the newly explored foreign territories.) Britain did not go to quite such extremes (although we did have the speculation fever of the South Sea Bubble in 1720), but the tulip established itself in this country as a firm favourite.

Sir Thomas Hanmer wrote that tulips were 'the Queene of Bulbous plants, whose Flower is beautifull in its figure, and most rich and admirable in colours, and wonderfull in variety of markings'. He went into great detail regarding the conditions the bulbs liked best and the soil which was required. Indeed, he knew of a florist in Paris – one of the ablest, apparently – who had made a great deal of money from growing and selling the plants, but to maintain the high standards 'he chang'd his habitation purposely every third or fowerth yeare, which he found infinitely better'd by varietyes of aire as well as Earth'.

Very quickly there were huge numbers of different varieties of tulip being listed by the cognoscenti. John Rea, a nurseryman from Shropshire during the 1670s, enumerated 174 varieties, and could have listed many more which 'would fill a considerable volum'. It was always the striped, feathered and marbled varieties which proved such a fascination, although there was a report in 1754 of a tulip with such a large, strong flower bowl that it would hold a pint of wine. By the end of the eighteenth century, another nurseryman, James Maddock (*c.* 1715–86), a Quaker who had established a business at Walworth in Southwark in London, was listing 665 different kinds. In the nineteenth century the tulip took on a new persona, that of a florist flower (see p.40) and again, because of rarity and exclusivity, the price of the individual bulbs shot up; indeed, armed guards were once employed in the Cambridge home of one Richard Headley to protect his show tulips.

Eventually different classes evolved as various forms, and new species were discovered which resulted in many different categories, including Early, Cottage, Double, Parrot,

Tulipa zomerschoon painted by Sally Crosthwaite. In her book *The Tulip*, Anna Pavord describes this bulb as perhaps the oldest tulip in cultivation, having been known since 1620.

Darwin, Lily-flowered, Fringed, Triumph and Mendal. Such botanical confusion reigned that, in an effort to unravel the history and breeding of the plants, a Tulip Nomenclature Committee was set up by the RHS in the early part of the twentieth century. This resolved itself into 'The Classified List of Tulip Names', which was published in 1948 and contained over 4,000 names; by the latter part of the twentieth century, however, only about 500 remained on the list.

There were other plants, and products, coming into the country as a result of the feverish desire for exploration, which was often assisted by the East India Company in particular. As has been seen (p.98), the company had been founded at the beginning of the century with that very reason in mind, and their ships played an important and, as time went on, an increasingly larger role in the transportation of seeds and plants.

The banana made its first appearance in 1633 when Thomas Johnson, an original member of the Apothecaries Society, displayed a hand of them in his window on 10 April, having received the fruit from his 'much honoured friend, Dr Argent, President of the Colledge of Physitions'. They had come from the 'Bermoothes' (Bermuda) and were in all probability the Horn Plantain, *Musa paradisiaca*, which was cultivated throughout the tropics for food. The name commemorates Antonius Musa, the physician to Emperor Augustus (63 BC–AD 14). The *Musa* name was later confirmed by Linnaeus; the second name reflects the Arabic name for the plant, the Tree of Paradise. By this time (1633) the word 'banana' had entered the vocabulary too; this had come from the Congo, where it was the local name for the fruit (see illustration on p.203).

Thomas Johnson kept his 'bananas' until June, when they were soft to the touch, and then cut them into small slices. He records that they had a pleasant taste but was surprised they had no seeds. Prior to that Johnson – who lived on Snow Hill in the City of London – had had them carefully drawn and engraved, and they appeared on the new frontispiece of the much enlarged edition of Gerard's *Herball*, which Johnson had just finished editing. Snow Hill, incidentally, acquired an unsavoury reputation during the eighteenth century, when gangs of young men – often aristocratic ruffians with not enough work to do – would seize elderly matrons, stuff them into barrels and roll them down the hill – all for fun, of course. The aficionados of these escapades were known as the 'Mohocks'.

Coffee, chocolate and tea had also begun to appear here.

Samuel Pepys drank his first cup of 'cha' (the Chinese name, and naturally brought here on an East Indiaman) in 1660, the same year that an import tax of 8d a gallon was imposed on the leaves. Chocolate was also expensive, as the imported cocoa too had a high duty put on it. The first chocolate house opened in London in 1657, but by 1698 the most famous was The Cocoa Tree Chocolate House in Pall Mall, in the centre of London. The '*chocolatl*' drink had been made by the Aztecs long before the Spaniards brought it back across the Atlantic in 1528. The beans came from the tree called 'food of the gods', *Theobroma cacao,* and although the tree was introduced to Britain in 1739, being noted as arriving from Trinidad, it cannot have survived, as it is truly tropical and no amount of 'plant against a south-facing wall' advice would have allowed it to thrive in the relatively cool and damp climate of even south-west Britain.

Theobroma cacao is in the *Sterculiaceae* family, which holds about 50 genera with some 750 species. Most of these are tropical trees, but just a few can manage to grow in favoured spots here. First are one or two of the flame trees (**Brachychitons**) from Australia; another is the beautiful Madagascan tree **Dombeya wallichii**, which arrived here in 1820. The family name of *Sterculiaceae* is named for Sterculius, the Roman god of privies (L. *stercus*, 'dung') – the wood and flowers of some of the trees stink; even Dutch and English pirates capturing Spanish ships with a cargo of the beans aboard called them 'sheep shit' and threw them overboard. It certainly makes one think twice about eating chocolate.

Despite the extraordinary diversity of plants being brought back from so far away, there were plants still to be discovered in Europe, especially from around the Mediterranean, and one of the most indispensable shrubs arrived from Portugal in 1648: the evergreen Portugal Laurel, *Prunus lusitanica*. It is a perfect hedge maker, and is especially favoured for game coverts. It makes the most beautiful specimen tree if allowed to develop, and the long white drooping flowers are sweetly perfumed. It is hardier than the Cherry Laurel, *Prunus laurocerasus*, which had arrived here in 1576.

Another plant of which we are still very fond, and which originated from the eastern end of the Mediterranean, is the very English Rose of Sharon, *Hypericum calycinum*, found by the young Sir George Wheler in 1676 when he was taking, in modern parlance, a year out before going to university. The British Ambassador's entourage, of which he was a member, had decided to explore some of the islands they were sailing

Agapanthus praecox subsp. *orientalis* was one of the first plants to arrive from South Africa in about 1650. It made its way to Europe courtesy of the Dutch East India Company.

past on their voyage from Venice to Constantinople. Unfortunately, on one of their jaunts the ambassadorial party left Wheler and a companion behind, so with time on their hands while they waited to be rescued they did a little gentle botanising and found this delightful St John's Wort. In due course the plant returned to England with Wheler, where it was called, well into the nineteenth century, 'Sir George Wheler's tutsan'. The word 'tutsan' means wholesome, and is derived from Old French and late Medieval English; it is the name usually applied to the native *H. androsaemum*, which for centuries has been known for its medicinal qualities. All the St John's Wort Hypericum spp were believed to have magical properties. The Greeks used the plant to ward off evil spirits, the Scots recommended it against witchcraft, and in England it was used as a preventative against lightning striking the house. Today St John's Wort is recommended as

an antidepressant or to relieve the stresses of modern life.

Just as the colonising of North America by the French and the British was reflected in the plant introductions from the New World during this century, so the same can apply to the Dutch exploration of South Africa. Plants from the Cape of Good Hope suddenly began arriving in Europe, and although most of them needed shelter from frost and winter cold, they were greeted with delight and a great deal of interest. One of the first to appear (noted by Parkinson in 1629) was the lovely blue lily *Agapanthus africanus*, or, if you prefer, the Love Flower from Africa. It comes from a small genus of about 10 species, all from southern Africa, and each one looks distinctly exotic. None of our native plants is remotely similar so they must have made a tremendous impact, with their dark blue to creamy-white star-burst flowers which bloom late in the summer. The one usually grown in England is *A. praecox* subsp *orientalis*; it is the toughest and has found a toehold in the sand dunes on Tresco, in the Isles of Scilly.

It was eventually realised that these architectural plants looked marvellous when planted in big tubs and pots, and the nineteenth century saw them much used to decorate terraces and driveways. It was not until the twentieth century that any hybridising took place – and that was by the then Treasurer of the RHS, the Hon. William Palmer (1894–1971); several of his 'Headbourne Hybrids' are available, including one named for his wife, Dorothy Palmer. Sometime during the 1980s, the 10 species comprising the Agapanthus genus were taken out of the *Liliaceae* family and put into the *Alliaceae* family, since they have several characteristics in common with the onion and the garlic.

A second plant to come from South Africa was the Stork's Bill, *Pelargonium triste*, thought to have been obtained by the elder John Tradescant in 1631. Twenty-five years later, Sir Thomas Hanmer remarked that 'the plant is very scarce in England' (it was then called *Geranium Triste*, the Sweete Cranesbill). It is surprising it has remained in cultivation, since its second name means 'dull' or 'sad'. It does have a rather self-effacing yellow and brown flower and as the apothecary Sir John Hill (1707–75) observantly recorded in 1756, 'it rejoiceth not by day' but comes into its own during the evening, when it gives off a most delicious scent – perhaps that is what has kept it flourishing. The following century was the time when the vast majority of geraniums and pelargoniums took the voyage north from South Africa. Between 1820 and 1828, the Devonian Robert Sweet

White Sweet Pea · · · · · · · · · · S R K

(1783–1835) published his huge opus *The Geraniaceae*, in five volumes, after which there was no stopping the Stork's Bills (erodiums) or the Crane's Bills (geraniums) developing into the modern hybrids we plant in our gardens today.

The Sweet Pea, *Lathyrus odoratus*, was one of the last plants to be recorded as arriving in Britain in the seventeenth century, and its entry in 1699 was once again due to the curiosity and generosity of friends exchanging plants and seeds. Dr Robert Uvedale (1642–1722), who studied divinity and then law, was widely recognised as 'a master of the greatest and choicest collection of exotic greens [plants] that is perhaps anywhere in this land'. He became headmaster of Enfield Grammar School and leased the Manor House both to take in boarders and to develop his exotics in some of the earliest hothouses recorded. He also must have learned of John Evelyn's advice on the planting of cedars, as he ensured that one was planted in the school grounds in 1670, where it became quite a landmark, remaining there until relatively recently (see p.125).

Like Reverend John Ray, Robert Uvedale was a tremendous correspondent (always in Latin), and he wrote to Father Franciscus Cupani of Sicily to congratulate him on the publication in 1697 of his book *Hortus Catholicus*, in which *Lathyrus odoratus* is described for the first time. It was a cultivated garden plant around the Mediterranean and had a very long pre-Christian history of being used for garlands and wreaths. The Sicilian monk responded by sending Uvedale some seed of his native *Lathyrus odoratus*, and of all the Sweet Peas which have been developed since then, none has such a sweet, pervasive perfume. The colour of the flowers could be purple, pale purple and red, or white and red. They were so popular and easy to grow that, by 1722, they were being recommended for use in London gardens – Enfield in Middlesex then very much being in the country.

Unlike the geraniums, which have been continually developed since their introduction, the 'sweet-scented Pea' remained untouched until quite late in the nineteenth century, when Richard Clarke, who hybridised cotton and begonias, also experimented with the raising of an all-blue Sweet Pea (although a pale blue species, *L. vernus*, was introduced much earlier in the seventeenth century). He is said to have created a white flower with a blue edge. There are two inventors of the modern Sweet Pea: one was a Scot, Henry Eckford (1823–1905), gardener to the Earl of Radnor in Berkshire; the other was Silas Cole (dates unknown), gardener to Earl Spencer at Althorp. The two men exhibited their blooms in successive years at the Crystal Palace in the first two years of the twentieth century. There was so much interest caused by the display of nearly 300 different varieties that almost on the spot the National Sweet Pea Society was formed.

One can, of course, have too much of a good thing, and the great horticulturist Edward Bowles (1865–1954) of Enfield, writing a little later of their all-pervading perfume, said: 'A dining-table decorated heavily with sweet peas spoils my dinner as I taste sweet peas with every course, and they are horrible as a sauce for fish, whilst they ruin the bouquet of good wine.' This heartfelt description makes them sound like the garlic equivalent of the floral world.

The change in ideas about gardening and flowers which took place during this century was certainly dramatic. John Aubrey, the antiquarian and folklorist, who had estates in Wiltshire, Herefordshire and Wales, wrote in 1691: 'There is now ten times as much gardening about London as there was in 1660… In the time of Charles II, gardening was much improved and became common.'

The formality of knot gardens and parterres became ever more complicated during the century, and each new plant introduction must have been greeted with excitement. Some of the new imports, like *Bulbocodium vernum* from the Pyrenees and two or three different Colchicums, all from Turkey and Greece (but not *Colchicum byzantinum*, which was too tall and not neat enough to be used in this way), as well as *Crocus serotinus* subsp *clusii*, collected from Spain in the 1620s, were quickly cultivated and taken into formal gardens. The design of the knots and parterres was made more open, so that the intricacy of the patterns could be viewed more easily from the house without detracting from the vista and *point de vue*. The edging of borders and beds was by now low and often of tile or stone, while the paths surrounding them were made of brick or 'a sparry marble' (so that it glittered) or even crushed coal – anything to get colour into delineating the increasingly complicated embroidery of the design. Gradually the spaces within the garden were remodelled and revitalised in the stylish French manner.

Left: The Sweet Pea *Lathyrus odoratus* is a Sicilian wild flower, the seed of which was sent to Britain in 1699. By 1782 the Leeds nurseryman, Thomas Barnes, could supply seed of four different colours: fine scarlet, white, painted lady and purple.

André Mollet, who worked in London during the first half of the century, was the first proponent of the new style; he revitalised St James's Park, which had earlier been heavily planted with copses of laurel (*Prunus laurocerasus*). Mollet began the huge alterations during 1660 with the digging-out of the half-mile-long canal, the straightening of The Mall and the planting of the double avenue of trees on either side of it – a layout which still, today, has such a triumphant feel about it. The radical changes to the royal park were the talk of the town, and Samuel Pepys, on 16 September of that year, strolled from Whitehall to view the changes.

It was André Le Notre (1613–1700), Louis XIV's *Controleur General des batiments du Roi*, who conceived the layout and implementation of the gardens of Versailles during the 1660s. He was the pre-eminent and most artistic landscape designer of this century. Although he never visited England, his influence here, as throughout Europe, was profound and had far-reaching effects.

Looking at the immense layouts of so many of the royal and great gardens which were being built or redesigned at this time throughout Europe and Britain and which were influenced, directly or indirectly, by Le Notre, one thought springs to mind, and that is the rigid symmetry, scale and absolute control that the design places on the outside space. In English terms, the scale of the designs appears to be out of harmony with the landscape of our countryside, which is on an altogether more intimate scale than that of France. Combined with what we now call the hard landscape – fountains and water features, grottoes, urns and statuary – these dominant Continental designs sit uncomfortably in our countryside. Having said that, the tragedy is that of all the effort and money which must have been poured into the creation of these gardens, so little remains for us to see today. Most, of course, were swept away without so much as a backward glance during the early years of the eighteenth century, since from 1720 onwards England was overwhelmed by a new philosophy and ideal. This was to welcome and embrace the natural world into our own gardened surroundings, and to re-create the landscapes of Italy and the ancient world in the English countryside. Thus, during the eighteenth century, the influence of such men as the writer and poet Alexander

Pope (1688–1744), the 'virtuoso grand jardinier' Charles Bridgeman (*c.* 1680–1730), the man full of 'capabilities' Lancelot Brown (1716–83), and Humphry Repton (1752–1818) with his *Red Books* was to obliterate all the great Jacobean and Renaissance gardens in order to accommodate *le jardin anglais* and its lax and lyrical ways. Sir Roy Strong (1935–) calls the loss tragic, and indeed it is, but it is why our countryside looks so inclusively landscaped and seems so welcoming today.

In a way, the seventeenth century horticulturally mirrors the twentieth. At the beginning of both periods, garden layout was well established, and paid its dues retrospectively to the Elizabethan reign and to the Victorian era. The change of emphasis during the earlier centuries from growing only food and medicinal plants to flower gardening was gradually reflected in the books published. The phrase 'flower-garden' appears to have been used for the first time by William Hughes in the book he wrote in 1672 entitled *The Flower Garden Enlarged.* John Parkinson referred to the style in his book of 1629 as 'the Garden of Pleasure', yet the word 'flower bed' with which we are so familiar is relatively modern and seems only to have been first used in 1873. The phrase accurately reflected the use of thousands of bedding plants in carpet bedding, which was as complicated as the earlier knots and parterres of the seventeenth century.

For the first time in the history of English gardening we are able, during the seventeenth century, to evaluate in an historic way the systematic recording of new introductions into both Britain and Europe, and to assess the impact the plants have had on our modern-day gardening life. There seems no doubt that the camaraderie and enthusiasm for new plants which exists between gardening people worldwide had its foundations in the apothecaries, herbalists, physicians and botanists of the seventeenth century. They, in their turn, were copying what the monastic orders had done for centuries previously: discovering and exchanging information about the new plants that they were growing. Suddenly in the seventeenth century, our modern world does not seem so far away, and we can relate nearly all our gardening ways and habits to the legacy and traditions created in this, the most floral of centuries.

Plant Introductions in the period 1600–1699

1601 *Colchicum agrippinum* Autumn Crocus. Probable hybrid between *C. autumnale* and *C. variegatum*. Unknown origin but possibly Turkey and the Near East.

Colchicum byzantium (*C. autumnale* '**Major**') Autumn Crocus. Unknown origin but possibly Turkey and the Near East

1603 *Abies alba* (syn. *A. pectinata*) European Silver Fir, Silver Fir. C. and S.E. Europe.

Tulipa schrenkii (originally *T. suaveolens*) Crimea, Turkistan.

1604 *Gladiolus imbricatus* C. and E. Europe, Latvia, Estonia.

1605 *Hyacinthoides italica* (originally *Scilla italica*) Bluebell. Portugal, Spain, S. E. France, N.W. Italy, Switzerland.

Zingiber officinale Ginger. E. Indies; see p.113.

1607 *Scilla peruviana* Cuban Lily. W. Mediterranean.

c. 1613 *Nicotiana tabacum* Tobacco. Tropical America. In Virginia, John Rolfe crossed seed from the local plant of Nicotiana with one from the West Indies, and it was this subsequent variety that produced the first tobacco crop; see p.113.

c. 1616 *Aesculus hippocastanum* Horse Chestnut. Albania, N. Greece, Macedonia.

1617 *Paradisea liliastrum* (syn. *Anthericum liliastrum*) St Bruno's Lily. Mountains of S. Europe. Named by the French in honour of the founder of the Carthusian Order.

1618 *Angelica archangelica* (syn. *A. officinalis*) Archangel. N. Europe.

Rosa acicularis (*Rosa moscovita*) Finland through Siberia to N. Alaska and Japan. Reintroduced (see p.215).

1620 *Larix decidua* European Larch. Alps and Carpathian Mountains. Collected by John Tradescant on his trip to Russia. It is a tree greatly suited to the highlands of Scotland. The 2nd Duke of Atholl is said to have planted twenty-seven million of them on his estate at Dunkeld in Perthshire.

Scabiosa atropurpurea Pincushion Flower, Sweet Scabious. S. Europe; see p.200.

c. 1620 *Crocus serotinus* subsp. *clusii* (syn. *C. clusii*) Autumn-flowering Crocus. Portugal, N.W. and S. Spain.

1621 *Ipomoea purpurea* (syn. *Convolvus purpureus/Pharbitis purpurea*) Morning Glory. Probably Mexico.

Pistacia terebinthus Turpentine Tree, Terebinth. Portugal to Turkey, Canary Islands, Morocco to Egypt.

Smilax aspera Green Briar. S. Europe, N. Africa.

Trifolium stellatum Starry Clover. W. Mediterranean. Reintroduced (see p.215).

1626 *Lobelia cardinalis* Cardinal Flower. E. Canada to S.W. USA.

1629 *Acanthus spinosus* Spiny Bear's Breeches. Italy to W. Turkey. Reintroduction (see p.34).

Agapanthus africanus African Blue Lily, Love Flower. Cape of Good Hope (see p.152).

Bulbocodium vernum Pyrenees, S.W. and W. C. Alps.

Helleborus niger Christmas Rose. S. Europe. Reintroduction (see p.33).

Lathyrus vernus (syn. *Orobus vernus*) Spring Vetchling. Europe, Turkey, Caucasus, Siberia.

Morus rubra Red Mulberry. E. USA.

Parthenocissus quinquefolia Virginia Creeper. E. USA.

Polianthes tuberosa Tuberose. Mexico; see p.103.

Ramonda myconi (syn. *R. pyrenaica*) Pyrenees and N. E. Spain. Reintroduced (see p.158).

Rhododendron ferrugineum Alpenrose. C. Europe, Alps. Reintroduced (see p.179).

Rhus typhina Stag's Horn Sumach. E. USA.

Tradescantia x andersoniana (syn. *Tradescantia virginiana*) Spider Plant, Spiderwort. E. USA.

1630 *Amberboa moschata* (syn. *Centaurea moschata*) Sweet Sultan. W. Asia, Turkey, Caucasus.

c. 1630 *Cedrus libani* Cedar of Lebanon. Lebanon to Turkey.

Robinia pseudoacacia False Acacia, Black Locust. E. USA.

1631 *Pelargonium triste* Western Cape, Northern Cape.

Platanus occidentalis American Sycamore, Buttonwood; see p.74.

Tiarella cordifolia Foam Flower. North America. See also p.158.

1632 *Rudbeckia laciniata* Cone Flower. C. and E. North America.

1633 *Ranunculus amplexicaulis* Pyrenees, N. and C. Spain. Genus named by Pliny in the first century, and derived from the Latin word for a frog – Ranunculus usually grow in damp places.

1634 *Cistus monspeliensis* Montpellier Rock Rose. S. W. Europe. Found by Sir Oscar Warburg and rediscovered in 1984 by Harold Read.

Saxifraga umbrosa var. *primuloides* (syn. *S. primuloides*) Pyrenees.

1635 *Actaea rubra* (syn. *A. erythrocarpa*) Red Baneberry. C. and E.North America.

1636 *Tulipa clusiana* (syn *T. aitchisonii*) Lady Tulip. Iran to Himalayas.

1637 *Aster tradescantii* Michaelmas Daisy. E. USA.

Lupinus perenne Lupin. E. USA. See also p.80.

Mimosa pudica Humble Plant, Sensitive Plant. Tropical America.

Sarracenia purpurea Pitcher Plant, Huntsman's Cup. Canadian Arctic to New Jersey.

1640 *Achillea ageratum* (syn. *A. decolorens*) Sweet Nancy. Portugal. Mediterranean; see p.200.

Achillea herba-barota C. Europe; see p.200.

Aquilegia canadensis Columbine. E. Canada to S. USA.

Astragalus tragacantha (syn *A. gummifera*) Gum Tragacanth. S. Europe, W. Asia; see p.157.

Carlina acaulis Stemless Carline Thistle. Alpine pastures, S.E. Europe.

Taxodium distichum Swamp Cypress. S. E. USA.

1646 *Menispermum canadense* Canadian Moonseed, Yellow Parilla. E. USA.

1648 *Citrus aurantifolia* Lime. Tropical Asia; see p.98.

Citrus limon Lemon. Asia; see p.98.

Citrus medica Citron. S.W. Asia; see p.98.

Prunus lusitanica Portugal Laurel. S.W. Europe.

1650 *Liriodendron tulipifera* Tulip Tree. E. USA.

c. 1650 *Agapanthus praecox* subsp *orientalis* (syn *A. orientalis*) South Africa.

Foeniculum var. *azoricum* Florence Fennel. S. Europe; see p.31.

1652 *Aster alpinus* Alps.

c. 1653 *Rhamnus alaternus* Italian Buckthorn. Portugal, Morocco, Mediterranean, Ukraine.

1656 *Gleditsia triacanthos* Honey Locust. C. and E. USA. Named for Johann Gleditsch (1714–86), Director of Berlin's Botanic Garden.

Hamamelis virginiana Virginian Witch Hazel. E. North America.

Juglans nigra Black Walnut. E. USA; see p.23.

Lonicera sempervirens Trumpet Honeysuckle, Coral Honeysuckle. E. and S. USA.

Rhododendron hirsutum Alpine Rose. C. Europe, Alps.

1658 *Citrus sinensis* Sweet or Portugal Orange. China; see p.99.

Cynara cardunculus Cardoon. S.W. Mediterranean, Morocco; see p.27.

Ledum palustre Marsh Ledum, Wild Rosemary. N. America. See also p.169.

1659 *Nerine sarniensis* Guernsey Lily. Northern and Western Cape S.Africa.

Pulsatilla alpina Alpine Pasque Flower. Mountains of C. and S. Europe.

1663 *Aruncus dioicus* (syn. *A. sylvester/Spiraea aruncus*) Goatsbeard. Europe to E. Siberia, E. North America.

Platanus x hispanica (*P. occidentalis x P. orientalis*) (syn. *P x acerifolia*) London Plane. Garden origin; see p.74.

1664 *Arisaema triphyllum* (syn. *A. atrorubens*) Jack-in the-Pulpit, Indian Turnip. E. and S. North America.

Juniperus virginiana Pencil Cedar. E. and C. North America.

Podophyllum peltatum Mayapple, American Mandrake. North America.

1665 *Erythronium americanum* Trout Lily, Yellow Adder's Tongue. E. North America.

Lobelia syphilitica Blue Cardinal Flower. E. North America.

1668 *Ananas comosus* Pineapple. Brazil; see p.201.

1670 *Asclepias syriaca* Milkweed, Silkweed. E. North America. Early colonisers used the down from around the seed to stuff their pillows.

Crocus pulchellus S. Balkans, W. Turkey.

1673 *Trillium cuneatum* (syn. *T. sessile*) Trinity Flower, Wake Robin. S. E. USA.

1675 *Chelone glabra* (syn. *C. obliqua* var. *alba*) Turtlehead. E. and S. USA. Reintroduced (see p.168).

Yucca filamentosa Adam's Needle. S.E. USA; See p.112.

1676 *Hypericum calycinum* Rose of Sharon, Aaron's Beard. N. W. and N. E. Turkey, S. E. Bulgaria.

1680 *Citrus bergamia* Bergamot Tree. Asia; see p.98.

Crocus angustifolius (syn. *C. susianus*) Cloth of Gold Crocus. S. Ukraine, Crimea, Armenia.

Ferula tingitana Tangier Fennel. N. Africa.

1682   *Cupressus lusitanica* (syn. *C. lindleyi*) Cedar of Goa, Mexican Cypress. Mexico to Guatemala.

1683   *Arum italicum* Lords and Ladies. Europe, N. Africa.

Cornus amomum (syn. *Swida amomum*) Dogwood. E. North America; see p.165.

Hyacinthoides hispanica (syn. *Endymion hispanicus*, *Scilla campanulata*, *S. hispanica*) Spanish Bluebell. Spain, Portugal, N. Africa.

1686   *Astrantia minor* Lesser Masterwort. Alps.

Tropaeolum majus Nasturtium, Indian Cress. Bolivia to Colombia; see p.110.

1688   *Acer negundo* Box Elder, Ash-Leaved Maple. N. America; see p.155.

Aralia spinosa Devil's Walking Stick. S. E. USA.

Magnolia virginiana (syn *M. glauca*) Sweet or Swamp Bay. E. USA; see p.162.

Rhus copallina Dwarf Sumach, Shining Sumach. E. North America.

1690   *Grewia occidentalis* Four Corners. S. Africa.

Liquidambar styraciflua Sweet Gum. E. USA.

Mertensia pulmonarioides (syn. *M virginica*) Virginia Cowslip. North America.

Thuja orientalis (syn. *Biota orientalis*, *Platycladus orientalis*) Chinese Arbor-vitae. N. and W. China.

1691   **Quercus coccinea** Scarlet Oak. E. North America; see p.155.

1693   *Sisyrinchium graminoides* (syn. *S. angustifolium/S. bermudiana/ S. birameum*) Blue-eyed Grass. E. North America.

1694   *Jasminum azoricum* (syn. *J. fluminense*) Jasmine. Madeira.

1696   *Abies balsamea* Balm of Gilead, Balsam Fir. North America extending into the Arctic; see p.155.

Coffea arabica Arabian Coffee. Ethiopia, Angola; see p.210.

1698   *Isoplexis canariensis* Canary Island Foxglove. Canary Islands.

1699   *Argyranthemum frutescens* (*Chrysanthemum frutescens*) Paris Daisy, White Marguerite. Canary Islands.

Buxus sempervirens '**Suffruticosa**' Common Box. Europe. N. Africa, Turkey; see p.29.

Lathyrus odoratus Sweet Pea. Sicily, Italy.

Passiflora caerulea Blue Passion Flower. C. and W. South America.

All around the World
1700–1799

Significant dates

So MANY EXCITING HORTICULTURAL EVENTS took place during the eighteenth century that it is not surprising we are still influenced by their effects today. First and foremost, a positive deluge of new plants rolled into England. Like a great tidal wave of exotica they just kept arriving to titillate the senses of the botanical world. Philip Miller (1691–1771), the illustrious Gardener to the Society of Apothecaries at the Chelsea Physic Garden, believed that the number of plants arriving here doubled during the middle period of the century – probably to somewhere between 6,000 and 10,000 species. It is from 1748 onwards that the flow of introduced plants becomes continuous. There are no gaps; every year at least one new plant (usually several, sometimes hundreds) was being recorded. And that has continued right up to the present day. Thus, for over 250 years Britain

has received a steady stream of some of the most beautiful horticultural gems, and with the arrival of the Antipodean collection in 1772, British gardens could be filled for the first time with plants literally from around the globe.

Although it was not realised at the time, an event of major political significance which took place between 1756 and 1763 had a profound effect on plant introductions into Britain, not only for the rest of the century but right up to the present day. This was the Seven Years War, which began as a conflict between France and Britain in North America, spread to Europe and became a global conflict between the two countries for supremacy. It ended with success for Britain, who then took on the mantle of the leading world power. This fundamental shift in the balance between the two European superpowers of the day had an immense and

Left: Banksia serrata is one of the 70 species of the genus *Banksia* named for Sir Joseph Banks which comprises the Australian branch of the *Proteaceae* family whose 'cousins' are the South African Proteas.

lasting effect on the horticultural life of this country.

In geographical terms, the conclusion of the war brought Britain vast new territories to explore and exploit, including Quebec and Florida in the Americas, several West Indian islands (St Vincent, Grenadines, Grenada, Tobago and Dominica) and large areas which had been under French influence in India, including Bengal, Bombay and Madras. These huge territorial gains, with their new governmental and commercial opportunities, reaped their horticultural dividends over the coming centuries. British botanical influence gradually spread throughout the world and, particularly during the nineteenth century, the combination of the powerful Royal Botanic Gardens at Kew and the development overseas of related botanic gardens brought to our gardens the richest of feasts. Thus do plants become the unlikely trophies of war. A few years after the end of the Seven Years War, when it came to the discovery of Australasia, circumstances were different and it was the superb seamanship of Captain James Cook which carried off the Antipodean prize, when it could just as easily have fallen to one of the many great French explorers on the same quest.

Two of the most influential men in the history of horticulture were both born early in the century: one was the landscape gardener Lancelot Brown (1716–83), who saw great 'capabilities' in the estates he was called on to remodel; and the other was a professor from Uppsala University in Sweden, Carl Linné, who is perhaps better known as Linnaeus (1707–78) and who attempted to bring order out of botanical chaos. Although they both made horticulture their life's work, they operated from either end of the gardening world: the first an essentially practical man who took the broad brush approach to his designs and manipulated the landscape accordingly; and the second who breathed order and logical structure into the classification of plants (as well as animals and minerals).

Humphry Repton, the follower of Brown, was born in 1752 – the same year as England belatedly adopted the Gregorian Calendar: we were 170 years later than the Continent in changing to it and by this time were so out of step with Europe that eleven days had to be 'lost' to catch up with the rest of the Continent. The eleven-day adjustment was taken in September, and caused a number of hiccups in the gardening world: for instance, the 'Michaelmas' daisy earned its name by appearing to bloom later, around the feast of St Michael the Archangel on 29 September, instead of at the beginning of the month.

Carl Linneaus, Professor of Botany at Uppsala University, was later ennobled as Carl von Linné.

If that were not enough, there were also the three exploratory world voyages of Captain Cook. During the second of these (1768–71), when he was accompanied by the young Joseph Banks (1743–1820), he discovered the Antipodean world and the continent of Australia. The Admiralty's instructions for the voyage were clear: Cook was first to observe the Transit of Venus from Tahiti and then, and more importantly, to 'Proceed to the southward in order to make discovery of the continent … until you arrive in latitude 40°' to discover the land which had hitherto been called Terra Australis Incognita – or Unknown Land Belonging to the South. (The Transit of Venus across the face of the sun is an astronomical event which occurs infrequently and at somewhat unpredictable intervals. It was of great importance in the eighteenth century in helping to calculate the fixed distances between the planets, stars and Earth. The first time the phenomenon was observed was in 1639; the next time it was seen was over 120 years later, in 1761, when more than fifty observations were made worldwide, thirty-two of them by the French. The Transit that Cook was sent out to record was in 1769. There was then another long gap of over a hundred years before two Transits came close together, one in 1874 and the second in 1882; none occurred during the twentieth century.)

Towering over all these events is the figure of Philip Miller and the book he wrote, *The Gardener's Dictionary*. To call his life's work just a book is a rather inadequate statement, for between 1731 and 1768 it was available in eight updated editions, and in several different versions. The volumes were translated into Dutch, German and French, and the seventh edition (1756–9) included the nomenclature changes brought about by Linnaeus in his *Species Plantarum*, published in 1753 (see p.16). In 1736, the great Swedish botanist had visited England and had called on Miller at the Chelsea Physic Garden. At first, Miller was not keen on using the changes, much preferring the Pitton de Tournefort method, but eventually he accepted the logicality of the Linnaean system.

He was not the only one to question the wisdom of the changes: the Earl of Bute, Lord Stuart (1713–92), who had botanic gardens at Luton Hoo in Bedfordshire and at Highcliffe in Hampshire and was an enthusiastic patron of botanical enquiries and of Kew Gardens, wrote to his fellow enthusiast Peter Collinson (1691–1771) regarding the new system: 'We shall have more confusion with order than we had formerly with disorder…' Despite his criticism of the new order of nomenclature, however, Stuart was honoured by Linnaeus with the naming of the genus Stewartia for him, even though he misspelled his name.

Philip Miller was in the same mould as the great Tradescants, a keen and observant botanist, a practical gardener and an experimenter with all that was new (especially exotic fruit). He took obvious pleasure in spreading around newly received plants, sending them to fellow gardeners both in England and further afield, to Holland, France and even America. It seems as if very little escaped his botanical or horticultural notice during his four decades at the Chelsea Physic Garden. What was remarkable about *The Gardener's Dictionary* and his other writings, like *The Gardener's Kalender*, was that they were continually being updated, so the most modern views and the newest plant introductions were constantly coming before the public. The two-volume *Kalender* was not, in fact, published during Miller's lifetime, but appeared almost twenty-five years later, between 1795 and

Peter Collinson lived at Peckham where he created a botanic garden growing many plants sent to him from North America.

1807. His *Dictionary*, a great opus of gardening scholarship linked with practicality, is still of value today, and the Miller thread can be seen influencing much of our modern garden writing. In particular, the Royal Horticultural Society's four-volume *Dictionary of Gardening: a Practical and Scientific Encyclopaedia of Horticulture*, published in 1951, is a most worthy successor. Each time one of the volumes is taken off the shelf and opened by those lucky enough to possess a copy, one should remember and be thankful for Gardener Miller.

During the 1700s, therefore, British gardens could be filled for the first time with plants from all the five continents. In the first five years of the new century alone there are records of plants from South Africa, South America, Asia and North America, all keen to make their horticultural mark. One of the first to arrive, in 1702, and one of the most impressive,

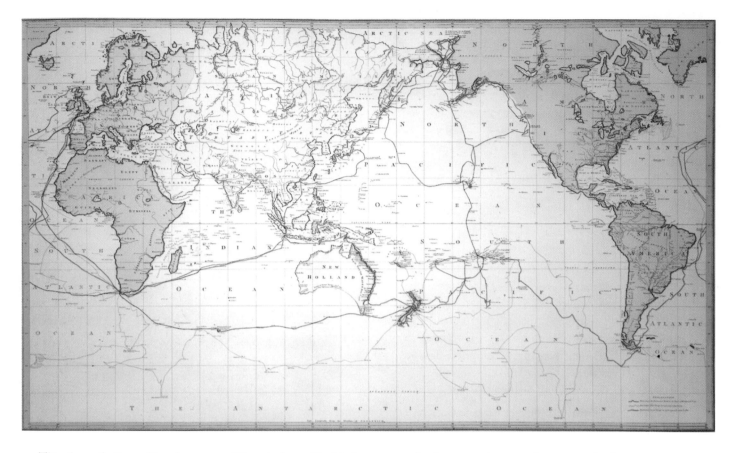

This chart of 1785 marking the routes of Captain James Cook's three voyages in the years 1769 to 1779 is amongst the first to include Australia and New Zealand, the discovery of which completed the cartographical picture and opened the flood gates to the Antipodian flora.

was a lily from South Africa, *Eucomis regia*, although another lily, *Agapanthus africanus*, had already made itself at home here, having come from the Cape in 1629 (see p.141). There are about 15 species of Eucomis in the genus and they are all native to southern Africa. The name given to them by Linnaeus comes from the Greek for 'a beautiful head', after the tuft of leaves they have on top of the flower spike. We know them as the Pineapple Lilies and they come into bloom in late summer. The second Eucomis arrived in 1760: *E. autumnalis*, so named as it can flower into early autumn. The third species to arrive, and the one with the most pronounced tuft of leaves or bracts at the top, was *E. comosa*, which was collected in 1790 by Francis Masson (1741–1806), who had been appointed by the Royal Botanic Gardens, Kew, to collect on its behalf in southern Africa. The most robust is the Giant Pineapple Flower, *E. pallidiflora*, which arrived here in 1887.

The next plant on the list was plucked from obscurity, coming across the Atlantic from Central and South America. It is the first of a species which, today, is immensely popular and which has proved a mainstay in our gardens. This is the Fuchsia, named to honour a sixteenth-century German physician and herbalist. The nomenclature of plants runs a strange course: Professor Leonard Fuchs, after whom this one was named, did not have the pleasure of seeing it or, indeed, of even knowing of its existence, since he died in 1566 while the first species did not arrive here until 1703, ten years after it had been collected from the West Indies by the Franciscan Father Charles Plumier (1646–1706). Plumier, on the other hand, certainly saw the genus named for him – the Plumeria, which is a native of the West Indies and includes the Frangipani *Plumeria rubra*, which was earlier called the West Indian Jasmine; that name was later given to *P. alba*. Both these plants have exquisitely perfumed flowers but, coming from tropical America, they can only be grown in England with protection.

The first fuchsia species to arrive here was what Father Plumier called *Fuchsia triphylla flore coccinea,* This was its early name, and later, when Linnaeus began his botanical renaming, he took this plant as the type species and gave it the name of **F. triphylla**. It can still be found for sale here but is a rather wimpish plant, with no staying power. Even when grown from seed by Philip Miller at the Chelsea Physic Garden in the 1730s it did not survive, and it was another 130 years before it made itself known again.

Problems of identification and of nomenclature began with the second and third fuchsia introductions. *F. coccinea,* which was found and described by another French missionary, Father Feuillée, in 1714, was probably not introduced to Kew until 1788. This plant looks very similar to the third species, *F. magellanica,* which has a date of introduction of 1823. They both come from southern South America and it seems as though there was a muddle over their names right from the start, with the two names being applied indiscriminately to one or other plant. Either *F. magellanica* had been described earlier or it had arrived here earlier than its supposed date of entry.

The problem was exacerbated in the early 1790s by James Lee (1715–95), who, in 1745, had set up as a nurseryman in a former vineyard at Hammersmith. A client whom he was showing around the nursery told Lee about a plant he had apparently noticed on a windowsill in a house in Wapping in East London which looked remarkably similar, though far superior, to *F. coccinea.* Being the knowledgeable plantsman that he was, Lee immediately took the equivalent of a fast car to Wapping, where he 'persuaded' the lady of the house – whose sailor husband was supposed to have brought it home from his travels – to part with her plant. This she did for a fee and the promise of two more plants when he had propagated them. By the following flowering season he was reported to have had three hundred plants to sell (history does not record whether the bargain was kept of two plants being returned to their Wapping home), and this is where the real confusion begins. The three hundred, which sold like hot cakes at a guinea a time, were all purchased under the name of *F. coccinea* (perhaps *F. wappingensis* did not sound quite right). When one of these blooms was presented to the Gardens at Kew, probably in 1793, it was identified not as *F. coccinea* – the name under which it was sold – but as *F. magellanica.* It was not the first time that plants had

Pub.ᵈ as the A.t direct: Oct.ʳ 1789 by W.Curtis Botanic Garden Lambeth Marsh

Fuchsia coccinea **is just one of a genus of over 100 species comprising trees, shrubs and perennials, whose original home stretches from central and southern America to New Zealand.**

been bought and sold with the wrong name attached to them, nor would it be the last. The real *F. coccinea* was thought to have died out and disappeared from the scene, until one day in 1867 Sir Joseph Dalton Hooker (1817–1911) found it growing quite happily when he was being shown around a greenhouse at the Oxford Botanic Garden and was able to identify it as the missing species.

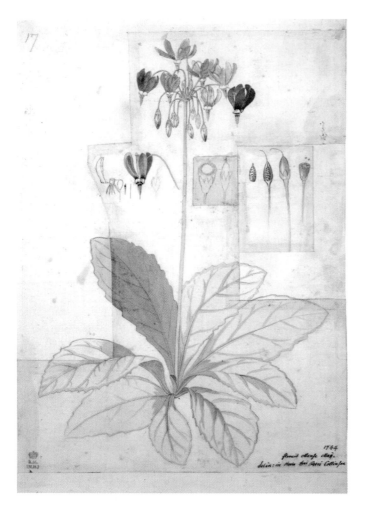

The American Cowslip *Dodecatheon meadia* is one of a genus of 14 species, all from the high alpine meadows or damp grasslands of North America.

By that time 20 or more fuchsia species had arrived. There are about 100 species in all, nearly all native to Central or South America but with one or two from New Zealand. It is *F. magellanica* which is the hardiest species and has the most variation – its home territory stretches an immense distance, from Peru to Tierra del Fuego – so consequently it has been much used in breeding.

Fuchsia blooms and florists' fuchsias took the public by storm in the nineteenth and early twentieth centuries, and in 1939, when the British Fuchsia Society was formed, breeding was set to go into orbit. Even *F. triphylla* came back into circulation and, as becomes the earliest entrant of such an admired flower, it has given its name to the Triphylla Group

of fuchsias (all single-flowered with very long tubes, usually having purple-backed leaves). There are now, at the beginning of the twenty-first century, some 8,000 hybrids and cultivars which are available to grow and nurture (If this all sounds too much to take in, fuchsia lovers should take heart; there are supposedly some 38,000 different camellias being grown.)

Another early arrival was the American Cowslip, seen and collected in Virginia by the missionary John Banister just before he was killed at the age of thirty-eight in 1692. The plant arrived in this country in 1704, and in 1709 Philip Miller saw it in flower in the garden of the Bishop of London, Henry Compton (1632–1713), the dendrologist and patron of botany. The bishop was responsible for sending missionaries and priests to the newly expanding colonies, and had originally dispatched Banister to spread the Christian message to the American Indians in 1685. The American Cowslip or Shooting Star he found was *Dodecatheon meadia*, named by Linnaeus and found throughout eastern North America. Its name derives from *dodek*, the Greek word for 'twelve' – because there are about that number of flowers in each umbel – and *theos*, meaning 'of the gods'. Pliny had first used the name for a primrose (the exact one is not known), and as the genus is in the *Primulaceae* family it seemed a quite logical name. There are 14 species in the family, all from North America. Unfortunately, this first one to arrive felt a little homesick and had to be reintroduced, this time by John Bartram, who sent it to Peter Collinson in 1745. Very soon after its second arrival it was being offered for sale by a number of nurseries. It cost 2s 6d from Anderson's, the Edinburgh nursery, in 1775, and later 1s 6d; by the turn of the century it was a bargain at just a shilling. John Galpine's nursery catalogue of 1782 considered *D. meadia* to be 'most beautiful'.

Another plant from North America, but one which showed not the slightest sign of wanting to return home after it arrived in 1710, was *Aster novi-belgii*, which became part of that huge tribe of *Compositae*, the daisy family. It was first described by a Belgian botanist, Professor Paul Hermann (1646–95): *Hermannia*, a family from South Africa, commemorates him. The seed of the Aster or Starwort must have been sent to him, for although he travelled to Ceylon (now Sri Lanka) and Batavia, it seems he did not visit America. There were already native asters in Europe. We had our own seaside species, *Aster tripolium*, and there was also a European native, *Aster amellus*, known as the Italian Aster, which had been grown here in our gardens since at least 1596 and which, according to both Sir Thomas Hanmer and John Parkinson, was quite an invasive species.

Some of the species which arrived from North America, where they are most numerous, have gone native, including *A. novi-belgii*. This is the plant with the strongest genes and has been the one most used in breeding our present-day Michaelmas Daisies. When grown well and looking happy, they add star quality to the late summer herbaceous border, but they are not as steadfast as some would like. They are vulnerable to attack by anything that moves and munches in the soil – snails, slugs, tarsonemid mites (minute sap-suckers), eelworms and aphids, as well as *Fusarium* wilt, leaf spot, powdery mildew and grey mould, and if that isn't enough, many of the taller varieties require staking.

A number of shorter and sturdier varieties were bred by

THE MISSIONARY BOTANISTS

It is remarkable how often the botanical nature of a country has been described and plants collected by the missionaries of the Gospel, in much the same way that the monks contributed to the medieval era. John Baptist Banister (1654–92) was a good botanist, who sent seeds of the American Cowslip not only to the Oxford Botanic Garden, which by then had been in existence some eighty years, but also, as we have seen, to his patron, Bishop Compton, and to the Revd John Ray (1627–1705), the eminent horticulturalist.

Henry Compton was Bishop of London for thirty-eight years and resided at Fulham Palace, where he was able to develop his passion for collecting and growing trees and plants. As Head of the Church for the American colonies, he was in a unique position to receive from his chaplains the new plant discoveries they were encouraged to search for and send home. The Bishop's gardening prowess was noted by Stephen Switzer, the nurseryman, who remarked in 1742 that 'This Reverend Father was one of the first that encouraged the Importation, Raising, and Increase of Exotics'. Switzer himself was believed to have eventually housed over 1,000 species under glass.

As well as the American Cowslip, Banister was responsible for introducing in 1688 *Acer negundo*, the Box Elder, and also the very first magnolia to be cultivated – *Magnolia virginiana*, the Sweet or Swamp Bay. In 1691 the Scarlet Oak, *Quercus coccinea*, made its spectacular entrance, followed in 1696 by the beautifully named Balm of Gilead, *Abies balsamea*.

Victor Vokes in the 1920s. He worked for the War Graves Commission and was interested in breeding Michaelmas Daisies specifically to plant in the war cemeteries. He crossed *A. novi-belgii* with *A. dumosus* (shrubby, bushy), which resulted in the named variety called 'Remembrance', which is often still in bloom on Armistice Day. It is widely available. It is not known when the species *A. dumosus* made its way across from America, but it is a war horse of a plant with good sturdy genes and, along with *A novi-belgii*, has been used extensively in breeding.

The Pelargonium of South Africa, like the Fuchsia of South America, is so familiar as to need no introduction, except to comment that we still like to call it the 'Geranium' even after a hundred years of its name having been changed. There are three different families within the *Geraniaceae* family. The first is the Cranesbill (Herb Robert), called in Classical Greek *geranion*, or 'crane', so geranium. Second is the Pelargonium, which comes from the Greek name for a stork, *pelargos*; and the third is the Erodium, from the Greek name for the heron, *erodios*. The differing shape of the three seed heads reflects the shape of the beaks of the cranesbill, stork and heron and thus determines the genera to which the plants belong. The confusion over the names began with one of the earliest of the plants to arrive, in August 1710, when it was called *Geranium Africanum*. It kept its original name for a considerable time but, as with all things in the nomenclature department, it was considered logical to alter it to *Pelargonium zonale*, because of the distinction of the leaf, which was banded or zoned in a different colour. It is the hearty parent of the pillar-box-red bedding 'geranium'.

All the pelargoniums derive from South Africa and the first one this century to put in an appearance arrived a few years prior to *P. zonale*. This was the rose-flowered *P. capitatum*, which was zonal-leaved (banded with the distinct colour pink). Mary Somerset, the first Duchess of Beaufort (c. 1630–1714), had been enthusiastically creating a botanic garden at Badminton in Gloucestershire, where she had formed an enormous collection of exotics, including some of the earliest pelargoniums. One of her introductions, in 1701, was *P. peltatum*, known as the Ivy-leaved Geranium, although it should be called the Shield-leaved since that is what the Latin implies. (The Duchess was commemorated in 1818 by the small Australian genus which includes the Swamp Bottlebrush, *Beaufortia sparsa*, exotic enough for anyone's taste.)

There are about 230 species of pelargonium, of which

about 20 have been persistently used for hybridising, and the thousands of resulting cultivars have been catalogued into six horticultural groups named Angel, Ivy-leaved, Regal, Scented-leaved, Unique and Zonal, some of which can be subdivided within their groups. Hybridising began almost immediately, and one of the earliest results was carried out by the nurseryman James Lee of fuchsia fame. He crossed *P. fulgidum* (meaning 'shining leaves'), which arrived in 1723, with *P. lobatum* (meaning 'lobed-flowered'), to produce in *c.* 1818 *P. x ardens*, the Glowing Geranium. It has staying power, for nearly two hundred years later it is still being grown.

In 1712, the most sumptuous lily arrived, also from South Africa. This was *Amaryllis belladonna*, the first name belonging to the shepherdess who was praised by both Virgil and Ovid. Shepherds are supposed to know their sheep, but in this instance the progeny has become hopelessly muddled, with the same name being used for the Hippeastrum from South America. Both are in the *Amaryllidaceae* family. There is just this single species of Amaryllis in the family and it was so quickly hybridised, with either *Brunsvigia* to make *x Amarygia* (syn. *x Brunsdonna*), or with *Crinum* (*x Amarcrinum* syn. *x Crinodonna*), that their parentage is endlessly complicated. The lily itself has the sweetest of perfumes, and the colour of the original Amaryllis can stretch from being a sugar-plum-fairy pink to a sturdy cherry-red.

We scuttle back to eastern Europe the following year to greet *Brunnera macrophylla*, a slightly taller-looking forget-me-not-like plant – its original name of *Anchusa myosotidiflora* bore that out – which occurs anywhere between the Caucasus and Asia. It is a hardy perennial (as it should be, being named for a Swiss botanist, Samuel Brunner 1790–1844) and grows from seed in the flower border or in light woodland.

In the same year, 1713, and from across the Atlantic came the very first of the Ceanothus. There are over 50 species in their family, the *Rhamnaceae*, which is a Greek word used for various spiny shrubs; the Buckthorn is in the same family. The Ceanothus name also means a 'spiny plant' and comes from the Greek word *keanothos*. *Ceanothus americanus* has a history very like the American Cowslip we have recently met, in as much as it was collected first by John Banister and sent to Bishop Compton to grow in his famous garden in Fulham; then it disappeared and was brought back into cultivation by Peter Collinson in 1751. It is a fairly small shrub or 'shrublet', only 1–1.25 m (3–4 ft) tall, but elegant, with dullish-white,

very dense flowers. It has a string of self-explanatory common names: Red Twig, Red Root and New Jersey Tea; the dried leaves are said to have been soaked as a beverage during the American War of Independence, the resulting 'tea' being given the name of 'pong-pong' which is an approximation of the Indian name. (Another plant from North America which was also used to make 'tea' re-entered England in 1763. This was *Ledum groenlandicum*, or Labrador Tea; see p.129.)

The Ceanothus themselves go under the collective title of the Californian Lilac, nearly all emanating from that western state. Most have blue flowers, although the first one to arrive was white-flowered; it was over hundred years before the first blue-flowered species arrived. *C. thyrsiflorus* (meaning 'having many flowers'), the Blue Blossom, made its spectacular entry from Monterey in California in 1837, collected as seed by Richard Hinds, surgeon-naturalist on board HMS *Sulphur* during its global voyage between 1836 and 1842. This Ceanothus is more of a tree than a shrub, growing to about 9 m (30 ft), and is a very hardy species.

It comes as a surprise that one of the next entries, in 1714, is included in the same family as the Heather and the Cassiope (mother of the beautiful Greek goddess Andromeda), the *Ericaceae* family. It is *Arbutus andrachne*, the Grecian or Cyprus Strawberry Tree. It is a rare cinnamon-barked small evergreen tree which was collected by William Sherard (*c.* 1658–1728) when he was 'our man' in Smyrna (Izmir). Sherard was born in Leicestershire, and before embarking on his travels to Turkey and Smyrna, in the early years of the century, he was engaged as tutor to Henry Beaufort (later the 2nd Duke), son of the Duchess of Beaufort, at Badminton, where no doubt he and the Duchess egged each other on in their mutual enthusiasm for botany and the growing of 'exotics'.

William's brother James (1666–1738) was an apothecary; he lived at Eltham, east of London, where he had a garden noted for its rare plants. Both brothers were responsible for a number of new introductions during the second quarter of the century. In 1725 they introduced *Coreopsis lanceolata*, a bright yellow daisy and one of the Tickseed family from North America; then in 1726 they were the first to cultivate *Verbena bonariensis* the South American Verbena brought back from Buenos Aires. The following year they procured seed of a pale yellow marigold from Mexico, *Tagetes minuta*, to grow in the Eltham garden. In 1730 Dr James Sherard's garden bloomed with the beautiful and sweet evening-per-

The painting of this nerine is by Daniel Frankcom who painted flowers from the botanic garden created by the Duchess of Beaufort.

In 1729, *Astragalus galegiformis*, one of the Milk Vetch or Goat's Thorn, came ashore from Siberia. It is in a genus with over 1,000 species in it, a number of which come from Siberia and Russia, as well as Serbia and the Mediterranean region. Several species were much earlier arrivals, in the sixteenth and seventeenth centuries. *A. tragacantha*, or Gum Tragacanth, was introduced in 1640 from southern Europe and western Asia. Many of them are spiny, scrubby plants, but they all have pretty pea-like flowers; some are grown for the rockery and others, including *A. galegiformis* with its handsome foliage, are taller and stout enough to look after themselves.

We must make more than a passing reference to the arrival in 1730 of the Weeping Willow, *Salix babylonica*, which, despite its name, is probably a native of central and northern China. It is not thought to be the willow mentioned in Psalm 137 of the Bible: that tree is believed to be *Populus euphratica*, which belongs in the same family of *Salicaceae*.

> By the waters of Babylon,
> there we sat down, yea we wept, when we remembered Zion.
> We hanged our harps, upon the willows in the midst thereof.

The true Weeping Willow, as becomes its mournful habit, is not quite hardy in Britain, and is relatively short-lived; it is more usual now to plant hybrids, which seem to be sturdier. The tree was noted and remarked on by Sir George Wheler (who brought St John's Wort back from Turkey) when he saw it on his travels in Asia in 1675: '... whose large branches were so limber, that they bend down to the ground ...and naturally make a curious shady bower around it'.

It was probably first seen in Europe when Pitton de Tournefort (1656–1708) brought it back from his travels at the end of the seventeenth century. According to Peter Collinson, the tree did not arrive in England until 1730, first brought here by a merchant trader, a 'Mr Vernon of Aleppo, Turkey' (Aleppo was then part of the Ottoman Empire, and is now in Syria), who planted the willow in his garden at Twickenham. However, there is evidence that the tree, and certainly its weeping habit, was well known in England at about the same time as its entry into France. Its shape certainly must have caught the eye of William Cavendish, 4th Earl of Devonshire and later 1st Duke, who in 1693 was having his extensive garden and waterworks created at

fumed *Phlox paniculata* from the eastern United States. Like the Michaelmas Daisy, this whole genus has become a most welcome addition to the herbaceous border, the main reason being not so much how they flower, but when they flower; in this case, like the aster, they bloom in late summer, giving an added touch of *joie de vivre* to our gardens. The one in Sherard's garden was the mauve-flowered *Phlox drummondii*, which commemorates Thomas Drummond (*c.* 1790–1836), who collected the seed from Texas and sent it home to England in 1835 (see p.206). It was an instant success, having such a wide and varied range of colour.

Chatsworth in Derbyshire by the Frenchman Monsieur Grillet, a pupil of Le Notre. Grillet, of course, could have seen the tree on its arrival in France and described it to the Earl. Lord Cavendish commissioned an elaborate life-size model to be made of the tree and placed in the garden. This novelty really did act as a Weeping Willow and was in the height of fashion, spraying water on friends and family alike most unexpectedly. What a jape to play on unsuspecting guests. The piece of water sculpture was remodelled by Joseph Paxton in the 1830s, again from copper and brass, with its 800 sprays of water depicting the weeping of the branches.

Celia Fiennes must have been familiar with at least the shape of the willow, too, as she writes in her journal after touring the Chatsworth grounds, that '…by the Grove stands a fine Willow tree, the leaves barke and all looks very naturall, the roote is full of rubbish or great stones to appearance, and all on a sudden by turning a sluce it raines from each leafe and from the branches like a shower, it being made of brass and pipes to each leafe and from appearance is exactly like any willow'. The journal entry was dated 1697, another pointer to the fact that the tree appears to have arrived here long before 1730.

There was a vogue for planting the real thing following the death of Napoleon Bonaparte in 1821 on the island of St Helena. He was buried under a weeping willow, and cuttings from the tree were brought from the island to England and planted, presumably by his admirers – history does not relate when that particular tree was planted.

There were two reintroductions in 1731, both from the seventeenth century. One was a plant first associated with John Parkinson in 1629, the little *Ramonda myconi* from the Pyrenees (see p.129), and the other was the Foam Flower, originally collected from North America by John Tradescant the younger in 1631. This is in the *Saxifragaceae* family and is called *Tiarella cordifolia* – *tiara* meaning 'a crown', and *cordifolia* meaning 'with heart-shaped leaves'; it is very pretty.

The year turned out to be a horticultural high spot for *Bletia verecunda*, the first tropical orchid to bloom in Britain. It had been collected from the Bahamas, probably by Mark Catesby (1682–1749), who visited the islands in 1725 and sent the plant home to Peter Collinson. Unlike most of the 35,000 Orchis species since discovered, this plant had hardly an ounce of flounce in it; its very name, *verecunda*, means 'mod-

esty', while the Spanish apothecary Luis Blet, who founded the Botanic Garden at Algeciras in the eighteenth century, is commemorated by the genus. Much later, in 1778, another Bletia, now named *Phaius tankervilleae* (in honour of Lady Emma Tankerville), with more of a flaunt about it, was brought into flower by both Peter Collinson and Dr John Fothergill (1712–1780). This orchid was imported from China and has red-brown flowers with a yellow throat and red lips – very fetching. The word 'orchid' is derived from the Greek name for testicle – which makes one view the flowers and rhizomes rather more closely, and which suggests at least one of the reasons why they were held in such regard. Records state that the ground-up roots of the orchid were made into a nourishing jelly or taken in a hot drink – which makes one think of Horlicks or Ovaltine. As the powder was imported from Turkey, however, there does not seem much chance of it being a soothing bedtime drink; it was probably imbibed for more exotic purposes (perhaps it was the Viagra of its day?).

One of the few orchids to lead a useful life is the Vanilla genus, of which there are about 70 species spread right around the tropical world, from Uganda to Demerara in Guyana to India. The plant which was developed for its vanilla flavouring was *Vanilla planifolia* (syn. *V. fragrans*), which was discovered in 1800 in the West Indies. The word *vanilla* is derived from the Spanish word for 'a small pod', and that is precisely from where the flavour is extracted.

To outsiders, the growing of orchids seems entirely mystical; even our native species look 'foreign', appearing to have an almost unearthly quality about them. One would not be surprised to discover them growing on the moon or Mars. The Chinese consider the symbolic nature of the orchid to be as a true gentleman, for its delicate fragrance in perfuming a room is only noticed once he departs. The writer Lesley Gordon remarked, more robustly, that the blooms are a 'mink-and-cleavage flower', which seems to aptly describe both their modern-day roles in life. Suffragettes may have had in mind the flowers' flaunting nature when in 1913 they attacked and nearly destroyed the Orchid House at Kew. (They burnt down the Tea Pavilion as well.)

We need to keep returning to the flora of South Africa, this time to welcome in 1731, from the watery margins of Lesotho, the serene Arum Lily, which always reminds me of

Right: The Willow Tree Fountain was first 'planted' in a dell at Chatsworth in 1692 and had been regularly restored since then.

Vanilla planifolia was used by the Aztecs as flavouring for chocolate. The pods of the orchid are picked unripe, packed tightly in boxes where the pods sweat, and turn from yellow to black and acquire their distinctive flavour.

a white swan gliding through the water. Its Latin name honours an Italian botanist, Francesco Zantedeschi, born in 1797, and thus we have *Zantedeschia aethiopica* in the *Araceae* family. Although not classed as a lily, it is sometimes called the Lily of the Nile or the Trumpet Lily.

During the following ten years, from 1732 onwards, huge shiploads of American plants crossed the Atlantic, plants which we have found of great decorative use in our gardens, such as the two sunflowers *Helianthus atrorubens* and *H. scaberrimus*, meaning respectively 'dark red' and 'rough'. There are about 80 species of sunflower, all of which come from the Americas and all of them cheery-looking and easy to grow. Children particularly like to grow *H. annuus* because it is a speedy and spectacular grower and brings out the competitive spirit in them, with each child trying to grow the tallest sunflower or the biggest flower-head. Competing for height in the growing of the sunflower is nothing new, and perhaps it brings out the child in all of us: there were stories in the sixteenth century from Spain, Italy and England (where it was described by Gerard in 1596) all boasting about the height the plants had reached. Alas, Gerard reported only a mere 'fourteene foote in my garden' (4.3 m), while Madrid reported 7 m (nearly 23 ft) and Italy topped the bill with a massive 12 m (39 ft). Today, of course, exaggeration is just not allowed and the RHS gives its imprimatur much closer to the height of John Gerard's sunflower, at 5 m (just over 16 ft).

The Black Snake Root, *Cimicifuga racemosa*, arrived in 1732 as well. The name means 'to drive bugs away', but that really refers to *C. foetida*, which arrived from Siberia a few years later.

A pair of Kalmias also arrived from America, one in 1734, the other in 1736, and were named by Linnaeus to commemorate Pehr Kalm (1715–79), his Finnish pupil and friend. Kalm, working as an agent for the Swedish government, was sent to America in 1748 to report on the continent's natural resources. The lovely evergreen but highly toxic Kalmias were known by a variety of names: Spoonwood was one (the Indians carved small tools from the root-wood) and Mountain Laurel another; it is still known by the latter name, and also as the Calico Bush – even though calico is not made from it. *Kalmia latifolia*, the first of the shrubs to be made welcome, had first been noticed by Mark Catesby when travelling in the Carolinas, collecting and recording the horticultural material there. He originally gave the low-growing shrub the Greek-derived name of *Chamaedaphne foliis*, a ground laurel 'full of leaves', which aptly describes the plant's habits; the name was later changed by Linnaeus to

Right: Subscribers to the 1727 plant catalogue of Robert Furber of Kensington Gore Nursery, London.

Kalmia. (Chamaedaphne as a genus now describes one plant which is in the same *Ericaceae* family.) In 1790, there were a large number of deaths in Philadelphia caused by the eating of both game which had been fed the Kalmia berries, and honey made from the flowers.

In 1736 *K. angustifolia* arrived: this too was poisonous to grazing animals, so confusingly it was called Sheep Laurel, as well as the more explicit Lamb-Kill and Poison Berry. There are only 7 species in the genus, all from North America except one which is from Cuba. Mark Catesby, Peter Collinson and the great Philip Miller all had difficulty in getting both the seed and the plant to grow. Propagation can be either by greenwood cuttings in late spring or semi-ripe cuttings which have to be taken later in the middle of the summer. They are fickle shrubs even now, but are great beauties.

An aristocrat with exquisite manners arrived in 1734 in the form of *Magnolia grandiflora*. Known as the Laurel Magnolia, with leaves which are a dark glossy evergreen and leathery in texture, it is one of about 80 species distributed throughout eastern North America and south-eastern Asia. This one was first discovered by the Quaker traveller and plant gatherer John Bartram (1699–1777) growing at Bulls Bay, South Carolina. For about thirty years he sent back to England seeds, cones, bulbs and corms, usually to Philip Miller, who then distributed them to collectors, who paid Bartram for his efforts. The species was given the name of yet another botanist, this time Charles Magnol (1638–1715), a French doctor from Montpellier who could well have seen the first of the genus to arrive from America in the late 1680s, one called *M. virginiana*.

From the floral point of view, the blooms are at the furthest end of the spectrum from orchids, being considered the simplest or most primitive of all flowers. To see the creamy serenity of the perfumed petals of *M. grandiflora* is a joy, although in America, the indigenous Indians believed that a fragrant bloom placed in the bedroom would result in the death of the occupant. All the magnolias, like the camellias, have great anticipation quality; their buds take an immense time developing, so they can be watched and drooled over in keen anticipation of the first sign of petals showing.

As can be imagined, the tree was quickly taken up by keen patrons. The Duchess of Beaufort made sure that it was planted at Badminton, Peter Collinson grew it in his Peckham garden in south London, and perhaps the first to plant it in the West Country, where it does so well, was Sir John Colliton, at Exmouth in Devon. The Parsons Green garden of Sir Charles Wager was reported to be where the beauty bloomed first of all, but Mark Catesby reckoned that honour should have gone to Exmouth. Both plants probably flowered in the same year, 1736, as we know that in the following year Georg Dionysius Ehret (1708–70), the brother-in-law of Philip Miller and a skilled botanical artist, walked the three miles from Chelsea to Parsons Green to watch the flower bloom and to paint it. (The genus Ehretia was later named for him.) This begs the question of either the date of its arrival in England or in what form it was sent here, as although it would not have been known at the time, seedling-raised plants of this species take about twenty years to come into bloom, whereas trees raised from cuttings (they can be easily rooted from layering) take only about five years. If there was any evidence of what form *Magnolia grandiflora* took when it arrived here, it has long since disappeared.

The Asiatic magnolias, like the American ones, are spread across the Tropic of Capricorn, and although they arrived on our shores a generation after the American trees, they had been in cultivation in the east for thousands of years. The first Chinese one to appear, in 1780, was an introduction by Sir Joseph Banks. It was familiarly called the Lily Tree, and in 1806 was described by Richard Salisbury, then secretary of the newly founded Horticultural Society, as looking like 'a naked Walnut Tree with a lily at the end of each branch' – a most apt description of *M. denudata* (syn *M. conspicua*).

The tree comes into bloom early in the year, from February onwards. The flowers have to make do without the protection of the leaves, which open later, so consequently are very susceptible to frost. Unlike *M. grandiflora*, however, it begins its flowering life both profusely and at the precociously young age of four or five. One of the best cultivators of the day, Sir Abraham Hume (1749–1838), had many successes in his grand garden, Wormleybury in Hertfordshire, particularly with Chinese plants, and in 1801 he planted a most famous Lily Tree. Twenty-five years later it was over 6 m tall (nearly 20 ft) and carried in excess of 900 flowers. In its native China , the tree can be found growing in both woodland and open areas.

Richard Salisbury (1761–1829) was a descendant of Henry Lyte, who, in 1578, had written *A Niewe Herball or Historie of Plantes*. He acquired the house and garden Peter Collinson had made famous at Mill Hill, then just outside London, and has the distinction of being remembered for *Salisburia adiantifolia*, a synonym for the much better known *Ginkgo biloba*, introduced in 1750.

Another Chinese magnolia, which arrived in 1790, was *M. liliiflora* with a flower which resembles a lily or a chalice, although its vernacular name is the Woody Orchid Magnolia. This tree had attached to it the added glamour of having supposedly been introduced to England by a prime minister – William Cavendish Bentinck, 3rd Duke of Portland (1738–1809), who was leader of the Whig Party and twice Prime Minister, in 1783 and again from 1807 until his death two years later. It had always been assumed that it was 'Yes, Prime Minister' for this particular magnolia. It may well have been politically correct for him to have claimed the introduction, but in delving a little deeper (as so often in politics), it is more likely to be 'No, Prime Minister' and a possible case of mistaken identity.

The real introducer could have been one of two people, neither one the Duke of Portland. The first could have been the 3rd Duke's mother, the Duchess of Portland, who was known to be interested in and knowledgeable about plants and who, like the first Duchess of Beaufort, made a 'botanic garden', and a museum too, at the family home, in Bulstrode in Buckinghamshire. Margaret, Duchess of Portland, was a lady of high intellect and the blue-stocking correspondent of Jean-Jacques Rousseau, who, incidentally, also created a 'botanic garden' at his home in France. The Duchess is commemorated by a small genus of Central American shrubs and trees, Portlandia, in the *Rubiaceae* family, first discovered in the nineteenth century and all requiring stove conditions to survive in England; none are found in nurseries today.

It seems more likely, however, that the introducer was Lord William Cavendish (1774–1839), a relation of the afore-mentioned Prime Minister. Cavendish was Governor of Madras, then Governor-General of Bengal and eventually of all India. He was extremely interested in the flora of India, and botanical matters in general, and his name was given to a genus of palm, the Bentinckia. There are only two members in this family: one, *B. nicobarica*, is a native of the Nicobar Islands, north of Sumatra; the other, *S. coddapanna*, comes from south India. Both palms are rare in the wild, having been collected very heavily, and consequently they are vulnerable to being lost altogether. Like Portlandia, neither of the Bentinckia species can survive in England. *M. liliiflora* has never been discovered growing in the wild; it was only ever found as a cultivated plant in both China and Japan. Its temple and garden use was documented for centuries. It therefore seems quite possible that the plant could have been brought to India at some earlier date, and cultivated to orna-

ment palace and garden alike. Once it was known how interested the Governor-General was in horticulture, the magnolia would no doubt have been brought to his attention, and what would have been more natural than for him to send it home to England in the next diplomatic bag. Prime ministerial influence may or may not have been brought to bear, but what is not disputed is that the first *M. liliiflora* to flower in this country did occur in the greenhouse at Bulstrode.

The horticultural writer J. C. Loudon (1783–1843) notes the year of entry for the Turkey Oak as 1735, and although there is no reason to doubt this, it is a surprise that the tree – or at least the acorn – did not catch the eye of an earlier traveller. *Quercus cerris* is a native of the Balkans, Syria and Asia Minor. It is not a garden tree at all, being much too big, but it does look outstanding growing in parkland. So self-assured is it that, except for its name, it could be a native, being quite at home in both woodland and hedgerow. Clipped as a hedge it makes the most superb wind break, especially against salt-laden tempests; when standing alone with the sun filtering through its dusky-silvery leaves, the Mediterranean world of its near relative, the Cork Oak, *Q. suber*, is brought to mind.

The following year, 1736, the Black Birch, *Betula nigra*, with its shaggy bark was noted arriving from the east side of America by Peter Collinson, as was *Tsuga canadensis*, Eastern Hemlock. This tree belongs in a small and very elegant genus in the *Pinaceae* family, although originally being grouped in with the larger *Abies* family. The 10 or 11 species are spread over the forests of most of south-eastern Asia, Japan and across to North America. The word *Tsuga* comes from the Japanese word for 'hemlock' but the tree has nothing to do with the drink that Socrates took; that was believed to have come from the common Hemlock, *Conium maculatum* (see p.29). The only connection between the two is that the crushed foliage of the tree hemlock smells similar to that of the plant.

A whole bouquet of European plants came into our gardens during 1739, including a member of the flax family, *Linum narbonense*. This is a lovely blue-flowered bushy perennial, first grown by Philip Miller in the Physic Garden at Chelsea. Flax belongs to a large genus of about 200 species found growing all over the northern hemisphere, including some British natives. *Linum flavum* came from Russia in 1793, and is a yellow-flowered species, while in 1820 the flamboyant *L. grandiflorum*, with its deep rose flowers, joined the group, arriving from north Africa.

The Woolly Mullein, *Verbascum phlomoides*, was collected from the Caucasus; *Artemisia valesiaca*, now called **S*eriphidium vallesiacum*,** came from the mountains of Switzerland in 1739 and is in the same family as tarragon. One of the Hare's Ear or Bupleurum species arrived from Central Europe in the same year; this member of the *Umbelliferae* family also has the name Thorow-wax attached to it. This rather strange word is a derivative of the 1548 'thoroughwax' and means that the branches of the plant seem to grow right through the leaves, and certainly that is what the flowers seem to do. At least 4 species arrived during this century, including *Bupleurum petraeum* from central Europe in 1768, and *B. gibraltaricum* in 1784, a native of southern Spain as the name implies.

Datisca cannabina, which was brought here from Crete in 1739 but is a native of the Middle East, is such a discreet plant that no one has ever been able to unravel the meaning of its first name, nor has it captured our imagination sufficiently for us to give it a vernacular name, even though it has a fine stature and good foliage. Although it is distantly related to the Begonia family, its shyness is joined by its exclusivity, as there are only 2 species in the genus and it is in the very small *Datiscaceae* family.

The Camellia arrived from Japan and was first noted in the same year, 1739, flowering in Lord Petre's (1713–42) greenhouse at Thorndon Hall in Essex. No one could have realised how Camellia-crazy the whole world would eventually become, but its early development was very much part of the gardening history of England. Like a number of other genera, for instance the Rhododendron, the Rose and the Clematis, the Camellia has become an international jet-setting star.

One of the reasons why it has become so popular and is viewed with such pleasure must surely be its habit of flowering when almost all else is in the deep slumber of winter. Its glowing good looks and vitality make it a perfect shrub (left long enough, the beautiful smooth silver-grey bark will thicken into a tree); it needs no pruning, although it can be clipped into a hedge; it can be grown as a standard, and it is evergreen. Some of the blooms are even scented – what more could one want? Perhaps a cup of tea provided by the leaves from *Camellia sinensis*. Tea had been brought from China by about 1652, and was gaining in popularity as a drink, but the plant itself arrived from China in 1768, when it was known as *Thea sinensis*. It can be grown outdoors in England given a mild climate and protection in the winter, or planted in a

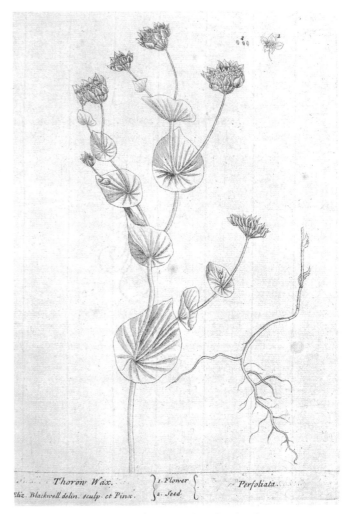

Thorow Wax. 1. Flower 2. Seed Perfoliata.
Eliz. Blackwell delin. sculp. et Pinx.

This unusual-looking plant is one of the family of *Apiaceae/ Umbelliferae* of the genus *Bupleurum*, Thorow-wax, in which there are about 100 species.

container and brought in for the winter. The white flowers are dainty, as are the leaves.

The Tsubaki, as the Camellia was called in Japan, meaning 'tree with shining leaves', was given its Latinised name of *Camellia japonica* in 1753 by Linnaeus, in honour of the Moravian Jesuit priest Father George Joseph Kamel (1661–1706). Kamel was an excellent botanist who wrote a history of the plants of the Philippines as well as carrying on his work there as a missionary. He died in the early part of the century, long before his name was botanically Latinised, and as we have seen so often before with the naming of plants, he never even had the satisfaction of seeing a

Camellia, for the Philippines were a Camellia-free zone.

Since then over 260 species have been discovered, including a number of the fabled yellow-flowered Camellias; these were kept such a closely guarded secret that up until the twentieth century, stories circulated claiming that there was no such plant. The two most important species are *C. japonica* and *C. saluenensis*, the first from Japan, and the second found and sent as seed from the Salween area of China in the early part of the twentieth century by the great plant hunter George Forrest. A recipient of *C. saluenensis* was J. C. Williams (1861–1939) of Caerhays Castle in Cornwall, one of the towering gardening figures of the Edwardian age. Sometime after its arrival in 1924, he crossed one of the *C. saluenensis* pink-flowered seedlings with pollen from a white-flowered *C. japonica*. The resulting hybrids, *C. x williamsii*, unlike so many offspring, took the best attributes of both parents and produced profusely flowering plants which could withstand the cold. In housewifely fashion, too, they also shed their old blooms once they have flowered, as opposd to *C. japonica*, whose flowers remain stubbornly clinging to the branch. In all, there are now well over 38,000 named Camellias, but the variety nearly everyone knows is *C. x williamsii* 'Donation', which was raised in Sussex probably in the late 1920s, according to the expert Jennifer Trehane, rather than the more usually stated 1941, which is when the RHS gave it an Award of Merit.

In 1741, two plants native to Siberia were noted for the first time. One was *Cornus alba*, the Red-Barked Dogwood, which was the first of the Asian Dogwoods to arrive, although the very first of the genus, *C. amomum*, came from the American continent in 1683. The second Siberian offering was a larkspur, *Delphinium grandiflorum*, which had been packed up and sent from St Petersburg by a Professor Amman to Peter Collinson. This plant is also a native of Mongolia, Japan and China, where it has the charming name of Herb of the Flying Swallow.

It was in 1703, at the instigation of Peter the Great (1672–1725), that the first great garden was begun at St Petersburg. Peter I, Emperor of Russia, was a most extraordinary character, ruling Russia from the age of ten and later bringing about many fundamental reforms. In 1698 he came to England for three months to gather information on an eclectic mix of subjects – some of which travelled back to Russia – including learning about horticulture and the design of our gardens. While here, he rented John Evelyn's house at Sayes Court in Deptford and caused (much to

Evelyn's annoyance) a great deal of damage to the garden and holly hedges by his riotous behaviour. However, his genuine enthusiasm for gardens led him to visit France and Holland as well, and he purchased books and engravings on the subject to take back with him to Russia. The Emperor founded St Petersburg in 1703, calling it his 'Window on the West', and filling the parks and squares he created with plants and trees brought in from the countries he had visited. He also arranged for gardening books to be sent from abroad in order to encourage the development of nurseries in his own country; horticultural and landscape design schools were established for the training of Russian gardeners.

Peter the Great's daughter, the Empress Elizabeth (who succeeded to the throne in 1741), and later Catherine the Great (reigning 1762–96) both continued and developed the Emperor's work. By the time Catherine showed interest in the English style of gardening, it had changed from the formality of Sayes Court to the free flow of the landscape movement. She was so taken up with this latter ideal that she wrote to Voltaire that 'I now love to distraction gardens in the English style… In a word, anglomania rules my plantomania.' This headlong admiration led her to commission the 'Table and Dessert Service, consisting of 952 Pieces, and ornamented in Enamel, with 1244 real Views of Great-Britain' – a *tour de force* of decorated china ordered from Wedgwood and Bentley in 1773. Known as the 'Frog Serviee' (Catherine's emblem), all the great and noble houses and gardens of the day were represented, some of them several times. The wide following and admiration that the English landscape park commanded in Russia led for several years to a bounty of horticultural presents being sent from Siberia in exchange.

A plant which Philip Miller grew around 1745 was called by him *Pisum americanum*, the Cape Horn Pea. It was discovered by Lord George Anson – or at least by his cook – who collected the seeds from the Straits of Magellan during his round-the-world voyage which began in 1740. The fragrant blue-flowered plant which Linnaeus retitled *Lathyrus nervosus* was well named, for although the *nervosus* part really refers to the strongly veined leaves, the plant has always been of an extremely nervous disposition in England, never quite able to make up its mind as to whether it likes it enough to stay. In addition, its identity was muddled with *L. magellanicus* for a number of years. Nomenclature has now resolved the situation by confirming the earlier Linnaean name as correct and *L. magellanicus* as the synonym.

In the same year *Gladiolius tristis* arrived from South Africa. That too was grown by Miller but was made of altogether sterner stuff, even though its name means 'sad' or 'dull'. Indeed, it is an unworthy epithet, since the flowers are delightful small trumpets of cream and green, with the added bonus that towards evening they exude a lovely perfume.

Although hardly a garden tree on account of its size, but one whose arrival just cannot be ignored, is the *Ginkgo biloba*, the single fossil leftover of the strange floral world of 350 million years ago. It is related to that other dinosaur among plants, the Cycad which will come into the story a little later (see p.189). The Ginkgo seems to have been a native of Zhejiang province in eastern China, and clung tenaciously to life for all those millions of years until noted within recorded history. It was found growing in association with monasteries in both China and Japan from at least the eleventh century; in fact, one or two specimens in China are reputed to be over 1,500 years old. In England, it has the vernacular name of the Maidenhair Tree because its leaves (which are shaped rather like a half-opened fan) look similar to our native Maidenhair Fern (*Adiantum capillus-veneris*).

The first the West learned about the Ginkgo (and the camellia) was as early as 1712, and came from the notebooks of Engelbert Kaempfer (1651–1715) a physician and botanist with the Dutch East India Company at their trading station at Deshima, Nagasaki in Japan. It is believed that seeds arrived first in Holland about 1730, but they took another twenty years to reach Britain. It is a noble tree, coming very late into leaf, but turning a glowing gold in autumn. Its one disadvantage, in the female tree, is the smell of its ripening fruit, which supposedly gives off the odour of vomit. The fruit hangs rather like cherries and looks like small plums; in China it is considered edible and is added to stews (one can buy it canned); in Japanese cooking it is the kernel which is mostly used.

Long after the tree's arrival in Europe it was given the name *Salisburia adiantifolia*, in honour of Richard Salisbury, whom we have already met and who in 1805 became the first secretary of the Horticultural Society. However illustrious Mr Salisbury was, his Latinised name gave no indication of the tree's weighty associations, and it has since reverted to its original long-held Chinese name, albeit with a European flavour. The word 'ginkgo' derives from the Chinese *yinxing* meaning a 'silver apricot', and is a reflection of the tree's antecedents as well as its looks. The small personal problem of the disagreeable odour certainly does not detract from its

The Maidenhair Tree *Ginkgo biloba* derives its name from both Chinese and seventeenth-century Japanese, in the latter it comes from *gin* (silver) and *kyō* (apricot).

beauty. The herbal medicine comes from the leaves of the tree, which contain oils believed to help increase the blood supply by dilating the blood vessels. So popular has this remedy become, and of such world wide proportions that fears are being expressed for the tree's very survival; considering its 350-million-year-old history, this is a trifling inconsiderate of the present generation.

Probably the oldest known Ginkgo in Britain is one which was originally planted in the 1750s on the Duke of Argyll's estate at Whitton Place in Twickenham, west of London (where James Lee, of fuchsia fame, had been employed for a short time). In 1762, however, the tree was transplanted via a

Thames barge to Kew Gardens, where it still flourishes. What an extraordinary sight that must have been – if only someone had recorded the event for posterity. The tree is very tolerant of pollution, but even so it comes as a delightful surprise to find it planted as a street tree in Fitzrovia in London, copying the use made of it in big cities in China.

The success of the introduction of plants or trees from abroad relied entirely on the ability to grow them to maturity and then propagate them in the host country. In England during most of the eighteenth century there were a large number of keen plantsmen and nurserymen eager to experiment with the flood of seeds and cuttings which were entering Britain. One of the ablest gardeners was James Gordon (c. 1708–80), who originally worked for both Dr James Sherard in his garden at Eltham, and also for the 8th Baron Petre of Thorndon Hall in Essex. Robert Petre was a very knowledgeable botanist and is best known as the introducer of *Camellia japonica*. He was very wealthy, and moved to Thorndon Hall in 1732, determined to make it a veritable hothouse of new plants and exotic species. He corresponded with Linnaeus and knew both Philip Miller and Peter Collinson. In the ten years before he died (in 1742), his tree-planting programme was prodigious. In one year alone, 1740, he is said to have planted 5,000 saplings of an eclectic mix, from Cedars of Lebanon to larches, the Tulip Tree, chestnuts and conifers. His importations from North America were particularly strong, many of them sent to him by John Bartram, and by the time of his death from smallpox – 'the greatest loss that botany or gardening ever felt in this island' – 50,000 more trees had been added to his plantations, clumps and thickets.

The appointment of James Gordon as gardener at Thorndon Hall proved successful. He was introduced to the foibles of the rare and unusual plants being gathered there, and the experience of looking after them and coaxing them to grow and bloom later stood him in good stead. Following Lord Petre's death, the family decided to auction his entire collection of greenhouse and nursery stock, some 200,000 different plants and Thorndon Hall itself was sold. In 1754, Horace Walpole (1717–97), the 4th Earl of Orford, visited Thorndon, obviously hoping to gain inspiration from his tour of the grounds. Walpole was the owner of the fabulous 'gothick' Strawberry Hill at Twickenham in Middlesex, and was at that time deeply involved in transforming his estate into the most fashionable picturesque ideal of landscape gardening. Between 1766 and 1772 Thorndon Hall's grounds

were much altered by Lancelot Brown – for which he was paid £5,059 2s. Later came the twentieth century's ubiquitous answer to the landscape park: a golf course was created and any remaining eighteenth-century planting was obliterated.

After the death of Robert Petre, James Gordon decided to open a nursery at Mile End, then on the eastern outskirts of London. Here he quickly gained wide renown for his ability to coax the most recalcitrant of seeds to grow. In 1768 Captain James Cook, when getting ready for his voyage to observe the Transit of Venus on Tahiti, visited the nursery to purchase a variety of vegetable seeds to take with him to plant on the island. Gordon's nursery was the first to grow the seedling *Ginkgo biloba* commercially, also the more difficult *Gardenia jasminoides* (now known as **G. augusta**); this arrived from the Far East in 1754, and Gordon managed to raise plants from a cutting.

Peter Collinson and Philip Miller both held Gordon in high esteem and later, after he had opened a seed shop at 25 Fenchurch Street in the City of London, often visited him to discuss the latest consignment of new and unknown recruits to be brought to flower in English gardens. A London merchant and importer of seeds and plants from America, John Ellis (c. 1705–76) in a letter to Linnaeus considered Gordon 'had more knowledge in vegetation than all the gardeners and writers on gardening in England put together, but he is too modest to publish anything'. It therefore seems entirely appropriate that his name was linked with the nurturing of the Ginkgo tree in English parklands.

James Gordon is also remembered horticulturally in the naming of a relative of the Camellia family, the small but elegant Gordonia genus. In 1763 he introduced one of the species, **Gordonia lasianthus**, or Loblolly Bay, but then spent twenty years in trials learning how to grow it successfully. The intriguing 'Loblolly' name is attached to a further two American imports: **Pinus taeda** (1741), which was known as the Loblolly Pine, and Loblolly wood, which came from **Blighia sapida** (syn. *Cupania edulis*), a tree named for William Bligh of *Bounty* and breadfruit fame (*Artocarpus incisus* see p.198). Although an import from the West Indies in 1793, the *Gordonia* was originally a native of Guinea on the west African coast, but at some time must have been taken to the Americas, where it became well established. It is a single species in its own genus in the *Sapindaceae* family. One would assume the word 'Loblolly' was a record of where all three were first discovered, but it appears not be so. The word itself was first recorded in use in England as early as

1597 and was derived from 'lob' meaning 'to eat or drink noisily', and 'lolly', a 'thick broth or stew'. Another choice from 1604 meant a 'rustic' or 'country bumpkin' but neither definition seems appropriate to any of the three trees, so the word association remains a mystery.

Both James Gordon and Peter Collinson were involved in the introduction from the other side of the world of the beautifully named Tree of Heaven, *Ailanthus altissima*, the seed of which was sent to Collinson in 1751 from China. He, in turn, passed some of it to Gordon to experiment with in his nursery. Belonging to the *Simaroubaceae* family, it eventually makes a handsome tree with leaves shaped like those of the ash. It is very tolerant of living in a polluted atmosphere and can be found planted along some of the streets in Shanghai. Another of its attributes is that it takes no offence at its branches being hard-pruned each year, the new ones producing bigger and bigger leaves.

John Bartram is thought to have been responsible for the arrival from America in 1752 of two plants from the same genus. One was *Chelone obliqua alba*, as it was then called, although it is now known as **Chelone glabra**; its vernacular name is Turtlehead because of the shape of the flowers. The second is now known as **C. obliqua.** The first species may well be a reintroduction, as there is evidence of *C. obliqua alba* arriving in 1675. All 6 species in the genus prefer partial shade or even boggy conditions. They are closely related to the Penstemons, both genera being in the *Scrophulariaceae* family. A third American arrival was **Chimaphila maculata**; this is commonly known as Prince's Pine, despite the fact that it is not a pine, nor indeed even a tree. It is an evergreen perennial with soft, rather holly-like leaves and small cup-shaped white or pale pink flowers, and is happiest being planted in a woodland garden. In the same year, 1752, *Lobelia erinus* from the Cape of Good Hope arrived. This was the forerunner of, and in large part responsible for, all the bedding and trailing lobelia that is planted each summer in our gardens. Philip Miller cultivated it in the Chelsea Physic Garden in 1759.

A southern belle in the form of **Daphne cneorum**, the Garland Flower, came from the mountains of central and southern Europe also in 1752. Like all members of the Daphne family, the shrub is exquisitely perfumed but highly poisonous, and this seeming dichotomy has not improved its character. The whole genus can be of a sulky nature and sometimes will just not get on and strut its stuff. Perhaps the chip on their shoulders is a result of the gods muddling their identity with the Bay Laurel, *Laurus nobilis*. The word

daphne is the Greek name for the Bay Laurel but was later transferred to this completely different genus – the two are not botanically connected and are not even in the same family. *Laurus nobilis* is in the (very small) *Lauraceae* family, while the larger Daphne genus, of some 50 species, is placed in the *Thymelaeaceae* family. However difficult it is to establish the latter in gardens, their exquisite fragrance forgives them all. The Daphne is also one of the relatively few species to have no vernacular name but is always known by its formal botanical name.

A plant which came to be closely associated with Victorian gardens arrived from Peru about 1757. *Heliotropium peruvianum*, the Heliotrope or Cherry Pie (because of the perfume of the flowers), is a member of the *Boraginaceae* family and has now been renamed **H. arborescens** on account of its woody bark. It is the main plant from which all the modern heliotrope hybrids descend. Its name is derived from, first, the Greek word *helios* for 'sun' and then *trope*, 'to turn', hence its early name was sometimes Turnsole, reflecting a plant whose flower-head is supposed to be constantly turned towards the sun. As John Gerard explained two centuries earlier regarding the European Heliotrope, it is not 'because it is turned about at the daily motion of the sunne, but by reason it flowereth in the summer Solstice, at which time the sunne being farthest gone from the aequinoctial circle, returneth to the same'. The rather mauvish violet colour of the flower came to be known as 'heliotrope'; it was a popular shade for clothes and accessories in Victorian times and was often associated with mourning.

You either love them or hate them; there seem to be no half-measures with reactions to the *Berberis* genus, of which there are about 450 species in total, although fewer than half of them are in cultivation. They began to attract the attention of plant hunters in America, and one of the first to put its spiny toe in the water arrived here in 1759. In fact, the early settlers had already exported our own native Barberry, *Berberis vulgaris*, to America, and in exchange we received their **Berberis canadensis**, although the two shrubs were not that different. *B. canadensis* is not a native of Canada, as its name might suggest, but was found growing in the Allegheny Mountains of Virginia and North Carolina. It had a certain popularity but by the middle of the twentieth century was rarely cultivated and has now entirely disappeared from the garden scene, overtaken by other more stylish Berberis. The name Berberis comes from the Arabic word which describes the edible fruit or berries. The derivation is somewhat anomalous, because although the plant is prolific, the Arabic world is singularly bereft of them.

The Berberis genus falls into two main groups: 280 species belong in the Asiatic Group (India, China, Japan, Tibet); and the rest, apart from a very few in Europe and Africa, to the South American Group, with 2 having slipped into North America. In England, as early as the sixteenth century, Thomas Tusser records the fruit being eaten as a conserve; it was then known as the Pipperidge Bush. The bark is a yellow-ochre colour and it was thought to be a cure for jaundice; indeed, two early names for it were Woodsour and Jaundice-berry. The Mahonia is a closely related cousin, and both genera are in the same family of *Berberidaceae*. The Berberis genus itself is notoriously difficult to study as the species exhibit such a wide variation in their distinguishing details.

In the same year (1759) another small clutch of plants arrived from Russia and Siberia. A creamy-yellow perennial, *Cephalaria gigantea*, or Giant Scabious, was one of them, as was the delicate and so aptly named Baby's Breath, although how this *Gypsophila paniculata* survives in Siberia one will never know, since the books say it is hardy except in severe winters.

The following year, 1760, saw the arrival of a tree which must be mentioned: the Purple or Copper Beech, *Fagus sylvatica* f. *purpurea*. Again, this is not a tree for a small garden, but one which has made a colossal impact wherever it has been planted. It is instantly recognisable, although the striking leaf shade is not to everyone's taste. There were at least three widely distributed sites on the continent, all noted in the seventeenth century where records show the tree was growing: the Vosges Mountains in eastern France; the canton of Zurich in Switzerland (in 1680); and the German state of Thuringia. Its arrival here was said to have been a seedling from the Thuringia group. The best form now is considered to be a grafted tree *F. sylvatica* 'Rivers' Purple', although in 1983 Graham Stuart Thomas (1909–), then the Garden Consultant to the National Trust, wrote that he found them an 'upsetting sight in our green landscape, and can set at nought the most carefully conceived schemes when seen at a distance'.

The first of the Gaultheria genus came to England in 1762 from America, where it was called Wintergreen or Checkerberry, a name which makes it sound like a square dance. It is *G. procumbens*, whose leaves were of use as the source of Oil of Wintergreen, methyl salicylate, which helps

to relieve muscular aches and pains. Another name given to it was Mountain Tea, and again the leaves were used to make a drink, although hopefully not perfumed with the 'wintergreen' smell. All of the 170 species are evergreen, and most are low-growing and even ground-hugging. One such is a little treasure *G. microphylla*, with its tiny white flowers and white or rosy-pink fruit, which arrived here from the Falkland Islands in 1762. It has a list of four synonyms, so many in fact that the poor thing has been quite smothered and has retreated back to its windswept South Atlantic homeland.

The whole of the Gaultheria race suffers from an identity crisis, having welcomed into its family all the x Gaulnettya and the Pernettya too. A Benedictine abbot, Dom Antoine Pernetty, is remembered in that genus: he sailed as priest-in-charge and natural historian with the navigator Louis Bougainville (1729–1811) on the French expedition to the Southern Ocean in 1763. He had an opportunity to explore the Falkland Islands with a view to founding a French colony, and while there he made a collection of plants to take home with him. As well as the little *Gaultheria microphylla*, the Abbot also discovered one of the tallest of the Pernettya family, a prickly 'shrublet' then called *Pernettya mucronata* but which because of the nomenclature changes within the genus is now known as *G. mucronata*. It grows to all of about knee height and has pretty but almost artificial-looking ornamental pink and white berries. Although the plant was brought to Europe in 1763, it was well over fifty years before it arrived in Britain, in 1828, and this was from seed which was sent – from Tierra del Fuego this time – to W. J. Hooker (Sir William Jackson Hooker 1785–1865) of the Glasgow Botanic Gardens.

The seed was sent by James Anderson (1797–1842), the botanical collector who sailed with the SS *Adventure* on the expedition to map the Straits of Magellan between 1825 and 1830. Anderson spent the best part of his working life in horticulture in the southern hemisphere, first as a plant collector and later as the superintendent of the Botanic Gardens in Sydney; these were established in the Farm Cove area as early as 1816, two years before the Glasgow Botanic Gardens. He was rewarded with the naming of a New Zealand sedge in his honour, *Carex andersonii*. Louis Bougainville, mentioned above, was also honoured when *Bougainvillea spectabilis* was named after him.

As we have seen, 1763 turned out to be a rather favoured year for plant entry to Britain, and it continued with plants from Alaska (the reintroduction of *Ledum groenlandicum*), and the sophisticated Golden Rain Tree (*Koelreuteria paniculata*) from China, introduced by the 6th Earl of Coventry. However, amongst all these treasures there is bound to be one that lets the side down, and this time it is the gladiatorial entry of *Rhododendron ponticum*. Found near the town of Pontus in north-east Turkey, this plant no doubt looked good in the glittering light and conditions of its native home, but having been given British citizenship, it threw caution to the wind and set about conquering the native population like the Vikings (or at least conquering the understorey of much of Britain's woodlands). As a bright young thing in its early days, it was regarded with favour and planted in all the smartest great gardens and estates. As it grew, however, it gradually revealed its true character – that of a killer, a smotherer, a choker-to-death of native woodland species and no plant for polite society. In its search for new victims it also spread along railway embankments, where its only merit is that one can sometimes see the wide variation of colour, from wishy-washy mauve to wishy-washy pink.

The Earl of Coventry (1721–1809) inherited Croome Court in Worcestershire in 1738 when he was seventeen, and developed over the years into a discerning plantsman and horticulturist. He commissioned Lancelot Brown to design and build a new house (in a dignified Palladian style) as well as to transform the landscape surrounding the house; the work, which began in 1751, was one of Brown's first house commissions. In 1760, Robert Adam was asked to undertake the design of an Orangerie. The whole scheme took twenty years to complete and eventually cost a staggering £400,000. It was obviously impressive and much admired, since in 1792, *The Gentleman's Magazine* could hardly restrain itself when it remarked: 'if there be any spot on the habitable globe to make a death-bed terrible, it is Lord Coventry's at Croome…'

The estate was known for the quality of its trees and exotic shrubs and was reputed to have been the first place in England to grow the tree that goes under the name Pride of India. In fact, *Koelreuteria paniculata* is a native of China, from where it was sent in 1763. In 1766 the Earl received, also from China, the aptly named shrub Wintersweet, *Chimonanthus praecox* (syn. *C. fragrans*), which he planted in his new conservatory where it lived quite happily for the next thirty years. By then plantsmen were experimenting with growing it in the open, and there is no doubt that it was (and is) tougher than was first thought – like so many other importations. Give it a sheltered spot and the reward will be

the wonderful fragrance that the flower emits during the depths of winter. J. C. Loudon, in the nineteenth century, considered it so desirable that 'no garden whatsoever should be without it'. *C. praecox* was for some time believed to be monotypic, but over the following two hundred years another 5 species of the genus were discovered – all native to woodland areas of China.

The clashing cymbals of 'Zenia Peruvia' began to arrive in England during the 1750s. The seeds of the first Zinnia, *Zinnia pauciflora*, with its red and yellow carnival colours, were delivered to Philip Miller at the Chelsea Physic Garden in 1753 from the Royal Gardens of Paris where they had arrived earlier from Mexico. There was some doubt as to its native home, as seeds sent to Linnaeus were labelled *Calthe de Brésil*, the Brazilian marigold. From a gardening point of view, the most important Zinnia species was *Z. elegans*, which sashayed across the Atlantic in 1796 and became the parent of the modern-day bubbly annual. It was called for some unexplained reason Youth and Age. The genus of about 20 species was named for Herr Professor Johann Zinn of Göttingen University in Germany.

This latter Zinnia had as company some seed of 'Cocoxochitl', and both were delivered by the good offices of Charlotte-Jane, Marchioness of Bute (d. 1800), the wife of John, 1st Marquess of Bute (1744–1814), the British Ambassador in Spain from where the seed travelled. 'Cocoxochitl' would have been a name to conjure with, instead of which the plant was named for Dr Anders Dahl, yet another pupil of Linnaeus (just like *Rudbeckia* for Rudbeck and *Kalmia* for Pehr Kalm). Dahl, a Swedish botanist, was quite sure that the Dahlia would become a new and trendy vegetable, but although edible it had not got that certain something that the cooks were looking for. Even cattle turned it down, and thus it failed to become the Mexican equivalent of the mangel-wurzel. It is perhaps fortunate that it never reached 'chef' status, as we may well have been denied the explosion of colour and form which the plant has since taken.

Plants noted in 1771 were the exotic Coral Tree, or Cock's Comb (*Erythrina crista-galli)* from Brazil; and two perfumed travellers from China, *Daphne odora* and *Osmanthus fragrans*, the latter being also known as the Fragrant Olive or Sweet Tea, from its Greek name. This plant seems not to have lasted the course, as it was reintroduced in 1856. From southern Europe and west Asia, came the pretty double-flowered Greater Celandine. Its Greek name of *Chelidonium majus* 'Flora Pleno' indicates that its flowering time (in Europe) coincides with the arrival of the swallows; indeed, one of its old names was Swallow-wort.

From a horticultural viewpoint the most important event of 1771 was the docking at three p.m. on Saturday 13 July of the barque *Endeavour* at Deal and the return of Captain Cook and Joseph Banks from observing the phenomenon of the Transit of Venus and then their epoch-making journey to Australia and the southern hemisphere. It was also the year that saw Banks become the horticultural and garden adviser to Kew, a post he held until his death in 1820.

By 1772, from the horticultural riches brought back by Cook and Banks, the very first two Antipodean plants were being grown. One of the first, called *Philadelphus aromaticus*, has always been on the cusp of winter tolerance here. It is a large shrub belonging to the Myrtle (*Myrtaceae*) family. All of the 80 species have very narrow seeds, a fact reflected in the Greek name they were given of *Leptospermum*; they are also known in the vernacular as the Tea Tree. This first introduction, *Leptospermum scoparium*, was collected from New Zealand, where it is also called the Manuka Tree. With its small white flowers which bloom in May or June, the shrub looks as if it has had a dusting of snow over it. A second *Leptospermum*, this time a native of Australia and Tasmania, joined it in 1774 – *L. pubescens* (now *L. lanigerum*) which, because it has soft hairy stems, is known as the Woolly Tea Tree. A third member of the family, *L. polygalifolium*, arrived in 1778 from New South Wales.

Joining the first *Leptospermum* in 1772 was another large shrub or tree, the New Zealand Laburnum, a name which gives some idea of the colour of the flowers and the shape of the leaves. This plant's original botanical name of *Edwardsia Macnabiana* was to honour a young and very talented botanical artist of the time, Sydenham Teast Edwards (1768–1819), while *Macnabiana* probably commemorates William Macnab (1780–1848), a most distinguished curator and principal gardener to the Edinburgh Botanic Gardens. Both names, however, were swept away when Linnaeus used the more apt Arabic name of Sophora, the 'tree with pea-shaped flowers', to describe the 50 or so species, and in this first instance, *Sophora tetraptera* means 'with a four-winged seed pod'. Its New Zealand name is Kowhai.

Certainly the Sophora and probably the Leptospermum were initially both given to the seventy-year-old Philip Miller to grow at the Chelsea Physic Garden. However, slowly but surely with the ebb and flow of new seeds and

Leptospermum scoparium was amongst the 1,500 plants that were brought back by Captain James Cook and Sir Joseph Banks from their epic world voyage which ended in 1771.

plants arriving over the next decade or so, it was Kew Gardens, under royal patronage and the younger leadership of Joseph Banks and William Aiton (1731–93), rather than the ageing Gardener Miller and Chelsea, that enjoyed the challenge of growing the new introductions from around the world.

Even though there was political upheaval in America during the latter part of the century – 1773 saw the Boston Tea Party and 1776 the Declaration of Independence – plants from across the Atlantic continued to arrive.

It was the ever-loyal John Bartram and his fifth son William (1739–1823) who, from their home in Pennsylvania, were still able to continue feeding the shrubbery and flower

Franklinia altamata is a single species genus in the *Theaceae* family. It was collected by John Bartram from near the Altamaha river in Georgia.

bed of England. They were responsible for introducing a substantial number of now well established plants. In 1773 it was *Hydrangea quercifolia*, the Oak-Leaved Hydrangea. Next on the list, in 1774, came a small tree or shrub of the most refined breeding – *Franklinia alatamaha*, which had first been seen by John Bartram in 1765 by the River Altamaha in Georgia.

It turned out to be a single species genus, although related to both the Camellia and the Gordonia. The small tree was very localised, so much so that within a very few years it had disappeared from its one known natural location. It has never been found in the wild again, and all the cultivated trees grown throughout the world derive from the seeds and seedlings from the Bartrams' garden in Philadelphia. It is named in honour of the statesman and scientist Benjamin Franklin. Because the tree needs plenty of hot sun for it to bloom in the late summer and early autumn, something Britain can rarely supply, it is a shy flowerer here. However, the leaves glow with autumn colour, and the thought that it has no near relations and is extinct in the wild gives an added reason for including it in our gardens.

One of a genus of 6 species in the *Liliaceae /Convallariacea* family was also collected by William Bartram, in 1778, and named (later) for a Governor of New York and the instigator of the Erie Canal, Mr De Witt Clinton. ***Clintonia borealis*** was originally called *Smilacina borealis*. It looks similar to a Solomon's Seal and in America it is named the Corn Lily.

A series of severe winters during the 1780s perhaps helped three plants which came from Siberia to feel at home. ***Bergenia cordifolia***, one of the huge *Saxifragaceae* family, arrived in 1779. The earliest of the tribe to appear was *B. crassifolia* in 1765 (also from Siberia); *crassifolia* meaning 'with thick stems', which it surely needs coming from such a cold home, while *B. cordifolia* means 'having a heart-shaped leaf', hence its common name of Elephant's Ears. The whole of the genus was named by Linnaeus to honour a physician and botanist from Frankfurt, Karl von Bergen. All the Bergenias seem to have a rather unrefined nature which considering the toughness of their home territory is not surprising, and appear to be particularly unfazed by the hordes of snails they secrete in their folds. If the plant is grown in sun rather than in shade or the 'odd corner', it does flower delightfully and in autumn its leaves glow a brilliant red. Perhaps that is why Bergenias were such a favourite of Gertrude Jekyll (1843–1932) although, conversely, Christopher Lloyd (1921–), in his book *The Well-Tempered Garden* (1970), writes that he finds their leaf colour rather liverish.

The second of the Siberian group to defrost into our gardens, a year later, in 1780, was the beautiful violet-coloured but deadly ***Aconitum volubile***, a twining Monkshood from eastern Siberia and the Altai Mountains of Manchuria. Ten years later it was joined by *A. japonicum* from Japan. The whole genus received its name from the Aconitus Hills in

Turkey, and all the species are poisonous, every scrap, even the pollen. The most deadly of all is *A. ferox*, which grows in the Himalayas but is not available in England. It is the source of the poison *bikh* which was used by the Nepalese army in the nineteenth century to doctor the water wells during the struggles against the British.

The Siberian trio was completed in 1780 by another familiar garden plant, **Sedum populifolium**. This belongs in the very large *Crassulaceae* family, which includes our own native Stonecrop. The hybrid *S.* 'Autumn Joy' (or more properly *S.* 'Herbstfreude') is a very familiar plant in our gardens, but strangely attracts only bees and not butterflies.

Centaurea ruthenica, a knapweed from Ruthenia in European Russia, made its way here in 1783. In the same year, the Spotted Laurel, *Aucuba japonica*, bowed in from Japan, its botanical Latinised name being taken from the Japanese word for the shrub 'Aukubi'. It is another plant which is similar to the Daphne and again, despite being called a laurel, bears no relation to *Laurus nobilis* and is in a different family – that of *Cornaceae*, the Dogwoods. The Spotted Laurel has some of the same characteristics as the Bergenia, a really tough cookie which will grow anywhere, put up with the most outrageous neglect and yet still come up smiling. The whole shrub is mildly poisonous but, looked at carefully, it has the most beautiful glossy leaves. One of its characteristics is that it is dioecious, meaning that the male and female flowers are borne on separate plants. Only the female plant was originally introduced to England, so to begin with none of the glowing red berries which some people think are the plant's only attribute were produced. It was not until Robert Fortune (1812–80) visited Japan in 1861, some eighty years later, that a male plant was aquired so that the berries could brighten up otherwise dull autumnal corners.

By 1786, another two Magnolias had arrived from America: *Magnolia fraseri* (parchment-coloured and with slightly fragrant flowers) and *M. macrophylla*, which has been described as 'an awe-inspiring small tree'. It has very big creamy and purple flowers and the largest deciduous leaves of any plant growing in Britain, hence its common name of the Umbrella Tree.

Even before the second millennium began, the next plant, **Paeonia suffruiticosa**, the Tree or Moutan Peony, had had a long and noble history in the gardens of China. So exquisite was its breeding, so refined its manners, that it was called by the Chinese 'King of Flowers'. Tree Paeonias were featured and revered by the mandarins in the Imperial Gardens of

China and later in the temples of Japan. It is difficult to imagine a plant of such stature growing in the hurly-burly of its natural surroundings of northern China, yet in 1698, a whole hillside of them was seen glowing red and scenting the air, an event described in a Chinese document about the Moutan Shan hills in Shensi province. When the plant collector William Purdom (1880–1921) visited that area in 1912, he could find no trace of such a heavenly sight. The luscious goblet-shaped flowers come in a wide range of white, pink, red or purple, with maroon markings in the base, and are sometimes scented.

The Tree Peony was introduced to Kew by Sir Joseph Banks in 1787, when a Dr John Duncan of the East India Company returned to England and presented him with a living plant. This was something of a miracle, because a number of attempts had already been made to bring the plant back to England. In China, commercial production took place along the Yangtze River near Shanghai, and as the Tree Peony is not native to Canton (now Guanghzou), the dormant roots had to be transported the four hundred miles to the port. These were then forced into growth, ready for buying from the markets. Rather as we now buy chrysanthemums or poinsettias in pots for instant decoration, the Chinese would buy their *Paeonia suffruiticosa* also in pots, and probably only keeping them for a year or two at the most.

The first Tree Peony to arrive at Kew did not live long, and a further consignment of seven of the plants arrived in 1794, this time probably procured by Daniel or Thomas Beale, two brothers who, for fifty years, were merchants in China and India. Thomas, in particular, cultivated a garden of renown at his home in Macao, and the brothers regularly sent plants home. This group of peonies could, though, have been eased through the tortuous Chinese regulations by the arrival the previous year of Lord Macartney's embassy to visit the Emperor, Qian Long, in his summer palace at Chengde (Jehol). They carried with them an ambassadorial greeting from His Britannic Majesty King George III, and a plea to ease the appalling regulations and restrictions that the Chinese insisted upon for all foreigners. Despite the grandeur and sophistication of the entourage, no relaxation of any description took place. (When the effort was repeated in 1816–17, this time led by Lord Amhurst in an even grander display, the result was the same.)

However they were gathered, the Moutan Peonies were packed up and sent to England aboard the East Indiaman

Triton. Unfortunately, the ship was battered by a bad storm in the Channel and two of the seven shrubs died. The others, as becomes such a high-caste plant, were distributed between the King and Sir Joseph Banks.

One further Tree Peony arrived in 1804, on the SS *Hope* with Captain Prendergass, and was given to Sir Abraham Hume to be grown at Wormleybury in Hertfordshire. Hume was a tremendous plantsman as we have seen, and this plant proved to be of a much tougher disposition. It had a different-shaped poppy flower, white with a delicious maroon splodge at its base. According to contemporary accounts, it grew strongly, and in 1826 was reported to be over 7 ft high (just over 2 m) with 660 buds on it. What a sight it must have been in flower. For a while this form had the separate name of *P. papaveracea* but after a few years it reverted to its original name of *P. suffruticosa*. It was fifty or so years after the first plant arrived in this country that Robert Fortune, travelling in north-east China in the mid-1840s, discovered the centre of their cultivation and was able to send home 12 or 13 new varieties.

Today, with instant images able to be beamed literally from anywhere on earth (and beyond) on to our television screens, it is hard to imagine the difficulties of growing a plant in the eighteenth century. However, the excitement of wondering how they would grow, large or small, what the flower would be like, what the colour, was it poisonous and, indeed, would it survive, was all part of the challenge of taking on the roller-coaster of responsibility for survival. Not knowing what a plant was going to look like or what its requirements were must surely have made gardeners and horticulturists exceptionally keen-eyed and observant.

Although collecting plants for decorating British gardens was always seen as the *raison d'être* of British exploration and travel, the system worked in reverse when a country was being settled: native plants from Britain, whether suitable for the climate or not, were exported and planted. The familiar plants helped to keep both homesickness and starvation at bay while native edible plants were found and cultivated.

From 1788 onwards, a whole raft of Australasian plants arrived on the gardening scene in Britain. New Zealand and Australia, despite being relatively close to each other in the southern hemisphere, are botanically poles apart. The majority of New Zealand's native flora, some 83 per cent, is unique to the country, while much of the flora growing in Australia can generally be found in the countries surrounding the Indian Ocean.

Sir Joseph Banks, because of his early Antipodean explorations on HMS *Endeavour*, always took a great interest in the development of both countries. On his return home he wrote a report for the British government on the suitability of developing an agricultural community in Australia – or New Holland as it was then called – using convict labour. Subsequently, the First Fleet of eleven ships set sail from Spithead on Sunday 13 May 1787 bound for New South Wales with about 750 male and female convicts aboard. In addition to its human cargo, the ships carried seed from England, and plants and seeds were also collected *en route* from Rio de Janeiro and the Cape of Good Hope, ready to be planted on arrival. When Australia was reached in January 1788, land was cleared to grow the essential vegetables and trees needed for survival.

The immigrants certainly chose a good spot, since, years later, in 1816, that first clearing became incorporated into the Sydney Botanic Gardens. Overlooking the harbour, the gardens still flourish today. In 1818, another botanic garden was created, also overlooking a harbour; this time it was 10 hectares (25 acres) at Hobart, the capital of Tasmania. Over the next sixty years a further nineteen Australian botanic gardens were established.

In 1788, as the First Fleet was docking in Australia, the first bottlebrush, *Callistemon citrinus*, was tentatively being grown here, probably in a greenhouse, since the whole genus of about 25 species, all indigenous to Australia, is really only suited to the most favoured and mild of sites. Despite its citrus name (which actually refers to the lemon-scented leaves when crushed), the flowers are a dense red colour. A second species, *C. salignus*, arrived on the same trip; this one is a little hardier and has pale yellow 'brushes' and narrow willow-like leaves. The Greek for 'beautiful' is *kalli*, and *stemon* means the 'stamen', so the name alludes to the great beauty of the flowers. They are part of the *Myrtaceae* family. Also in 1788 came another pair of typically Australian introductions, this time members of the *Proteaceae* family and named to honour Sir Joseph Banks, *Banksia serrata* and *B. integrifolia*. Although their leathery leaves make them appear tough, they are frail in our climate.

Sir Joseph's first attempt to send plant-hunters to Australia was in 1789, but this ended in disaster when, on its way to Botany Bay, HMS *Guardian* was wrecked and two young Kew gardeners, James Smith and George Austin, were among those who lost their lives. Sir Joseph Banks then commissioned David Burton (d. 1792) to collect seeds and

Araucaria heterophylla, the Norfolk Island Pine, flank the west side of Sydney Bay, Norfolk Island, south of New Caledonia, in 1790, in which HMS *Sirius* was shipwrecked.

plants for him. Burton did this successfully even though, at the time, he was Superintendent of Convicts at Parramatta, Sydney (then called Port Jackson).

One of his introductions was a delightful shrub with pink sweet-scented flowers, *Boronia pinnata*. All the ninety five Boronia species are native to Australia and Tasmania so it is a quirk of fate that that the entire family was named in honour of Francesco Boroni, the manservant of Dr John Sibthorp (1758–96), Professor of Botany at Oxford University. Sibthorp was a traveller and plant collector who appears to have had no connection with the Antipodes. He spent his whole life working and cataloguing the native plants of Greece and the eastern Mediterranean. His opus magnus, the huge, ten-volume *Flora Graeca*, took nearly forty years to publish and was completed forty-four years

after his death in 1840. Four more Boronias were eventually sent from Australia in 1823 and 1824, none of which found conditions here for growing quite what they required. However, two of the genus did settle and are still available to us – *B. megastigma*, the Brown Boronia from Western Australia, and *B. serrulata*, the Sydney Rock Rose from New South Wales, and they both arrived in Britain in the 1870s.

A presentation of at least four Pittosporums was received here during the last twenty years of the eighteenth century, three of them coming from Australia; but the first to be recorded, however, in 1783, was *Pittosporum coriaceum*, from Madeira. It must have been gathered by Francis Masson, the Kew collector, who, having recently returned from the Cape, then made a two-year trip to Portugal and Tangier. The Australian contingent began with *P. ferrugineum* in 1787. It

can grow into a tree measuring over 15 m (50 ft); it was grown in this country as a shrub and needed greenhouse treatment, but is no longer available here. That was followed, in 1789, by *P. undulatum*, which can survive outdoors (just, in sheltered sites) and has perfumed creamy-white flowers. In 1795 another tender member arrived, *P. revolutum*, so named because the edges of the leaves curl upwards. A further cluster of Pittosporums arrived between 1804 and 1806, including *P. tenuifolium* from New Zealand (from which nearly all the hardier hybrids have been developed), *P. tobira* from China and Japan and *P. viridiflorum* from South Africa. This last species has leathery but lustrous dark green leaves and clusters of jasmine-scented creamy-white flowers, and is another of the Pittosporums which needs special TLC in the greenhouse to survive in Britain. There are about 200 species in all and they form their own family of *Pittosporaceae*. The name aptly describes in Greek the resinous coating on the seeds – *pitta* meaning 'pitch' and *sporum* 'seed'. They are sometimes known as Parchment Bark or Australian Laurel.

A group of herbaceous lilies with which we are now very familiar began arriving from Japan and eastern Asia towards the end of the eighteenth century. In the vanguard, in 1790, came the kimono-clad beauty *Hemerocallis coerulea,* with wonderful leaves but a rather sickly flower. It changed its kimono to *Funkia ovata* quite soon and then, still not feeling comfortable in that set of clothes, changed again in the 1950s to the European style, with the name *Hosta ventricosa* (to honour the Viennese physician Nicolaus Host, 1761–1834). All 80 Hosta species come from the East: Korea, China, eastern Russia and Japan. It was in Japan that Philipp van Siebold (1791–1866), the German doctor to the Governor of the Dutch East India Company, discovered *H. plantaginea* and *H sieboldii (*syn *H. albomarginata)* between 1826 and 1830. Today, one is almost overwhelmed by the success of the hybridisation, and there are many beautiful Hostas to chose from, but unless one is on full Hosta alert from February each year, the slugs will have them first.

Norfolk Island, a windswept and uninhabited small island south of New Caledonia, was discovered and named by Captain Cook in October 1774. He also noted 'most prominently' great spruce pines, seeds of which were collected at the same time. What he saw is one of the most architectural trees in the world and instantly recognisable as a silhouette. The genus is a very ancient one with many fossilised remains. Part of the *Araucariaceae* family, its botanical name is ***Araucaria heterophylla***, and in its vernacular tag – the Norfolk Island Pine – it carries the name of that remote island all around the world; it is also known as the Australian Pine. As becomes the tree's isolated provenance, it is part of an ancient and very small genus of 18 species of evergreen coniferous trees, all found in southern hemisphere tropical rainforests. It arrived in this country off the ship HMS *Resolution* in 1794, and can only really be grown as a conservatory plant.

Two years later, *Araucaria araucana* – or, as it is better known, the Monkey Puzzle Tree – began to preen itself ready for its front-of-house position in Victorian gardens. It came from Chile and Argentina and is, alas, the only one of the genus which is England-friendly and can be grown outdoors. (The native Chilean word for the trees is *araucanos*, and they grow naturally in the Arauco province.) The Monkey Puzzle Tree was so named because the branches curl into the shape of a monkey's tail. The trunks are so prickly that they were supposedly proof against being eaten by dinosaurs; now they are planted as street trees as proof against vandals.

The nut of the cone is edible and was first noticed in 1796 in a dessert bowl of fruit and nuts by Archibald Menzies at a dinner at St Jago in Chile at which he and his fellow naval officers from HMS *Discovery* were being entertained by the Spanish viceroy. Not having seen anything like them before, he slipped at least five into his pockets – which must have been voluminous, since the cones are quite large, or perhaps it was just the seeds from the cones that he took. HMS *Discovery* with Capt. George Vancouver in charge, was nearing the end of its five-year round-the-world voyage, and Surgeon-Naturalist Menzies managed to get the seeds to germinate by the time he arrived back in England. One of the tree seedlings was planted at Kew and grew there for nearly a century, until it died in 1892.

The extraordinary nature of the tree meant that for a long time it was considered a curiosity, and it was not until fifty years after Menzies' discovery that William Lobb (1809–64), the great plant collector, procured sufficient pine kernels for them to be distributed on any scale.

A bouquet of herbaceous plants bedded themselves into

Left: The Monkey Puzzle *Araucaria araucana* is the only one of the 18 species to be successfully grown throughout Great Britain, being frost-tolerant. Here the tree is seen in its natural habitat in Chile.

our gardens during the last few years of the eighteenth century. Bleeding Heart, *Dicentra formosa*, came from the western side of America in 1796; an earlier species, *D. cucullaria*, or Dutchman's Breeches, had arrived from the eastern seaboard in 1731; and a few years later in 1816, *D. spectabilis* arrived from Siberia and Japan. The same year, 1796, saw another stalwart of our gardens, *Verbascum phoeniceum*, the Purple Mullein, come from the heights of the Altai Mountains in Central Asia. Although sometimes overgenerous with its seed, the whole genus of some 350 species makes a spectacular statement in our gardens.

From Mexico in 1799 came two near relations of the Dahlia: *Cosmos bipinnatus* (with leaves arranged like a feather) and *C. sulphureus* which was originally named *Coreopsis parviflora*. The seed had first arrived in Spain, and as with the Zinnia had been sent to England by the Marchioness of Bute. A further pair crossed the Atlantic in 1835, *C. diversifolius* (syn. *Bidens dahlioides*) and one that shows how simple it is for plants to drift away out of fashion and out of nursery catalogues unless they are continually loved and nurtured: *C. atrosanguineus*, the deliciously chocolate-scented dark maroon annual from Mexico. The seed was received in 1835 by William Thompson (1823–1903), who had earlier founded a nursery at Ipswich (which later became the world-famous firm of Thompson and Morgan). The plant made an immediate impact, with its dramatic deep maroon colour and was widely grown, but despite being admired – and commented on by such plantsmen as E. A. Bowles (1865–1954) – and receiving an RHS Award of Merit in 1938, it fell out of favour. It was only at the very end of the twentieth century that it was 'rescued' and recovered its self-esteem to flourish again in our English gardens.

Finally, it was an Australian which grabbed a toehold for itself in 1799 – *Bracteantha bracteata*, the Golden Everlasting Strawflower, a member of the sunflower family. We now know these as the easily dried 'everlasting' flowers, which have rusty papery bracts and brightly coloured flower-heads.

Today, having just turned the corner into the third millennium, we in Britain are reaping the horticultural rewards of the inward investment made to our gardens and landscapes during the eighteenth century. The result of the combination of the explosion of new plants and the talent of so many good and great gardeners and plantsmen who were endlessly fascinated by the newcomers has proved invaluable to today's gardens. Almost untrammelled by thoughts of conservation and worries about planning laws, the men who masterminded the new 'landskiping' of vast tracts of estates and clothed them with some of the newest of arrivals altered the face of Britain almost beyond recognition. Suddenly, the adornment of the countryside, and its foreign elements of both trees and decorative buildings, became quite normal. Gardening became a national obsession. Of all the thousands of seeds, nuts, plants and cones that were tirelessly sent home from around the world, only a fraction could have survived both the journey and the new conditions. Those that did have enriched our gardens, most of the time without our even realising how far from home they travelled and how alien their new surroundings must have seemed to them. However, the introductions of the eighteenth century are as nothing compared with the forthcoming hundred years. If this century experienced a deluge of new plants, the nineteenth century saw a positive Noah's flood of horticultural delights enriching the cultivated land of Britain.

Plant Introductions in the period 1700–1799

nsd = no set date

nsd   *Populus nigra* ssp *betulifolia* Manchester Poplar. Europe; see p.75.

c. 1700   *Aubrieta deltoidea* Sicily to Asia Minor; see p.122.
Picea mariana Black Spruce. Canada, N.E. USA; see p.71.

1701   *Pelargonium capitatum* South Africa.
Pelargonium peltatum Ivy-leaved Geranium. South Africa.

1702   *Eucomis regia* Pineapple Lily. South Africa.

1703   *Fuchsia triphylla* South America, West Indies.

c. 1703   *Plumeria alba* West Indian Jasmine. Puerto Rico, Lesser Antilles.
Plumeria rubra Common Frangipani. Mexico to Panama.

1704   *Dodecatheon meadia* (syn. *D. pauciflorum*) Shooting Star. N. W. USA. Reintroduced 1744.

1705   *Pinus strobus* Eastern White Pine, Weymouth Pine. E. North America.

1710   *Aster-novi-belgii* Michaelmas Daisy, New York Aster. North America. See also p.135.
Pelargonium zonale South Africa, see p.205.

1711   *Aesculus pavia* (syn. *A. splendens*) Red Buckeye. E. USA.

1712   *Amaryllis belladonna* Belladonna Lily. South Africa

1713   *Brunnera macrophylla* (syn. *Anchusa myosotidiflora*) Caucasus.

     Ceanothus americanus New Jersey Tea. E. USA.

1714   *Arbutus andrachne* Grecian Strawberry Tree. S.E. Europe, Turkey, Lebanon.

     Pelargonium inquinans South Africa; see p.205.

     Rudbeckia hirta (syn. *R. gloriosa*) Black-Eyed Susan. C. USA.

     Veronicastrum virginicum (syn. *Veronica virginica*) Culver's Root. N. America.

1719   *Erythrina herbacea* Coral Bean. S.E. USA, Mexico.

1720   *Aloe variegata* (syn. *A. ausana, A. punctata*) Partridge-breasted Aloe. South Africa.

1722   *Catalpa bignonioides* Indian Bean Tree. S.E USA.

     Citrus grandis Shaddock, Pomelo. Polynesia; see p.99.

1723   *Pelargonium fulgidium* South Africa.

     Pelargonium x ardens The Glowing Geranium. Garden origin.

1724   *Wisteria frutescens.* S. E. USA.

     Rosa x centifolia Moss Rose, Provence Rose. Garden origin. France.

1725   *Coreopsis lanceolata* Tickseed. C. & S. USA

1726   *Verbena bonariensis* (syn. *V. patagonica*) South American Verbena. Brazil to Argentina.

1727   *Tagetes minuta* Mexico.

1728   *Hermannia althaeifolia* Western Cape of South Africa.

1729   *Astragalus galegiformis* Milk Vetch. Siberia, Caucasus, Asia Minor.

1730   *Phlox paniculata* Perennial Phlox. E. USA.

     Salix babylonica Weeping Willow. N. China.

1731   *Bletia verecunda* Modest Orchid. Bermuda.

     Dicentra cucullaria Dutchman's Breeches. E. North America.

     Ramonda myconi (syn. *R. pyrenaica*) Pyrennes, N. E. Spain. Reintroduction (see p.129).

     Tiarella cordifolia Reintroduction (see p.145).

     Zantedeschia aethiopica Arum Lily, Lily of the Nile. South Africa, Lesotho.

1732   *Cimicifuga racemosa* Black Snake Root. E. North America.

     Helianthus atrorubens (syn. *H. sparsifolius*) Dark-eye Sunflower. S.E. USA.

     Helianthus pauciflorus (syn.*H. scaberrimus/H. rigidus*) W. to C. USA.

1733   *Ipomoea japala* (syn. *Batatas japala*) Mexico, S. USA; (see p.113).

     Turnera ulmifolia West Indian Holly, Sage Rose. West Indies, Mexico; see p.101.

1734   *Kalmia latifolia* Calico Bush, Mountain Laurel. E. USA.

     Magnolia grandiflora Bull Bay. S.E. USA.

1735   *Quercus cerris* Turkey Oak. C. and S. Europe.

1736   *Betula nigra* Black Birch, River Birch. / E. USA.

     Kalmia angustifolia Sheep Laurel, Lamb-Kill. E. North America.

     Tsuga canadensis Eastern Hemlock. E. North America.

1738   *Lilium superbum* American Turkscap Lily. E. USA.

1739   *Camellia japonica* Japan.

     Datisca cannabina Middle East.

     Linum narbonense Flax. W. and C. Europe.

     Parkinsonia aculeata Jerusalem Thorn. S. USA, Mexico; see p.129.

     Rhododendron ferrugineum Alpenrose. Reintroduction (see p.129).

     Seriphidium vallesiacum (syn. *Artemisia valesiaca*) Switzerland.

     Verbascum phlomoides Woolly Mullein. Caucasus.

1740   *Magnolia acuminata* Cucumber Tree. E. USA. So-called because of the shape and colour of the fruit. Discovered by John Bartram.

1741   *Cornus alba* (syn. *Swida alba, Thelycronia alba*) Red-barked Dogwood. Siberia, N. China to Korea.

     Delphinium grandiflorum (syn. *D. chinense*) Siberia, Mongolia, China, Japan.

     Pinus taeda Loblolly Pine. N. America.

1742   *Stewartia malacodendron* S.E. USA.

1743   *Pinus rigida* Pitch Pine. S. E. USA. The first one in the country was planted at Woburn.

c. 1744   *Lathyrus nervosus* (syn. *L. magellanicus.*) Lord Anson's Blue Pea. South America.

1745   *Gladiolus tristis* Western Cape of South Africa.

1746   *Betula papyrifera* Paper Birch, Canoe Birch. North America.

1749   *Helianthus decapetalus* Thin-leaved Sunflower. C. and S.E. USA.

1750   *Anagyris foetida* Mediterranean; see p.107.

     Ginkgo biloba (syn. *Salisburia adiantifolia*) Maidenhair Tree. S.China.

c. 1750   *Tradescantia spathacea* (syn. *Rhoeo discolor/R spathacea*) Moses in-the-cradle, Three-men-in-a-boat. Central America; see p.133.

1751   *Ailanthus altissima* Tree of Heaven. China.

1752   *Chelone glabra* (syn.*C.obliqua* var. *alba*) Turtlehead. E. and S. North America. Reintroduction (see p.146).

     Chelone obliqua Turtlehead. C. and S.E. USA.

     Chimaphila maculata Prince's Pine. E. North America.

     Daphne cneorum Garland Flower. Alps, S E. France N. Italy.

Lobelia erinus Cape of Good Hope, South Africa.

1753 *Alstroemeria pelegrina* (syn. *A. gayana)* Lily of the Incas. Peru; see p.210.

Rhododendron ferrugineum Alpenrose. C. Europe, Alps. Reintroduced (see also p.129).

Zinnnia pauciflora (syn. *Z. lutea*) Mexico.

1754 *Gardenia augusta* (syn. *G. florida, G. grandiflora, G. jasmi noides*) Gardenia, Cape Jasmine. China, Taiwan, Japan.

1755 *Acer rubrum* Red Maple, Scarlet Maple. E. North America.

1756 *Euonymus atropurpureus* Burning Bush. E. and C. USA.

1757 *Heliotropium arborescens* (syn. *H. peruvianum*) Cherry Pie, Heliotrope. Peru.

Ornithogalum thyrsoides Chincherinchee. Western Cape of South Africa; see p.95.

1758 *Boltonia asteroides* C. and E. USA.

Populus nigra var. *italica* Lombardy Poplar. Italy; see p.75.

1759 *Berberis canadensis* Barberry. E. USA.

Cephalaria gigantea (syn. *C. tatarica, Scabiosa gigantea S. tatarica*) Giant Scabious, Yellow Scabious. Caucasus, N. Turkey.

Gypsophila paniculata Baby's Breath. C. and E. Europe, Siberia.

1760 *Eucomis autumnalis* (syn. *E. undulata*) South Africa.

Fagus sylvatica f. *purpurea* Copper Beech. C. Europe to Caucasus.

1761 *Mitchella repens* Creeping Box,Partridge Berry. E. and C. North America. Introduced by John Bartram.

1762 *Gaultheria microphylla* (syn. *G. antarctica, Arbutus microphylla, A. serpyllifolia, Pernettya seryllifolia*) Falkland Islands.

Gaultheria procumbens Checkerberry, Wintergreen. E. North America.

Ledum palustre Marsh Ledum. Reintroduction (see p.129).

1763 *Gaultheria mucronta (syn.Pernettya mucronata)* Chile, Argentina, Falkland Islands.

Gordonia lasianthus Loblolly Bay. S. E. USA.

Koelreuteria paniculata Golden Rain Tree,Pride of India. S.W. China, Korea.

Ledum groenlandicum Labrador Tea. Greenland, Alaska, Canada south to N. USA; see p.129.

Rhododendron ponticum Turkey. Armenia, Caucasus, Lebanon, Spain, Portugal; see p.122.

1764 *Aesculus flava* (syn. *A. octandra*) Sweet Buckeye, Yellow Buckeye. E. USA.

1765 *Bergenia crassifolia* Elephant's Ears. Siberia.

1766 *Chimonanthus praecox* (syn. *C. fragrans, Calycanthus praecox*) Wintersweet. China.

1767 *Kalmia polifolia* Eastern Bog Laurel. Canada and N. E. USA.

1768 *Anchusa azurea* (syn. *A. italica*) Alkanet. S. Europe, N. Africa, W. Asia. Cultivated by Philip Miller at the Chelsea Physic Garden.

Bupleurum petraeum C. Europe.

Camellia sinensis (syn. *Thea sinensis*) Tea Plant. China.

1769 *Erica australis* Spanish or Portuguese Heath. Portugal, W. Spain, Tangier.

1770 *Pistacia vera* Pistachio Tree. W. Asia; see p.131.

Zinnia multiflora Mexico.

1771 *Chelidonium majus* 'Flora Pleno' Greater Celandine. S. Europe and W. Asia.

Daphne odora. China and Japan.

Erythrina crista-galli Coral or Cock's Comb Tree. Brazil, E. Bolivia to Argentina.

Osmanthus fragrans Fragrant Olive, Sweet Tea. Himalayas, China, Japan. Reintroduced (see p.218).

1772 *Leptospermum scoparium* Tea Tree, Manuka. New Zealand, S.E. Australia.

Sophora tetraptera New Zealand Laburnum, Kowhai. New Zealand.

1773 *Calceolaria pinnata* Pouch Flower, Slipper Flower. Chile, Brazil, Jamaica, Bolivia; see p.209.

Hydrangea quercifolia Oak-leaved Hydrangea. S. E. USA.

Strelitzia reginae Bird of Paradise, Crane Flower. South Africa. Introduced by Sir Joseph Banks.

1774 *Buddleja globosa* Orange Ball Tree. Chile, Peru, Argentina.

Gladiolus carneus (syn. *G blandus, G. blandus* var. *carneus*) Western Cape of South Africa; see p.204.

Leptospermum lanigerum (syn. *L. pubescens*) Woolly Tea Tree. Tasmania, Queensland, Victoria.

Franklinia alatamaha (syn. *Gordonia alatamaha*) Georgia.

1775 *Androsace alpina* (syn. *A. glacialis*) Alpine Rock Jasmine. Switzerland.

Campanula cenisia Mount Cenis Bellflower Italian Alps. Collected by Thomas Blaikie.

1776 *Alstroemeria caryophyllea* Brazil.

Gentiana andrewsii Bottle Gentian. E. North America.

1777 *Aitonia capensis* South Africa; see p.185.

Calceolaria fothergillii Falkland Islands, Patagonia; see p.209.

Cimicifuga foetida Bugbane. Siberia to N. Mongolia.

Encephalartos altensteinii Prickly Cycad. Eastern Cape of South Africa; see p.189.

Genista tenera var. *virgata* (syn. *Cytisus virgatus*) Madeiran Broom. Madeira. Collected by Francis Masson.

1778 *Clintonia borealis* (syn. *Smilacina borealis*) Corn Lily. E. North America.

Leptospermum polygalifolium (syn. *L. flavescens*) New South Wales, Queensland, Lord Howe Island.

Phaius tankervilleae (originally *Bletia*) C. China N. India, Sri Lanka, S.E. Asia to Australia.

1779   *Bergenia cordifolia* Elephant's Ears. Siberia.

1780   *Aconitum volubile* Monkshood. E. Siberia, Manchuria, Japan.

Buxus balearica Balearic Box. Spain, Sardinia (see p.29).

Magnolia denudata (syn. *M. heptapeta*) Lily Tree, Yulan. China.

Sedum populifolium Siberia.

c. 1780   *Citrus reticulata* Clementine, Mandarin, Tangerine. S.E.Asia; see p.99.

1781   *Betula pendula* 'Laciniata' (syn. *B. pendula* 'Dalecarlica' of gardens) Swedish Birch. Named for the country's province of Dalecarlia.

1782   *Stachys byzantina* (syn. *S. lanata*, *S. olympica*) Lambs' Ears, Lambs' Lugs, Lambs' Tails, Lambs' Tongues. Caucasus to Iran.

1783   *Aucuba japonica* Spotted Laurel. Japan.

Centaurea ruthenica Cornflower, Knapweed. E. Europe.

Pittosporum coriaceum Madeira.

1784   *Bupleurum gibraltaricum* S.Spain.

Chaenomeles speciosa (syn. *C. lagenaria, Cydonia speciosa, Pyrus japonica*) China; see p.61.

Paeonia lactiflora (syn. *P. albiflora, P. japonica* of gardens) E. Siberia, China. Reintroduced (see p.215).

Veronica gentianoides Speedwell. Crimea, Caucasus, C. Turkey.

1785   *Decumaria barbara* S. E. USA.

1786   *Brunia nodiflora* South Africa; see p.240.

Dicksonia antartica Man Fern, Soft Tree Fern, Woolly Tree Fern. E. Australia, Tasmania; see p.185.

Galax urceolata (syn. *G. aphylla*) Wandflower. S. E. USA.

Magnolia fraseri S. E. USA

Magnolia macrophylla Umbrella Tree, Great-leaved Magnolia. S. E. USA.

1787   *Paeonia suffruticosa* Moutan or Tree Peony. China, Tibet, Bhutan.

Pittosporum ferrugineum Australia.

1788   *Banksia serrata* Saw Banksia. New South Wales, Victoria.

Banksia integrifolia Coast Banksia. Queensland to Victoria.

Callistemon citrinus Crimson Bottlebrush.New South Wales, Victoria.

Callistemon salignus White or Yellow Bottlebrush, Willow. Bottlebrush South Australia, New South Wales.

Fuchsia coccinea Chile.

1789   *Pittosporum undulatum* Australian Mock Orange, Cheesewood. Queensland, Tasmania.

1790   *Aconitum japonicum* Monkshood. Japan.

Eucomis comosa (syn. *E. punctata*) South Africa.

Hosta ventricosa (syn. *Funkia ovata*) China, N. Korea, Japan.

Magnolia liliiflora (syn. *M. quinquepeta*) Woody Orchid Magnolia. Central China.

c. 1790   *Penstemon barbatus* (syn. *Chelone barbata*) Beardlip Penstemon. W. USA to Mexico; see p.196.

1791   *Berberis ilicifolia* Holly-leaved Barberry. Tierra del Fuego, Chile.

1792   *Chamaecytisus purpureus* (syn. *Cytisus purpureus*) Purple broom. S.E. Europe; see p.107.

Macleaya cordata (syn. *Bocconia cordata*) Plume Poppy. China and Japan.

1793   *Blighia sapida* (syn. *Cupiana edulis*) Akee Tree. Guinea.

Artocarpus incisus Breadfruit. Malaya, Pacific Islands; see p.198.

Linum flavum Golden Flax, Yellow Flax. C. & S. Europe.

Lupinus arboreus Tree Lupin. California; see p. 80.

Olearia tomentosa S.E. Australia; see p. 213.

1794   *Araucaria heterophylla* (syn. *A.excelsa* of gardens) Norfolk Island Pine. Norfolk Island.

Boronia pinnata Tasmania, Australia.

Ligustrum lucidum Chinese Privet. China, Korea, Japan. Introduced by Joseph Banks.

Lupinus nootkatensis Lupin. N. W. America; see p.80.

1795   *Grevillea linearis* New South Wales; see p.185.

Pittosporum revolutum Queensland, New South Wales.

1796   *Araucaria araucana* (syn. *A. imbricata*) Monkey Puzzle Tree, Chilean Pine. Chile, Argentina.

Dicentra formosa (syn. *D. exima*) Bleeding Heart. W. USA.

Diospyros kaki Chinese Persimmon, Japanese Persimmon, Kaki. China; see p.113.

Verbascum phoeniceum Purple Mullein. S. Europe, Altai Mountains, N. Africa to Central Asia.

Zinnia elegans Youth and Age. Mexico.

1797   *Ipomoea batatas* (syn. *Batatas edulis*) Sweet Potato. Tropical America. Reintroduced (see p.112).

Malva alcea Hollyhock Mallow. S. Europe.

1798   *Arabis caucasica* (syn. *A. albida, A. alpina, A. billardieri*) Rock Cress. S. Europe; see p.40.

Digitalis lanata Foxglove. Italy, Balkans, Hungary.

Digitalis parviflora (syn. *D. kishinskyi*) N. Spain.

1799   *Bracteantha bracteata* (syn. *Helichrysum bracteatum*) Golden Everlasting Strawflower. Australia.

Cosmos bipinnatus Mexico.

Cosmos sulphureus (syn. *Coreopsis parviflora*) Mexico.

For the Improvement of Horticulture
1800–1899

Significant dates

1800–15	Napoleonic Wars with France		1858	Government of India transferred from East India Company to British Crown
1801	General Enclosure Act		1859	*The Origin of Species* published by Charles Darwin
1802	Founding of Liverpool Botanic Garden, the first botanic garden to be open to the public		1861	Prince Albert tours Horticultural Society Gardens at Chiswick; the appellation 'Royal' is bestowed on the Society
1804	Founding of the Horticultural Society		1861–5	American Civil War
1806	End of Holy Roman Empire		1864	Hilliers' Nursery founded at Winchester by Edwin Hillier (1840–1929)
1808–14	Peninsular War		1869	Opening of Suez Canal
1811	Prince of Wales installed as Regent (1811–20) due to George III's insanity		1870	John Ruskin (1819–1900), author and art critic, creates his garden Brantwood at Coniston in Cumbria
1815	Battle of Waterloo, final defeat of Napoleon Bonaparte (1769–1821)		1877	The Society for the Protection of Ancient Buildings founded by William Morris (1834–96)
1820	George IV (1820–30)		1879	Zulu War in South Africa
1821	The Horticultural Society leases land from Duke of Devonshire in Chiswick		1880–1	First Boer War
1822	*Encyclopaedia of Gardening* published by John Loudon (1783–1843)		1883	First Garden Suburb begun, at Bedford Park in west London, by Norman Shaw (1831–1912)
1829	Westonbirt Arboretum in Gloucestershire begins to be planted by Robert Stayner Holford (1808–92)		1888	Establishment of Arts and Crafts Exhibition Society by William Morris
1830	William IV (1830–37)		1891	RHS Conference held on Aster species
1831	Charles Darwin (1809–82) begins his voyage on the *Beagle*		1895	National Trust founded by Octavia Hill (1838–1912)
1837	Victoria (1837–1901)		1899	International Hybridisation Conference held by RHS
1840	*Gardening for Ladies* published by Jane Wells Loudon (1807–58)			The Garden City Association founded by Sir Ebenezer Howard
1845	Potato famine in Ireland		1899–1902	Second Boer War
1851	Great Exhibition held in Crystal Palace, designed by Joseph Paxton (1803–65)			
1854–6	Crimean War			
1856	China opens to European trade; Japanese trade agreement			
1857–8	Indian Mutiny			

Left: Amherstia nobilis painted here by Marianne North (1830–90) hangs in the gallery she built at Kew Gardens as part of the collection of her life's work which she bequeathed to the nation.

A BOOKSHOP SEEMS AN UNUSUAL VENUE for the beginning of a great horticultural movement, and yet in 1804, at 187 Piccadilly, London, where Hatchards the booksellers traded – and still do – a group of seven friends met and formed an organisation '. . . for the improvement of horticulture'. An appealing ideal, it would seem, for within two decades, there were over fifteen hundred Fellows – as members were called – among them three kings (of the Netherlands, Denmark and Bavaria) and one emperor (of Russia). By 1821 the expanding Horticultural Society had leased land from the 6th Duke of Devonshire to create its own demonstration gardens adjacent to Chiswick House on the western outskirts of London. On 5 June 1861, on what became one of Prince Albert's last public engagements before his death that December, he visited the gardens and conferred the appellation Royal on the organisation.

Prince Albert took a very lively interest in the Society's affairs, and in Queen Victoria's journal for 1859 she remarked that her husband busied himself so much with it that he could not find the time even to take a walk with her. Her opinion of the Society does not seem to have been high, although four years after the Prince's death, her first appearance in public was to visit a flower show in those very same gardens. Over the following two centuries, the Society grew and grew until, during the twentieth century, it became a truly international gardening organisation though still with the same 'mission statement' of 1804 at its heart.

Five of the seven friends who originally founded the Society were each honoured by having a newly discovered genus named after them. The Banksia genus was named for the greatest grandee of horticulture and science, Sir Joseph Banks (1743–1820). Dicksonia, the tree fern, commemorated James Dickson (1738–1822), a Scottish nurseryman who owned a seed and herb shop in Covent Garden, not far away from the bookshop. William Forsyth (1737–1804) was another Scot, for whom the yellow spring-flowering Forsythia was named; he invented a 'secret' poultice to enable trees to recover from broken or diseased limbs – it later proved to be of no help at all. He was gardener to the Duke of Northumberland at Syon House in west London until 1771, then took over from Philip Miller at the Chelsea Physic Garden. By 1784 he had been appointed Gardener Royal to George III at both Kensington and St James's Palaces. He wrote two books on fruit, the second being published in 1802 and entitled *A Treatise on the Culture and Management of Fruit-Trees*. The fourth plant to commemorate a founder member of the Society was

Grevillea are all part of the *Proteaceae* family. One of the earliest to arrive, in 1822, from New South Wales, was *G. rosmarinifolia*.

Grevillea. This is a more exotic shrub from Australia, whose leaves resemble rosemary although the two are not related. It was named for the Honourable Charles Greville (1749–1809), an introducer and grower of rare plants in his 'botanic garden' at Paddington. The fifth founder member was Richard Salisbury (1761–1829), whom we met as the introducer of the renamed *Ginkgo biloba*.

It somehow seems an oversight that neither of the two other founder members of this great Society has ever been commemorated in such a way. John Wedgwood (1766–1844) was the eldest son of the more famous Josiah. It seems particularly hard that he received no horticultural appellation because the forming of a horticultural society was probably his idea in the first place. He wrote as early as 29 June 1801 to Sir Joseph Banks that he had 'been turning [his] attention to the formation of a Horticultural Society'. He became its first treasurer and was obviously interested in plants, since in 1812 he wrote a paper for the Society entitled 'Culture of the Dahlia'.

The final founder member, who also remained 'nameless', was William Aiton (1766–1849), the 'Mr Kew' of the age as his father had been before him. He remained in charge of the Royal Botanic Gardens until 1845. William Aiton Senior (1731–93) was the first Superintendent at Kew, from 1759 until his death, during which time he catalogued the entire plant collection then growing there, some 5,500 plants. In 1789, four years before his death, the results were published as *Hortus Kewensis*. Unlike his son, he did have a genus named in his honour, an evergreen shrub collected from South Africa in 1777 and called *Aitonia capensis*.

It is interesting to note that all five of the genera named for the Society's founders came from the southern hemisphere, and that three of them, the Banksia, Grevillea and Dicksonia, were from the newly discovered Antipodes. They were the plants of the moment, with everyone clamouring to grow them. Five different Banksia – Australian honeysuckle – had arrived in the previous fifteen years, and *Grevillea linearis* came from New South Wales in 1795. The first tree fern had been introduced, also from Australia, in 1786. A second species, *Dicksonia arborescens*, had been sighted on St Helena by a French botanist and was later collected and brought to England in 1801. The Ginkgo, or Salisburia as it had been first called, had already been discovered in China and had arrived in Britain in 1750. The only late arrival was the Forsythia species, also a Chinese import, which did not arrive until 1844, when *F. viridissima* was among a package of plants sent home to the Horticultural Society by Robert Fortune (1812–80).

Collecting plants from China and Japan was quite different from plant hunting in most of the rest of the world. Not only was access most trying and restrictive, but the sophistication of the gardens and the cultivation and training of plants had been taken to an altogether higher plane and practised for much longer than was the case in Europe. Consequently the horticultural imports were not collected from the wild but almost uniquely were plants bought from nurseries or seen in gardens. If it had not been for the earlier botanical work carried out by Jesuit missionaries working in China, almost no information would have filtered through to Europe until well into the nineteenth century, when both China and Japan began to open up their horticultural borders. Early in the century, China's floral exports amounted to only about fifty plants, and those from Japan a mere ten.

The frustration of being confined to Macao – the island across the Zhu estuary from Hong Kong, with an area of only 16 sq km (6 sq miles) – with only occasional visits to Canton (Guanghzou) when the merchant fleet was in, lasted until 1842. In 1816, a second ambassadorial visit was made by Lord Amherst to the Chinese Emperor; the first had been in 1792 and led by Lord Macartney. This time, the entourage was seventy-two strong and included Mr Clarke Abel, a surgeon after whom *Abelia chinensis* was named in 1844, and James Hooper, a gardener from Kew who, despite the restrictions, managed to collect 300 packets of seeds. In spite of Lord Amherst's best endeavours, no relaxation of the rules occurred and confinement of all Europeans continued to be restricted to the coast and the various large ports of China until 1860. Tragedy even struck the seeds which had been so carefully collected; they were all lost from HMS *Alceste*, which was wrecked on its homeward journey. Thankfully no lives were lost.

However, thirteen years prior to the ill-received overture of friendship from Britain, there had arrived in Canton the very first plant collector to reside in China for more than just a few weeks. William Kerr (d.1814), a gardener who had been trained at the Royal Botanic Gardens, Kew, stayed eight years in China through the good offices of the East India Company. He began sending material back as soon as he could, and one of his first introductions, in 1804, was a shrub which now bears his name, *Kerria japonica*; it is a genus of a single species with bright yellow blooms. He also sent back the Tiger Lily, *Lilium lancifolium*, the colour and markings of the flower being spectacularly similar to those of the big cat. Sir Joseph Banks (1743–1820) and William Aiton Junior were both so excited at seeing the lily that they hastened to Kew to examine it as soon as it arrived in August of that year.

We have Kerr to thank for the fragrant *Begonia grandis*; also the Japanese Spindle, *Euonymus japonicus*, and the China Fir, *Cunninghamia lanceolata*. The last of these was named for James Cunningham (d. *c*.1709), who probably first discovered it in 1702 on Chusan (Chou-shan) island off the coast of Shanghai, during the time he was surgeon to the East India Company there. The tree's fragrant and durable wood was highly prized by the Chinese for making coffins. The perfumed *Rosa banksiae* 'Alba Plena' followed in 1807, and was named for the wife of Sir Joseph Banks.

Until 1843, travel within Japan was even more restricted, and had been since 1639 when all foreigners, including missionaries, had been banished. Grudgingly, an enclave was allowed for foreigners, but that concession was only for members of the Dutch (not the English) East India Company. It consisted of a tiny artificial walled island in Nagasaki Harbour measuring about 236 by 82 paces from which once a year the Europeans were allowed to make a carefully controlled journey to Tokyo (called Edo until 1868) to pay their respects to the Emperor. On the other hand, the

THE TRANSPORTATION OF SEEDS, PLANTS AND TREES

Importing live plant material around the globe from one country or continent to another was always a hazardous business, and of the three elements involved in a successful introduction – collecting, transporting and then growing – the most vulnerable was always the transportation. Long sea voyages are not conducive to the well-being of most of the horticultural kingdom. The Romans faced no such problems during the first century ad, when they imported from the Continent fruit trees, vegetables and plants (including the vine) for growing here. Almost all their introductions were cultivated plants, as was most of the material arriving with the monastic orders and the returning Crusaders during the eleventh, twelfth and thirteenth centuries.

Difficulties started to arise when serious exploration and long sea voyages began to be undertaken. Not only were the plants collected from the wild, so there was no known history of their specific cultivation, but they had to face a salt-water baptism of some length, two months or more in the case of the early Atlantic crossings. Without doubt, corms, bulbs and seeds must have been the easiest to transport, taking up the least amount of room in the tiny ships and, more importantly, requiring no watering. It was much more complicated when it came to the transportation of actual plant material, roots or seedlings: the packing needed was more bulky, the plants probably required water and someone had to be responsible for opening the boxes or barrels to let in the light and generally look after them.

The problem of transporting plants, whether from this country to aid the survival of newly arrived settlers (see p.174) or for importation into England, was obviously being thought about, since Sir Thomas Hanmer, in his *Garden Book* of 1659, has a very practical paragraph on 'How to Packe up Rootes and Send them to Remote Places':

The firbrous sorts must bee made up alone by themselves, with moist mosse or grasse about them tyed fast, which is better than earth, which will dry and fall away, unless the voyage bee short, but all the other kinds must bee packt up dry without earth in papers, and soe boxt up that they shake not, nor have any heavy thing to presse them hard together, and thus they may bee convey'd saffely very farr.

Twenty-seven years later, in 1686, John Evelyn was given the opportunity of receiving some plants from America and was busying himself with writing a letter to Samuel Pepys, then Secretary to the Admiralty, detailing instructions to the sea captain responsible for transporting them: '. . . the trees in Barills their rootes wraped about mosse. The smaller the plants and trees are,

the better; or they will do well packed up in matts; but the Barill is best, & a small vessell will containe enough of all kinds labells of paper tyed to every sort with ye name'. He advised packing seeds into paper and 'their names written on them and put in a box'.

In the following century, paper was again recommended for use as seed packets, this time by Peter Collinson (1691–1771). In his eagerness to support plant collectors and make sure their precious cargoes arrived home intact, he sent to John Bartram (1699–1777) parcels of paper to make up into bags. The problem with transporting seeds was, and still is, how to keep them dry. Putting the bags in sand was one idea, while Mark Catesby (1682–1749), when sending horticultural material back to England, conscientiously packed dry seed in labelled paper packets and then put those into a dried and sealed gourd husk. Wax was also tried, for the larger seeds; it was recommended that they be pushed into malleable wax only, and not covered with melting hot wax. Collinson thought that watertight ox bladders could be used to pack orchids and more delicate corms and roots, since these would keep them fresh, and he recommended that tree roots should be bound with damp moss and then packed close together in a cask for safe travel.

Detailed instructions were written specifically for sea captains to read; after all, these men were the key players in bringing home the horticultural goodies safely. In 1770, the London seedsman John Ellis published *Directions for Bringing over Seeds and Plants from the East Indies and other Distant Countries in a State of Vegetation* This was followed by an article written in a similar vein by John Fothergill, who had a celebrated garden of newly imported American plants at Upton, West Ham, east of London.

When the emphasis of exploration moved to the Far East and the rest of the southern hemisphere, the resulting sea voyages were greatly lengthened, lasting six months or even more. Consequently, during the latter part of the eighteenth century and beyond, the packaging and storage of horticultural material became more systematic, but the same problems still beset its transportation – either not enough light or too much salt water. Understandably, no sea captain liked herbaceous clutter on deck or anywhere else; sometimes the Great Cabin was commandeered and a fire lit to keep the plants warm, but much depended on whether the captain was interested or not. One or two were exceptionally helpful and gained such a good reputation that a newly arrived plant would occasionally be named in their honour – for instance, *Camellia welbankiana* was named after Captain Welbank (see p.188).

There were, however, often problems on board. Archibald Menzies, who brought home the first Monkey Puzzle Tree seeds (see p.177), requested that a servant be allotted to him when he was

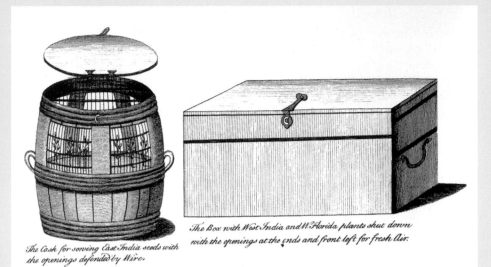

The Cask for sowing East India seeds with the openings defended by Wire.

The Box with West India and W. Florida plants shut down with the openings at the ends and front left for fresh Air.

The Box with divisions for sowing different seeds in earth & cut moss from the southern Colonies and the West Indies.

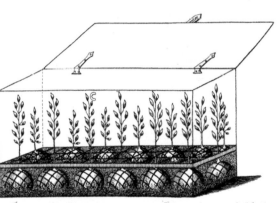

The Inside of the box shewing the manner of securing the roots of W. Florida and W. India plants surrounded with earth & moss tied with packthread and fastened cross & cross with laths or packthread to keep them steady.

Methods of carrying plants from one country to another or one part of the globe to another is as old as civilisation; even the ancient Egyptians went on expeditions to collect plants and trees. Nursing plants home across long sea voyages tested the ingenuity of both collectors and sea captains.

travelling on board HMS *Discovery*, commanded by Captain George Vancouver, on the voyage up the western coast of America during the 1790s. The servant, who was a member of the crew, was detailed to look after Menzies' collection of plants. Captain Vancouver was unsympathetic to Menzies' work and at the first opportunity the servant was transferred back to nautical duties. When Menzies protested, he was promptly placed under arrest. Tempers eventually subsided, and Menzies was allowed to return to his plants, but he had to work alone.

Sir Joseph Banks, too, had his problems. He had had designed small glasshouses known as 'plant cabins' or 'garden hutches',

which weighed, when full, up to three tons, and instructed these to be placed in specific positions on deck. A quantity of these plant cabins, packed with fruit trees, were travelling to Australia from England on board HMS *Guardian* during 1789–90 when the ship was holed by an iceberg. In order to lighten the load, the first items to be jettisoned were the plant cabins and fruit trees.

Experiments to improve the transportation of horticultural material continued, including bedding seeds in coarse brown sugar or currants, and hermetically sealing them in bottles and boxes. It was not until the invention of the Wardian Case in 1830s (see p.202) that seeds and plants stood a better chance of survival.

Chinese were permitted to trade with the Japanese, so despite the difficulties even that produced, a number of plants and seeds did manage to come into the West via that route. (Perhaps that is how, in 1804, *Euonymus japonicus* was introduced.)

John Livingstone (d.1829), another surgeon employed by the East India Company, was sent to China in 1793. A methodical and scientific man, he experimented with and wrote about extending the viability and survival of seeds and plants on the long journey home, and made some useful suggestions to the Secretary of the Horticultural Society, Joseph Sabine (1770–1837), as well as calculating the high cost of each plant that survived. The average cost of 'postage and packing' at that time was 6s 8d per plant, and perhaps out of a thousand dispatched only one would survive. There is at least one record of a consignment of plants aboard a ship, the *Winchelsea*, sailing from the Far East early in the century – arriving with not a single one still alive. It was almost certainly not the only shipment to end in such disaster.

Livingstone was, quite unusually, more interested in the vegetables than the flowers he saw, and stated in a letter he wrote to the Horticultural Society that the Chinese cultivated 'a very small variety of flowers, comprehending only showy or odoriferous plants, shrubs and trees'. In 1821, he is recorded as having dispatched a packet of vegetable seeds from Macao to the Horticultural Society, including a variety of the Chinese Cabbage or pe-tsai, *Brassica pekinensis*.

Some years earlier, in 1805, Livingstone introduced rice paper to Britain. This, in fact, is a misnomer, for it has nothing whatsoever to do with rice. The edible 'paper' was probably produced from *Tetrapanax papyriferum* – a shrub or small tree from China and Formosa which, just like the Canna plant, can be grown here to use in subtropical bedding. On the other hand, Japanese rice paper was usually manufactured from the shrub *Edgeworthia chrysantha*, named for a Mr Edgeworth, a botanist in the service of the East India Company. This paper was used both in cooking and by artists. The plant arrived here in 1845 (from China) and was described as only able to be grown in 'favoured positions'.

John Reeves (1774–1856) was also employed by the East India Company, as an Assistant Inspector of Tea. He arrived in China in 1812 and, like John Livingstone, kept up a useful scientific correspondence with the Horticultural Society back in England, only returning home twice in twenty years. (His son, also John Reeves, spent thirty years living in Canton.)

Reeves Senior, like his colleague, was keen to succeed with a more effective way of transporting plants back to Britain, and there is no doubt that, following his arrival in China, he instituted improvements which resulted in a far better survival rate. In 1816, in preparation for one of his trips home – via Java, Cape Town and St Helena (the usual route taken) – he potted up months before he sailed the hundred plants he wanted to bring to England so that they would be well established and able to withstand the rigours of a double crossing of the equator. The East Indiaman the *Cuffenels* docked with ninety of the original plants still alive, a most remarkable feat, although it should be said that the ship's captain, Welbank, already had a high reputation for the transportation of camellias.

A number of Reeves' potted-up plants were, in fact, camellias, and one which arrived safely on the *Cuffenels* was named in Captain Welbank's honour: *Camellia welbankiana*, a double-flowered white japonica which flowered in 1819. One of the other survivors was a climber which John Livingstone had seen growing in the garden of a merchant in Canton and had urged him to propagate. This was *Wisteria sinensis*, a native of the northern province of Fo Kien – it also grew in hedgerows on Chusan island – which had been brought to Canton by the nephew of the Chinese merchant, Mr Consequa. Both the wistaria and the camellia were grown by Charles Turner at his home in Surrey. By spring 1818 the climbing wistaria was flowering, and by 1819 it was having its portrait painted for *The Botanical Magazine*. A second plant, which was presented by Reeves to the Horticultural Society, flourished lustily, but the wistaria which he presented to Kew did even better. This was eventually placed in Sir William Chambers' Great Stove, the largest conservatory at Kew, being 34.5m (114ft) long. The building was erected in 1761 and was demolished a hundred years later when the contents, except the mighty wistaria, were transferred to the new Temperate House. The gnarled and wizened wistaria still survives and is supported by the corset of an iron frame.

Japan, too, produced a wistaria, *Wisteria floribunda*, collected in 1830 by Philipp von Siebold (1791–1866). Like its Chinese cousin, the fragrant blooms hang in racemes, though while the Chinese flowers are pale mauve, the Japanese are violet blue. Both species disport velvety seed pods, but their leaves differ slightly from each other and *Wisteria sinensis* is a more robust grower. There is, however, one fundamental difference between them, which seems

Encephalartos altensteinii is a native of the Eastern Cape of South Africa, having very striking cones, and which grow to a height of between 4–7 m (12–22ft).

quite inexplicable. They twine in opposite directions: the Japanese *W. floribunda* climbing clockwise and the plant from China, *W. sinensis*, in an anti-clockwise direction. One wonders if the Professor of Anatomy at the University of Pennsylvania, Caspar Wistar, after whom they are named, put his mind to such an extraordinary phenomenon – although it was not unique, see p.84.

In relative terms, the number of plants arriving from the Far East in the early years of the century must have seemed small compared with all the horticultural introductions which were being sent to Britain from the rest of the world. However, it was the fact that they were cultivated plants already well established in oriental gardens which made these rarities so sought after on their arrival. The number of Chinese plants rose steadily rather than spectacularly so that in the second edition of *Hortus Kewensis, 1810–13*, edited by

William Aiton Junior, 120 species were credited as recent introductions.

In the competition for novelty, the Cape Province of South Africa was proving to be a veritable cornucopia of flowers, thus justifying Kew Gardens' decision to send their first plant collector, Francis Masson (1741–1806), there in 1772 – the same year as HMS *Endeavour* returned from her epoch-making voyage. By the early years of the nineteenth century, Masson had introduced nearly a thousand plants from his various South African expeditions including 183 species of Erica, 175 of Mesembryanthemum, 102 of Pelargonium, 57 Oxalis and 42 Stapelia. None was more important than the Cycad, *Encephalartos altensteinii*, which he had sent home in 1777. In 1819 Banks paid what was probably his last visit to Kew, to experience the first 'flowering' of this plant. It remains in the Palm House at Kew today, where it is affectionately known as 'the oldest pot plant in the world'.

Over a century later, Ernest Wilson was to write that 'The Cape flora is astounding in both quantity and variety.' The area covers less than half the size of England but supports over 5,500 native species, surely the highest density of different genera anywhere in the world.

It was from the continent of Australia – following the surveying and botanical exploration of the entire coastline between 1801 and 1805 by Captain Flinders and his team of scientists on board HMS *Investigator* – that the flow of seeds and plants really got into its Antipodean stride, although it was not until 1829 that the whole of the vast continent was claimed as a British dependency. Robert Brown (1773–1858), a naturalist with Flinders' Australasian expedition (and later President of the Linnaean Society 1849–53), was responsible for selecting about 300 different plants to be sent to Kew, and it was there, just before the expedition's return, that a building was established to house some of the smaller plants; plants from the Cape were also grown there. Sir Joseph Banks, because of his early exploration of Australia, was always keen to support and promote the botanical material arriving at Kew. A form of plant exchange was established: 'plant cabins' (tubs and casks of living plants and seeds) were shipped to Australia, where the English botanical material was exchanged for the native vegetation to be sent back to Kew. The first nurseryman to set up business in Sydney, in 1826, was Thomas Shepherd (d. 1834), who had previously run a nursery in Hackney, East London, for over twenty years. *Flora Australiensis*, a seven-volume catalogue of

Australian vegetation, was undertaken and completed by George Bentham (1800–84), then President of the Linnaean Society, for Kew between 1863 and 1878.

India was being groomed as an horticultural 'hotspot' too, although its gardening background was quite different from that of China or Japan. Domestically, gardens were rarely seen, the struggle to grow food and water those crops proving enough for rural India to cope with. The two greatest influences on the earlier princely gardens – Buddhism, with its restrained quietness, and Hinduism, with its rainbow opulence of colour – had somehow combined during the reigns of the first six Mogul emperors (1508–1707). Between 1632 and 1654 the finest of all tomb gardens, the Taj Mahal, was created for Mumtaz Mahal, the much-loved wife of Emperor Shah Jahan. Later, with the increasing influence of the British – in the form of the East India Company – gardens and landscaping became quite the vogue amongst the no doubt homesick precursors of the British Raj. Similarly, the gardens of the French residency in Pondicherry and of the Dutch governor's official residence in Sadras both reflected their own European form of garden.

The British East India Company, which by now was already over two hundred years old and virtually governed India, employed a series of botanists to assess the economic and commercial value of a variety of domestic crops and trees that were growing or would grow in India. One economic move which Sir Joseph Banks in 1788 considered viable and of 'the greatest national importance' was the establishment of the tea industry in the subcontinent. Eventually, long after Banks had died, 20,000 *Camellia sinensis* were transported from China to India to form the nucleus of the tea industry. The Saharanpur Botanic Gardens in the North-West Provinces, with Surgeon-Major William Jameson (1815–82) as Superintendent, became the centre for the successful experiment, and by the 1850s tea was being grown commercially in Assam. In 1880 well over 20 million kilos (45,370,000 lbs) of tea were exported from India to Britain. Jameson collected plants in India and Burma during his tenure of office which spanned thirty-three years, beginning with his appointment there in 1842.

The Saharanpur Botanic Gardens had been founded in 1817 and were used as the base for all the plant-hunting expeditions which took place. Even earlier than Saharanpur, Calcutta was able to boast a Royal Botanic Garden from 1787, and over the next one hundred years at least twenty other botanic gardens were created on the subcontinent. India was set to play an increasing role in the gardening life of the home country.

The Indian Mutiny of 1857 brought to an horrific end the East India Company's rule of India. Queen Victoria, reflecting popular opinion, wrote in her journal in November of that year: 'India should belong to me.' Peace was finally declared in July 1858, and in August of the same year legislation was passed which transferred Indian administration from the East India Company to the Crown. Victoria had her wish. Long live the Queen-Empress.

The Earl of Amherst, who had earlier travelled as the representative of His Britannic Majesty to China, had been created Governor-General of India in 1823, and from then until 1828 both he and in particular his wife, Sarah, Countess of Amherst (1762–1838) collected plants to send home. During a tour of the remote Northern Provinces in 1826, Lady Amherst found the plant she is best remembered for introducing, **Clematis montana**, and the autumn-flowering white anemone, **Anemone vitifolia**. She has the distinction of having the one and only species of a spectacular Burmese tree named in her honour, **Amherstia nobilis**.

The 'Amherst' tree obviously fired ducal imagination, because when the 6th Duke of Devonshire saw a painting of it in 1835, he dispatched John Gibson (1815–75), a twenty-year-old gardening apprentice from Chatsworth, to India to collect the tree and other rare seeds and plants. Gibson was attached to the staff of the newly appointed Governor-General of India, Lord Haytesbury, and travelled with him via Madeira and the Cape of Good Hope, botanising and collecting all the while. The young man must have grown up considerably in the two years he spent in India, where he was based with Dr Nathaniel Wallich, the Danish superintendent of the Calcutta Botanic Garden. He came home laden with horticultural delights to be distributed to Kew Gardens, the East India Company Headquarters and, of course, to his employer, the Duke of Devonshire. Of *Amherstia nobilis* Gibson wrote that the tree 'was unequalled in the flora of the East Indies and perhaps not surpassed in magnificence and elegance in any part of the world'. Then came the let-down, because even Joseph Paxton (1803–1865) himself, who became the great Sir Joseph, Head Gardener of Chatsworth, could not entice it to flower.

In 1846, therefore, three more seedlings travelled to England from the subcontinent and were given to Kew and the Horticultural Society, but this time the third specimen

Anemone hupehensis var. *japonica*, the Japanese Anemone, was originally discovered in China in 1844 by Robert Fortune on the first of his five journeys to the Far East when he was collecting on behalf of the Horticultural Society and the East India Company.

went not to the Duke of Devonshire but to the Duke of Northumberland at Syon House in Middlesex. It made no difference: *Amherstia nobilis* steadfastly refused to bloom, and it was not until 1849, in the hothouse of Ealing Park, the Middlesex home of Louisa Lawrence (*c.* 1803–55), that it was coaxed into flower. The Amherst tree continued to flower at Ealing Park until 1854 – the year before Mrs Lawrence died – when it was dug up and transferred to Kew. There it was painted by Marianne North (1830–90), Flower-Painter-in-Ordinary to Queen Victoria. Its travels did not end there, however, as three years later it was transferred to the specially altered Palm House, but this move proved to be one too many. It did not settle into its new abode and died. (There is another *Amherstia nobilis* at Kew which also flowers infrequently, the last record being in 1978.)

Mrs Lawrence had a long reputation for the quality of her exotics and hothouse plants, and by 1838 had already won over fifty awards from the Horticultural Society. She created her first garden, at Drayton Green in west London, during the 1830s, and later made an even more sumptuous garden at nearby Ealing Park, to which she and her husband moved in 1840. The designs of her gardens were modern and created in the up-to-the-minute gardenesque style, with rustic arches, urns, and both a French and an Italian garden; Ealing Park was decorated with at least 500 roses, 140 pelargoniums and 227 orchids. No wonder J. C. Loudon (1783–1843), the founder and proprietor of *The Gardener's Magazine*, wrote following his visit to Drayton Green on 27 July 1833 that 'The brilliancy of the flowers, the immense numbers of statues and vases, and the sparkling waters of the various cascades, produced an effect that was perfectly dazzling.' Mrs Lawrence's style and exquisite taste were much copied, even as far away as Australia, where George Macleay in the 1840s made a

similar beautiful garden at Brownlow Hill near Sydney.

There must have been something in the air in 1849, because that was also the year that the two Dukes, at Chatsworth and at Syon, also finally gazed on their flowering *Victoria amazonica* (originally named *V. regia*), the gorgeous Amazon Water Lily, the seed of which had originally been brought to England by Thomas Bridges from Brazil in 1845. Just like the Amherstia, however, it refused to flower, and new seed had to be obtained before it finally revealed its beauty. Examples of blooms from both the *Victoria amazonica* and the *Amherstia nobilis* were presented to Queen Victoria.

Governor-Generals of India seem to have been particularly helpful to keen botanists and plant collectors: just as John Gibson had been helped earlier by a newly appointed Governor-General, so Joseph Dalton Hooker (1817–1911) was encouraged by James, 10th Earl of Dalhousie (1812–60), the youngest Viceroy ever sent to India. Perhaps the Earl had been influenced by his mother from a young age: Christina Ramsay, Countess of Dalhousie, had earlier collected plants in Nova Scotia as well as in India and Penang, during the time her husband, the 9th Earl, had been successively Governor-General of Canada and then Commander-in-Chief in India. Their son, the new Viceroy, with immense energy, oversaw the birth of the modern Indian transport

Mrs Louisa Lawrence's home and garden at Drayton Green, Middlesex, was described at the time as a 'model of its particular kind... the humblest and most economical possessor of a villa residence of two acres may take a lesson from Mrs Lawrence's taste'.

The *Victoria amazonica* is perhaps one of the best-known water lilies. It was the veining of its huge leaves that gave Sir Joseph Paxton his inspiration in designing the Crystal Palace for the Great Exhibition of 1851.

system, since it was during his term of office that the sub-continent's vast railway network was planned and begun. Two thousand miles of dust tracks were metalled and so made into all-weather roads, bridges were constructed, and the Ganges canal was built. This wholesale modernisation of the general infrastructure greatly assisted the exploration of the interior of India, and as a consequence the opening up of what had previously been remote and inaccessible areas helped commerce and agriculture, the amazing growth of the tea industry from the 1850s onwards being a prime example.

The two young men, Joseph Hooker and James Dalhousie, became well acquainted on their journey out to India during the winter of 1847, but it was another year before Hooker, who had planned to explore Sikkim, actually received permission from the Rajah to do so. His travels, explorations and plant collecting in India and the Himalayas were to influence him for the rest of his life (just as the earlier Sir Joseph Banks had been dominated by his travels to Australia and New Zealand). Hooker's Rhododendron and other plant introductions from the Himalayas showed the opportunities there were for altering the landscape of Britain. The way had, in fact, already been indicated years earlier by Major-General Thomas Hardwicke (1755–1835) of the Bengal Artillery, who was the first to bring the Himalayan Rhododendron to England: he also collected plants from South Africa, St Helena and Mauritius. Sir Joseph collected 43 different species of rhododendron from Sikkim during his two-year tour between 1848 and 1850, and within a relatively short space of time the seeds and plants were being distributed both by him and from Kew.

Charles Darwin received twelve assorted rhododendrons (as well as the newly arrived *Berberis darwinii* from Chile, which we shall meet later). HRH Prince Albert was also on the distribution list for plants and seeds arriving from abroad; his allocation was requested for the newly developing gardens at Osborne House, designed by Prince Albert himself, on the Isle of Wight. Nurseries were obvious recipients; one of the biggest at that time was Standish & Noble of Bagshot in Surrey. Seeds and plants were also packed up and sent from Kew to botanic gardens like Glasnevin in Ireland, and Edinburgh; many of Cornwall's gardens too including Tremough, Carclew, Scorrier, Killiow, Menabilly and Heligan, were happy receivers of these Himalayan giants. Plants literally went around the world, packed in the modern Wardian Case (see p.202): they were sent to the botanic gardens at Dijon in France, Florence in Italy, to Canterbury in New Zealand, Sydney in Australia and Baltimore in North America, to Jamaica, and to St Helena in the South Atlantic.

Sir Joseph Hooker's botanical interest in the Indian sub-continent remained with him all his life. His taxonomic work included his *Rhododendrons of Sikkim-himalaya* (1849–51), and over many years (1872–97) he produced the seven volumes of the *Flora of British India* which perhaps remains as the summation of his devotion to India.

The young United States was also being systematically opened up, and enlarged as well when, in 1803, the remaining French territory in North America entered the new Union in what became known as the Louisiana Purchase. These botanically rich lands were eagerly looked upon to enhance the American garden, but they also stimulated the demand for American plants in Europe, so much so that 'American gardens' became all the rage in early nineteenth-century England.

In the following year, 1804, an expedition sponsored by the United States government set off from St Louis to traverse the Pacific Ocean. Of the two men chosen, both British, neither Captain Merriwether Lewis (1774–1809) nor Captain William Clark (1770–1838) was either a botanist or a naturalist, but during the two years of the successful exploration they both discovered new plants. To compliment them on their great achievement of the first successful coast-to-coast crossing, the two captains each had a genus named for them – Lewisia, of which there are about 20 species, and Clarkia, which is related to the Fuchsia and of which there are some 36 species. The latter has species growing in both North and

South America, a wider botanical spread than the Lewisia, which are all found in North America.

The two men were also responsible for finding the first *Mahonia aquifolium*, the Oregon Grape. This they discovered in Oregon Country, the huge area now comprising the modern states of Idaho, Oregon and Washington, where the expedition over-wintered in 1805–6. The Oregon Grape took its time arriving in Britain, as the first record of its introduction is not until 1823. After finding conditions to its liking here, it began to venture out of the garden gate into the countryside. Now, at the beginning of the twenty-first century, the shrub is well established in dry woodland areas.

As well as continuing their interior exploration of North America, the British had also been pressing ahead with the discovery of the western coastline of the continent. Just as earlier the voyages of Captain Cook had opened Pandora's botanical box of the southern hemisphere, so Captain George Vancouver (*c.* 1758–98), on board HMS Discovery during 1794–5, along with the surgeon-naturalist Dr Archibald Menzies (1754–1842), did the same for the whole of the west coast of America, in particular California – or New Albion as it was then called.

The Horticultural Society was keen to recruit someone to collect the plants which Menzies had identified and so well described. A young Scot, David Douglas (1799–1834), was chosen. His first journey, however, during 1823, was along the eastern seaboard of North America, through the states of Pennsylvania and New York, and then up the Canadian coast. On this occasion Douglas brought little home that was new, as the area had been botanically both shaken and stirred over the past two hundred years. However, he did find and introduce a yellow-leaved honeysuckle, *Lonicera hirsuta*, and, rather fortuitously for the Society, returned with a large number of cultivated fruit trees including twenty-one varieties of peach. These were planted in the recently acquired 13.35 hectares (33 acres) of land at Chiswick, of which about half was set apart for fruit and vegetables. The Society praised his efforts, saying that he had 'obtained many plants which were much wanted, and greatly increased our collection of fruit trees'.

The following year, Douglas embarked on an altogether different voyage, this time to the almost unexplored west coast of America. The Hudson's Bay Company (which had been founded in 1670) was prepared to sponsor a Horticultural Society nominee, but warned that it would need to be someone very resourceful, since he would discover 'the fare of the country rather coarse and be subject to some privations'. Douglas's adventures in America can only bear out the accuracy of the statement, but then all plant collectors and explorers suffered hardship of one sort or another; it was all part and parcel of being a plant hunter. That warning, together with Sir Joseph Banks's diktat (adhered to even after his death) that 'Collectors must be directed by their instructions not to take upon themselves the character of gentlemen', makes one wonder whether there was much competition for the post.

The twenty-five-year-old Douglas was keen enough and sailed from Gravesend on board the *William and Mary* on 25 July 1824 together with a surgeon-naturalist friend from his Glasgow days who had been recruited for the voyage. Dr John Scoular (1804–71) was then only nineteen, but went on to become the distinguished Professor of Geology, Zoology and Botany to the Royal Dublin Society.

They journeyed around Cape Horn and looked earnestly at Mas-a-Tierra (one of the islands of the Juan Fernandez group, where Alexander Selkirk had earlier inspired the persona of Robinson Crusoe), and then sailed on to the Galapagos Islands. The islands' high humidity wrecked almost all of the botanical specimens Douglas had so far collected. Although he and the doctor noted the oddities of the animals and the plants, there was not the slightest hint of the evolutionary ideas which were to galvanise Charles Darwin ten years later. (On the Galapagos, Darwin made a collection of plants to send to his friend Sir Joseph Hooker to identify.)

Finally, after glimpsing Cape Disappointment on the west coast of America, the *William and Mary* dropped anchor in the estuary of the Columbia River, Oregon, in April 1825. David Douglas could hardly wait to inform the Horticultural Society about what he had found and wrote to them with glee about *Gaultheria shallon* which was the first plant he took in his hands and which he recognised because 'Mr Menzies correctly observes that it grows under thick pine-forests in great luxuriance and would make a valuable addition to our gardens'. This first plant duly arrived in England the following year, and with its tiny pink and white flowers and purple berries has become a popular and very useful American representative in our gardens.

The name of David Douglas is usually linked with conifers, and he would have had no conception of the visual impact his resinous stream of dark green 'fir' introductions was to have on our landscape. That first year he collected

seeds of what has became known as the Douglas Fir, its botanical name being *Pseudotsuga menziesii*; Menzies had first described the tree in 1792. It was originally called *P. taxifolia*, then later *P. douglasii*, but whatever its name it is one of the stateliest of conifers, and a most important timber tree. The same year, Douglas collected the Western Yellow Pine, *Pinus ponderosa*. Then, in 1827, followed the collection of seed from the largest pine of all, *Pinus lambertiana*, named for Aylmer Lambert (1761–1842), who was a botanist and, more importantly, wrote *The Genus Pinus* in 1804. It was Lambert who encouraged Douglas to collect all the pines he could and send them to him for inclusion in a proposed new edition of his book. The tree's vernacular name is Sugar Pine, a reference to the sweet exudation which oozes from its trunk or from a broken or sawn branch; this was often used as a sweetener in place of sugar or honey.

In 1830 *Abies grandis*, the Giant Fir, was collected, and this was followed the next year by *A. procera*, the Noble Fir. The Sitka Spruce, *Picea sitchensis*, was another tree which Menzies had noted, and which Douglas collected in 1831; it was to prove of great economic importance, although its all-enveloping cloak on our uplands is now less appreciated. At the same time, he collected *Pinus contorta*, commonly known as the Beach Pine for its ability to grow in sand dunes or light stony soil. In 1832 a tree with edible seeds, the Digger Pine, *P. sabiniana*, was gathered up and named for Joseph Sabine who, as Secretary of the Horticultural Society, had been instrumental in helping the young David Douglas prepare for his American journeys. The same year, *P. coulteri*, the Big-Cone Pine, was discovered; its name is certainly apt, as the cones are indeed large – up to 35 cm (14 in) long and weighing anything up to 2 kg (4½ lb) – although the Sugar Pine's cones can be even bigger, up to 50 cm (20 in) long. In 1833 Douglas collected the Monterey Pine (*P. radiata*) from the salt-laden windswept peninsula of the same name, and so the list went on.

Enthusiasm for planting these dark green, startlingly shaped trees, particularly in Scotland, generally outstripped demand. The fashion was also fed in part by the publication in 1831 of the complete listing of all the known exotic conifers in cultivation, which was followed in 1838 by *Pinetum Britannicum*, both publications being written by Charles Lawson, head of the famous Seed and Nursery Company of Edinburgh, which had been founded in 1770. Much later, in 1854, a Perthshire plant collector, following in the footsteps of Douglas, sent home some conifer seed to the nursery: the

Pinus ponderosa, the Western Yellow or Ponderosa Pine, shown here in Zion National Park, Utah, USA, was collected by David Douglas in 1826 when he was exploring the Rocky Mountains.

conifers, a selection of his many herbaceous plant introductions includes an (unnamed) member of the Evening Primrose family, Oenothera, as well as that jazzy Californian orange poppy which was introduced in 1825 and named in honour of a botanist called Eschscholtz, hence its somewhat awkward name of *Eschscholtzia californica*. In the same year, while travelling downstream to the mouth of the Columbia River, Douglas discovered the starchy lily bulb that the Chinook Indians ate and which apparently tasted much like baked pears. The Indians called them *quamash*, meaning 'sweet', but Douglas found they had the anti-social habit of producing flatulence. Indeed, on one occasion he complained of being almost blown out of his sleeping quarters in a Chinook lodge 'by the strength of the wind', but whether this was of his own making or that of his companions was not made clear. The Latin name is *Camassia quamash*, and the flower looks similar to a hyacinth – indeed, it is related to the Scilla.

We also have to thank David Douglas for at least 18 different species of Penstemon which he collected from 1827 onwards. He found most of them around the Columbia River and the Cascade Mountains. The earliest Penstemon (from the Greek for 'five', pente – the plant having five stamens, one of them sterile) to arrive in Britain was collected long before Douglas was born, and came from northern Mexico. It is a carmine beauty whose home territory stretches from Mexico to Colorado, and it must have arrived in Britain in the early 1790s, since by 1793 *Penstemon barbatus* (syn. *Chelone barbata*) was growing at Kew. None of the family is long-lived, but all are charming and uncomplicated. Reginald Farrer (1880–1920) described them as having 'a crowded hour of glory rather than a longer existence of mere usefulness'. In the Blue Mountains of Oregon, Douglas discovered the one and only peony to be found in North America and named it *Paeonia brownii*, in honour of the brilliant botanist Robert Brown.

Whilst in the Blue Mountains in British Columbia, Douglas found a flowering currant which he named *Ribes munroi*, to commemorate a Scottish gardening friend, Donald Munro, Head Gardener of the Horticultural Gardens at Chiswick. At the same time he discovered a yellow lupin, one of the 21 distinct forms of Lupin that he found during his travels. This one proved to be a cause of some embarrassment when it was sent back to the Horticultural Society. The notes accompanying the specimen detailed where it had been found, the type of soil and

resulting seedlings were given the name of *Chamaecyparis lawsoniana*, or Lawson Cypress; this is the conifer that is the most familiar and has probably given rise to the largest number of coniferous offspring, nearly 200 at the last count.

Lest it be thought that David Douglas sent home only

The *Camassia quamash* is called Camas Grass or Bear-grass. Its two Latin names are derived from the word *Kamas* which comes from the language of the North American Nootka Chinook tribe.

terrain, the botanical details and the name Douglas proposed to give it. He had chosen to honour John Turner, the long-serving Under-Secretary to the Society who had been so helpful to him in preparing for his travels. However, during Douglas's absence, and unbeknown to him, Turner had been exposed as an embezzler and dismissed from his post, and was now being hunted by police forces on the Continent where he was believed to have fled. The plant was therefore held discreetly until Douglas returned to England in October 1827, when he enjoyed a hero's welcome 'in consequence of his great services'. He obligingly and quietly renamed the lupin *Lupinus sabinii* in recognition of the work of the Society's Secretary, Joseph Sabine. Even that name has since been altered: and the lupin is now known as *Lupinus polyphyllus*, meaning 'many-leaved', and is the ancestor of all our garden lupins.

Along the Willamette River (which drains the Cascade Mountains and runs into the mighty Columbia), Douglas found a beautiful Erythronium lily which, because of its unusual height for the genus, 15–30 cm (6–12 in), was given the name *Erythronium grandiflorum*. Later, in the early part of the twentieth century, a taller Erythronium was discovered, *E. californicum*, which reaches a height of 35 cm (14 in). No doubt if this had been discovered prior to *E. grandiflorum* the names might have been reversed. All the Erythronium species from Europe to Japan and North America are small bulbous plants and because the tiny corm looks similar to a dog's tooth they are sometimes known as either the Dog's-tooth Violet or the Trout Lily; most of the species do not exceed 15 cm (6 in) in height.

A genus containing one species and belonging to the Lauraceae family was found by Douglas in 1829 and was called *Umbellularia californica*, the California Bay or Laurel. The pungent aroma of the crushed leaves is so strong that another name for it is the Headache Tree. It had the reputation of being difficult to establish, even though Douglas had hoped that it would become a valuable addition to the garden; several old gardens in Cornwall and Sussex grow it, and before the First World War there was a fine specimen in the gardens of Osborne House. Nowhere, however, does it ever seem to reach the height it attains in the wild, 18 m (60 ft). *Dendromecon rigida*, the gorgeous Tree Poppy, with its bright yellow flowers, was also discovered by Douglas but not introduced until 1854, when William Lobb (see p.206) sent it home. The shrub needs the protection of a south-facing wall if it is to survive our cold winters.

In all, David Douglas spent ten years exploring and searching for new plants and trees, mainly in and west of the Rocky Mountains. His total tree and shrub introductions amount to about 200 different species, which, as they have matured and multiplied, have made an enormous impact on the landscape and gardens of Britain. He was returning home to England in 1834 when the ship on which he was travelling called at Hawaii. True to his exploring instincts, he went off botanising and walking in the mountains and on 12 July met with a terrible and fatal accident, falling into a deep pit dug to catch the local ferociously wild cattle. At least one animal had already been trapped in the pit; Douglas stood no chance as he was trampled and gored to death. He was only thirty-five.

The combination of Sir Joseph Banks and then Sir William Hooker (1785–1865) as successive directors at Kew established the Royal Botanic Gardens as the *Hortus primo* of the all-encompassing British Empire, if not the world, and their paternalistic overview confirmed Kew (and Britain) as the premier influence on the burgeoning horticultural economy of the Empire. Botanical exploitation and the movement of plants from one continent to another were undertaken with breathtaking sangfroid. It was the Royal Botanic Gardens that was the taxonomic spider at the centre of this worldwide web of plants, overseeing what came in, assessing their value and deciding where best to send these horticultural heavy-weights in the Empire. This process had already begun in the previous century, when Sir Joseph Banks started experimenting by introducing the easily grown and nutritious breadfruit from the Pacific islands into the West Indies. As the nineteenth century progressed, it was the commercial benefits of the intercontinental traffic in plants which would have a lasting economic and social effect on both Britain and the Empire then approaching its apogee.

Artocarpus communis is the botanical name for the bread-fruit, the staple starchy food of south-east Asia and the Pacific. Joseph Banks had first seen and investigated this plant growing on the island of Tahiti when he was there with Captain Cook to view the Transit of Venus in 1769. Banks believed that the plant should be able to grow in the West Indies, thus helping to relieve the constant food shortage endured by the thousands of slaves involved in the produc-

tion of sugar. Eventually plants were collected from Tahiti by Captain William Bligh of HMS *Bounty* in 1787. 'Breadfruit Bligh', as he came to be known, had with him an experienced botanical collector David Nelson (d. 1789), who had sailed with Captain Cook on his third and last voyage. Bligh's orders from the Admiralty were to transport breadfruit from Tahiti to the West Indies to see if it would grow there. During the six-month stay on the island, 800 breadfruit saplings were potted up ready for the journey, as well as 700 assorted plants to go to Kew Gardens. However, none of the plants ever arrived at their destination, since they were all thrown overboard during the famous mutiny.

On his return to England, Captain Bligh was dispatched in 1791 aboard HMS *Providence* back to Tahiti. He was accompanied this time by the botanist James Wiles (fl. 1790s–1800s) and an assistant, a Kew gardener named Christopher Smith (d.1807) who later became the Superintendent of the Botanic Garden in Penang. This time the mission was safely accomplished, and by 1793 breadfruit trees were flourishing in Jamaica and other West Indian islands. The fruit, which looks like a large melon, is very versatile; it can be cooked in all imaginable ways including being ground into a flour for baking bread. It is also very easy to cultivate. The cookery writer Tom Stobart remarks that because of the tree's easy providing nature and nutritious ways, 'it seems to make possible that fortunate state of complete idleness', something the industrious Joseph Banks would not have approved of.

In 1859, Richard Spruce (1817–93) – the spruce tree was not named for him – and Robert Cross (1836–1911), another Kew-trained gardener, collected from the high Andes in Ecuador plants and seeds of *Cinchona succirubra*. The bark of this genus was the source of quinine, and the Kew authorities were anxious to get the healing properties of the tree to India and other tropical countries. Plants were eventually taken from Ecuador and Peru to Kew Gardens, where a few of the saplings were retained for experimentation, with the rest being transported onwards to the Nilgiri Hills in southern India. The exercise was successful, and by the end of the century there was a government cinchona factory at Mungpoo.

Robert Cross, when he was in Brazil in 1876, also collected the Rubber Plant, *Hevea brasiliensis*, which was sent, via Kew, to Sri Lanka and south-east Asia, where it grew and prospered as a commercial crop. A year earlier, Cross had

THE BARK OF THE CINCHONA TREE

PERUVIA COLLECTA NOVIS CHINCONIUS ORIS
ACCIPIT A SERVO PHARMACA FEBRIFUGA

A *trompe l'œil* painting showing the Countess of Chinchon taking cinchona to ease her malarial fever.

The Cinchona was named in honour of the Comtesse de Chinchon, the wife of the Governor of Peru in 1638, Chinchon being the name of the Comtesse's estates in Spain. She was cured of a fever by drinking a concoction of the ground-up bark. In a haphazard manner, the bark's healing properties had been known about for centuries, in the same way as 'the fever' or 'swamp fever' was known about. (The word 'malaria' was not coined until the 1740s; it comes from the Italian for 'bad air'.) Because of the Jesuit presence and influence in South America, it was the Society of Jesus which came to be associated with the making of the powder from the bark; it eventually became known as Jesuits' Bark. In the Protestant world of northern Europe, this meant the powder was not very welcome. Oliver Cromwell – who suffered from the fever all his life – absolutely refused to take any medicine with such apparent Catholic overtones, and is believed to have died of the fever.

been in Panama collecting a different Rubber Tree, *Castilla elastica*; *Manihot glaziovii* is another tree from the same area which was also a source of commercial rubber. None of these rubber plants can be grown in British gardens but were to prove economically valuable elsewhere. (The house plant we grow and know as the Rubber Plant is *Ficus elastica*, a member of the Fig family, which came to Britain from tropical Asia in 1815.)

As we have seen, this development of natural resources was a way of helping newly explored countries to reach their full economic potential via the 'mother country', and it is the laying-down of the self-confident horticultural expertise during this period by Kew Gardens, together with the far-sightedness and entrepreneurial spirit of private nurseries and the Horticultural Society, that has kept Britain as the main contender to be 'gardener to the world' in the twenty-first century.

Commercial interest was certainly being shown by a number of nurseries and seedsmen all eager to accept and trial newly introduced plants. Both the quality and quantity of the plants offered for sale was increasing. For instance, the catalogue of the premier London nursery of Loddiges of Hackney early in the century carried an ever-lengthening list of plants; by 1836 they were advertising over 1,500 different roses for sale. The firm had been founded in the 1740s by John Busch from Hanover, who later achieved fame as the gardener to Catherine II, Empress of Russia, laying out for her in 1772 the gardens at Tsarskoe Selo. In 1771 Busch sold the nursery to his compatriot Conrad Loddiges (*c.* 1739–1826), who carried on the business himself and then in partnership with his son George (1784–1846). The nursery also helped to pioneer the Wardian Case for better transportation of plants both to and from this country (see p202).

By the 1830s, the Hackney firm was considered to be the largest nursery in Europe, even having its own arboretum. New seeds and plants were constantly being sought, and there were several freelance collectors, such as Allan Cunningham (1791–1839), who travelled throughout Australia and New Zealand and sent packages directly to Messrs Loddiges. James Veitch (1815–69), grandson of the founder of the great Veitch Nursery, went one stage further and commissioned plant hunters to work exclusively for him. Over a period of sixty-five years – beginning in 1840 with William Lobb (1809–64) and including in 1905 the staggering introductions from China by Dr E. H. Wilson, (see p.226) – twenty-two men travelled on behalf of Veitch.

In all this colonial enterprise which took place so far away, it must not be forgotten that Europe in the nineteenth century could still offer botanical treasures of its own. In 1802 an expedition led by Count Apollon Mussin-Puschkin, a former Russian ambassador to London, and his friend Baron von Bieberstein set out to tour the Caucasus and Mount Ararat for almost three years. They found a number of garden-worthy plants, some of which were sent to Loddiges Nursery for propagating. One of them was the Mourning Bride or Pincushion Flower, *Scabiosa caucasica*, a close relation of our native Devil's-bit Scabious (*Succisa pratensis*) and belonging in the same family that teasels inhabit, the *Dipsacaceae* family. The Latin word *scabies* means 'an itch', and the leaves of the whole family are so scratchy that they are supposed to relieve any irritation. This particular species is the one which has been mainly used to develop the many cultivars there are on offer. A much earlier species to arrive was the Sweet Scabious, *S. atropurpurea*, from its native Italy in 1620, when it was confusingly known as Indian Scabious. It also had the name of the Mournful Widow because it was used in wreaths, particularly in Portugal and Brazil.

The evergreen perennial yarrow *Achillea filipendulina* was found by Mussin-Puschkin and von Bieberstein to join *Achillea herba-barota* from central Europe, and Sweet Nancy, *A. ageratum*, from around the Mediterranean, both of which had arrived many centuries earlier and had been used, like the leaves of Feverfew (see p.33), according to Gerard in his *Herball*, to 'easeth the pain of the migraine'. A native yarrow, *A. ptarmica*, is known as Sneezewort (yarrow and wort are both Anglo-Saxon words), and was widely used as a type of snuff.

Other plants which came from the Caucasus as a result of the ambassadorial searchings were *Geranium ibericum*, with its lovely iris-blue flowers; a catmint, *Nepeta mussinii* (now *N. racemosa*) and a pretty bellflower, *Campanula sarmatica*. This last plant, although collected from south-east Europe, was named after an area in eastern Poland, Sarmatia. A Turk's-cap Lily of deep yellow, *Lillium monadelphum*, and the dainty Lebanese Squill, *Puschkinia scilloides*, which was named in the Count's honour, were also recorded.

As we saw in the last century, Siberia had proved to be a rich source of plant introductions, and the area contributed about 220 species during the early years of the nineteenth century. This was largely due to the fact that from 1823 onwards, Dr Ernst Ludwig Fischer was in charge of the newly restored St Petersburg Botanic Garden and carried on an enthusiastic correspondence with Sir William Hooker. Hooker, at that time was Professor of Botany at Glasgow, and from 1841 was Director of the Royal Botanic Gardens, Kew. In addition to written correspondence between the two men, there was much exchange of plant material, hence the Siberian influence on our garden flora. Under Dr Fischer, a formidable botanical library of about 24,000 volumes was established at St Petersburg, although this did not compare with Kew, which had well over 100,000 volumes by the end of the century.

From the other side of the Atlantic early in the century, about 1804, from 'down Mexico way', came the Dahlia – or Georgina as it was first called. This had flounced its way into continental affections in the latter part of the previous century, although it had been flirting around the edges of European society from as early as 1651 when it was first described and illustrated by Francisco Hernandez, botanist to Philip II of Spain, in his posthumously published book *Thesaurus*. Hernandez was in Mexico from 1571 to 1577, and it was during this time that he first named the Dahlia 'Cocoxochitl'. The mini-skirted twirls of barbaric colour described by Hernandez were not immediately apparent in the three species which Baron Friedrich von Humboldt (1769–1859) brought back to Spain after travelling for five years in Central and South America. These were the single red *Dahlia coccinea*, *D. pinnata*, a double purple, and *D. rosea*, a delicate single. That same year, 1804, Lady Holland, who was in Spain with her husband, sent some of the tubers back to her London home, Holland House, where they were successfully nurtured into blossom.

The booming bloomers were set to go into orbit. Over the next twenty years there was an extraordinary explosion of dif-

The Dahlia is a native of the mountains of Mexico and Central America.

ferently coloured and shaped blooms, and the Dahlia was quickly adopted as a florist flower (see p.40). Many new varieties were developed, especially in France, with the coming of peace after the defeat of Napoleon at Waterloo in 1815. There were at least 1,500 different varieties during the Dahlia mania (John Wedgwood alone was said to have grown over 200 named varieties). J. C. Loudon wrote that 'they were the most fashionable flower in this country', and even Sir Joseph Paxton of Chatsworth found time to write enthusiastically about them. It will be recalled that Linnaeus had honoured one of his Swedish pupils – Dr Dahl from Upsala University – in the naming of the Dahlia.

The ending of hostilities with Napoleon had consequences all over the world, not just in France. In 1799, Malta had been captured from the French by Admiral Horatio Nelson, so at the end of the war in 1815 the island's botanic garden, which had been founded in 1676 by the Order of St John of Jerusalem (the Hospitallers), became available for use by Sir Joseph Banks and Kew Gardens. In South America, there was a certain amount of relaxation of nationalistic tensions when Spanish rule was rejected by Chile, Peru and Argentina, and even Portuguese Brazil was a little less hostile. The imperial rise of Britain with its concomitant exploration and plant gathering in the New World, could now be put on a rather more formal basis than the 'hit-and-run raids' which Banks and the crew of the *Endeavour* had had to endure when they ventured into the Brazilian port of Rio de Janeiro in 1769. Thirty years later, before Spanish rule had been overthrown, Humboldt was granted very special permission to explore Mexico, and the Spanish possessions in South America, known as 'New Spain'.

Banks, eager to garner the horticultural riches of South America, dispatched two Kew-trained twenty-year-olds to Brazil in 1814: James Bowie (*c.* 1789–1869) and Allan Cunningham. They each received a salary from the Treasury of £180, and despite receiving a lecture on 'frugality' and their employer's exclusive right to the plants collected, both men devoted their whole lives to working for Kew and botany in different parts of the world. Bowie went on to an horticultural life in the Cape Province of South Africa, while Cunningham sailed from Brazil in 1817 for Australia and New Zealand. He returned to England for a time but could not settle and in 1837 went back to Sydney, where his brother Richard had been Superintendent of the city's botanic gardens.

The two men's floral introductions from their three-year stay in Brazil were, however, just a taster of what was to come later. They were the first to send home the lusciously leaved *Gloxinia speciosa*, the name reflecting the eighteenth-century French botanical writer Benjamin Gloxin. However, this small genus of just 6 species was later subsumed into the larger Sinningia genus (named to honour Wilhelm Sinning, the head gardener at Bonn University). There are about 40 species in this genus, in the *Gesneriaceae* family, all from Central and South America and 8 of them were found growing all in the same area. The Gloxinias we buy as conservatory plants today have all been bred from *Sinningia speciosa*, and are rather confusingly known as Florists' Gloxinias.

A number of tropical bromeliads and orchids were dispatched to Kew at the same time. Bromeliads are that multifarious family of normally rain forest plants, ranging from air plants (epiphytes) to terrestrial plants. Some of them look similar to the pineapple (*Ananas comosus*), which is, in fact, a member of the vast *Bromeliaceae* family. There are some 2,000 species in the family and they are all widely diverse. The one thing they do all like is heat, so none succeed in the open in England but have to be grown in the greenhouse or conservatory – like *Bromelia balansae*, the Heart of Flame, which arrived in 1873.

The opening-up of South America encouraged traders, engineers, merchants and even diplomats to seek new opportunities, and they collected seeds and plants wherever they went. This was in spite of discovering that everyday life could be trying and uncomfortable for plant collectors, just as David Douglas was finding in North America. On the route between Argentina and Chile in the 1820s was a series of post-houses from which food and accommodation were available. John Miers (1789–1879), engineer and botanist, when writing home made them sound like the original 'flea-pit': 'In these dreary receptacles is the incredible number of fleas, bugs and even still more disgusting vermin. The fleas breed in the very earth . . .' So disgusting must these post-houses have been that one feels it might have been preferable to have sailed between the two countries, risking the weather off Cape Horn, rather than submitting to the unpleasantness of the land route.

The Horticultural Society was also very keen, once the danger of the Napoleonic menace had passed, to gather the flowering spoils of the newly emerging nations, and found a way of securing the maximum botanical return for the least amount of monetary outlay – always a consideration where plant collecting was concerned. Berths for men employed by the Society were found on ships calling at numerous ports on their long voyages. The more calls the ship made, the better it was, since it gave the collector the widest range of opportunities to gather plants. It is known that this fast-track method of plant collection took place around the coasts of west and east Africa, Brazil, Chile, Peru and Hawaii.

Gardeners trained at Kew also joined expeditions. David Lockhart (d.1846) took part in Captain Tuckey's exploration of the River Congo in 1816, and went on to become the first Superintendent of the Botanic Garden in Trinidad. George Barclay (fl. 1830s–40s) joined HMS *Sulphur* on travels which lasted five years from 1836 and took him to, amongst other places, Chile, Peru, Panama, Fiji and Hawaii. A few years later HMS *Herald* sailed up the west coast of America and into the Arctic with Berthold Seeman (1825–71) on board, specifically employed to collect plants on behalf of Kew. The same ship went off a year later, in 1852, with William Grant Milne (d.1866) aboard, on a voyage to Fiji which lasted until 1856.

The numerous new and exotic plants which crossed the oceans of the world and streamed into an eager Britain had an amazing effect on the nineteenth-century garden. At the same time not only were there new scientific ideas, but new inventions, horticulturally friendly legislation and gardening magazines all coming together to give the eager gardener every possible assistance. Just as in the eighteenth century gardens spread out and jumped over the ha-ha to create landscape parks, so in the first half of the nineteenth century gardening made a further leap over the far-off park boundary and brought horticulture into the suburban world of villas and town houses. No longer was the creation of gardens the prerogative of the landed gentry; growing plants and displaying them became a preoccupation and almost an obsession of the burgeoning middle classes. When J. C. Loudon launched his *Gardener's Magazine* in 1826, his first essay remarked that 'landscape gardening about a century ago was as much the fashion as horticulture is at present'. It was the beginning of modern gardening.

One of the most fundamental changes came about with both the repeal of the Glass Tax in 1845 and the invention of the sheet glass process in 1847. Glass in all its forms had been variously taxed from 1695 onwards but, specifically, a Window Tax of 1810 and a complete Glass Tax in 1812 had severely restricted the building of greenhouses, hothouses and conservatories. All those orchids and exotics now arriv-

ing from around the world had to go somewhere – and with glass becoming more widely available, orangeries, pineries (for the growing of pineapples), vineries and ferneries all became part of the garden scene.

Industrial techniques made the use of knives, scissors, clippers and secateurs (which arrived from France in 1881) more common. However, it was a bigger piece of equipment which must have very quickly changed the look of nearly all gardens from around 1830 onwards: this was of course the mowing machine. Since lawns are an integral part of English gardens, it seems only natural that it took an Englishman, Edward Budding, to come up with such a miraculous machine. He took out a patent on his invention and in 1833 sold it under licence to Messrs Ransomes of Ipswich. He was justly proud of his idea, and believed that 'Country gentlemen may find, in using the machine themselves, an amusing, useful and healthy exercise.' A most worthy ideal for earnest Victorian gentlemen to follow.

The Wardian Case has been mentioned briefly earlier. It was the well-thought-out invention of Dr Nathaniel Bagshaw Ward (1791–1868) for transporting plants around the world. To start with, Ward successfully grew a number

The technique of using a sealed box – the Wardian case – for the transportation of seedlings and plants greatly enhanced the chances of survival. Later it was found that small holes bored at either end of the case increased the number of living plants arriving at their destination.

of plants in sealed containers in his own home, which, incidentally began a vogue for highly decorative and elaborate glass containers in both Britain and, especially, in America. He then developed this into a hold-all for moving plants around; it was, in effect, a portable mini greenhouse. In 1835 a selection of plants was sent to Australia in a Wardian Case and arrived in far better condition than anything that had previously been exported. Inevitably, the same happened in reverse: plants from Australia were delivered to Kew in a sealed and airtight box which again had a high success rate. Dr Bagshaw Ward's box was, if used correctly, a timely invention of immense significance. He had found that the trick was to get the balance of the light correct, so that photosynthesis could take place. It meant that the water vapour condensing on the glass kept the soil damp, thus keeping the plants not only alive but in tip-top condition. The cases were not difficult to make: they looked rather like hen coops. Despite being somewhat cumbersome, they revolutionised the survival rate of plants during transportation.

Kew, in particular, benefited from using the Wardian Case when, in just one year, 1851, plants were sent to New Zealand, Tasmania, India, Trinidad, Jamaica, Sierra Leone and British Honduras (now Guyana). The invention was also used by Robert Fortune to dispatch the 20,000 or so small tea plants (*Camellia sinensis*) from China to the foothills of the Himalayas (see p.190). It had been employed earlier to transport the Chinese banana around the world. As we have seen (p.140), the first bunch of bananas had arrived in London as early as 1633, but it was not until 1829 that the first plant of the Musa genus arrived from China and did so without the benefit of Dr Ward's special case. When it was successfully grown at Chatsworth for the 6th Duke of Devonshire, it was named *Musa cavendishii* in 1829 in the Duke's honour (although it was later renamed **M. acuminata**). A few years later, several plants were transferred to Kew, where they were grown on before being packed in Wardian Cases and successfully transported to the islands of Samoa, Fiji and Tonga. In 1962, Fiji became the last recipient of a delivery of plants by Wardian Case from Kew Gardens.

The remark made by John Loudon about horticultural improvements becoming the fashionable cause of the day was to prove to be remarkably pertinent during the middle years of the century. A positive surfeit of newly arrived plants was eagerly devoured by the cognoscenti. Novelty was the order of the day, and the voracious appetite of the gardener led inevitably to botanical indigestion. Even Joseph Paxton,

The banana's native habitat is at forest margins in the light woodlands of N.E. India and Bangladesh and from S.E. Asia to Japan and Northern Australia.

in 1838, warned against such behaviour when he wrote: 'There is a fashion in all things, and novelty as respects flowers is now a complete mania'. It really made no difference, however; it was the fashionable mixture of newness and exclusivity which made the growing of all those new arrivals so attractive. The inevitable result was that if the plant did not perform well in the greenhouse or garden it was thrown out and another new introduction taken up.

A bonus of this wonderful floral bonanza was that gardeners began to show great interest in artificially developing bigger, brighter and better blooms from both existing plants and new introductions. Hybridisation – a word coined in 1845 – was the art of cross-breeding two distinct species and was taken up with such tremendous enthusiasm that in 1899 the Royal Horticultural Society held an International Hybridisation Conference ('hybrid' comes from the Latin word *hybrida*, and by 1775 its horticultural meaning was 'the offspring of two plants of different species'. A much earlier meaning denoted a child born of a Roman father and a foreign mother). Botanically the idea was not completely new: in 1717 Thomas Fairchild (1667–1729) had begun experimenting. (He was a near contemporary of Nehemiah Grew, whom we met in the seventeenth century, and was the first person to describe in his book *Anatomy of Plants* (1682), how pollen was needed to fertilise seeds.) Fairchild describes how he took the pollen from the Wild Carnation, *Dianthus caryophyllus*, and put it on to the stigma of the Sweet William, *Dianthus barbatus*, which had arrived here in 1573. Both plants are from the same family and, with hindsight we know that the experiment worked, but at the time it must have been a most exciting experience to watch the development of 'Fairchild's mule', as it was called.

Almost unbelievably to us, hardly anyone took any notice of Fairchild's experiments, and it was not until early in the nineteenth century that the Honourable and Revd William Herbert (1778–1847), later Dean of Manchester, wrote about his own experiments, and of how cross-breeding gave him 'an endless source of interest and amusement'. He thought long and hard about all aspects of the plant he wanted to create, its brilliancy of colour, its perfume, hardiness, and 'profusion of blossom'. Herbert was not only a very good botanist but was also an acknowledged expert on bulbous plants, in particular the Gladiolus species, *Gladiolus carneus*, which Francis Masson had so assiduously collected from South Africa in 1774.

There are at least two plants with *spofforthianus* as a second name (Gladiolus and Camellia), which commemorates the parish in Yorkshire where William Herbert spent twenty-six years as rector, but he was also honoured, while Dean of Manchester, by his name being used for a small genus of corms in the *Iridaceae* family native to Chile, southern Brazil and Texas, the blue- and purple-flowered Herbertia, the first two of which arrived respectively in 1827, *Herbertia pulchella*, and 1839 *H. drummondiana* (named for one of the two Drummond brothers, whom we shall meet shortly). Herbert's adventures in horticultural experimentation caused a great deal of interest and there is no doubt that his ideas sparked off

the explosion of cross-breeding during the 1830s and 40s. The arrival of so many novel plants on our shores, combined with the knowledge and power to be able to 'create' new plants, caused a complete change in the gardening scene. Gone were the discreetly refined ideas of the naturalising of the landscape and in came the exuberant artificiality of 'bedding out'. Ornamenting gardens in such a way can be likened to different styles of music – the classically inspired landscape parks seem to be filled with the sound of Mozart or Haydn whereas there is a 'big band sound' to many of the ribbon and carpet beds that were being developed. Between the classical orchestra and the brass band, both horticulturally and musically, there was a certain *frisson* – both are wonderful in their own way but not necessarily compatible.

The layout of flower beds in the middle of the nineteenth century took on the look of an earlier age, with inspiration coming from Tudor knots and Jacobean parterres. Later in the century, during the 1860s and 70s, the contrived national characteristics of gardens from Italy, Holland and France were all the rage. Also in the 1860s, subtropical bedding (frost-tender plants grown to ornament summer displays), carpet-bedding, ribbon bedding and embroidered parterres all had their moments of glory. It was the crisp contrast of bright colours and immaculate grooming which was so appealing – and still is.

It was this immaculate grooming, combined with 1870s high fashion, that catapulted the Malmaison Carnation to centre stage, both as a buttonhole for frock-coats and as decoration for elegant salons and chic dining tables. The flower began life as a seedling 'tree carnation' in France in the 1850s, and because its blooms resembled those of the then new rose 'Souvenir de la Malmaison' (raised in France in 1843) it took the same Malmaison name. The flowers were gloriously blushed, fragile-petalled beauties, and it needed every ounce of the gardener's skill to grow and produce the heavily perfumed blooms – which, of course, was part of their charm and appeal. Of these early varieties, one was named 'Old Blush', another 'Nell Gwynne' and yet another 'Princess of Wales' (after the future Queen Alexandra). Technically, they are half-hardy evergreen perennials which, given some protection, flower sporadically throughout the year.

As with all fads and fashions – combined with the fact that they were difficult to grow – the Malmaisons only had a few years of glory. They then lost favour and gave way to other equally delightful fashionable blooms – for example, Parma Violets – so that after a few years the sparkling Malmaison

Bedding to celebrate the Millennium at Waddesdon Manor, Buckinghamshire. Commissioned by Lord Rothschild, this creation was designed by the painter John Hubbard and uses a revolutionary process of growing the bedding plants. The old system of bedding out the 50,000 plants would have taken an army of gardeners several days to complete. Here, a computer-generated programme devised by Kernock Park Plants of Cornwall enabled the modern design to be laid out in one day.

ceased to exist. It was only after much research and an intensive breeding programme during the 1970s and 80s that the flower was once again a delight for us to drool over. Due to modern techniques, they are now easier to cultivate and are therefore much more widely grown. Let us hope that this time the fashion for them lasts longer, and that we shall be able to decorate our homes and indeed our buttonholes for many years to come.

Back in the garden, the flowers developed for the increasingly complicated bedding patterns were almost without exception imports from the earlier part of the century, and came mainly from South America – Verbenas, Calceolarias, Petunias and (later) Begonias – or from South Africa with its Gazanias, Lobelias and, most popular of all, Pelargoniums. *Pelargonium zonale* and *P. inquinans*, which arrived in this country in 1710 and 1714 respectively, had both hung around patiently waiting to be found a useful role. Their chance came in 1844, when the combined charms of their hybridisa-

tion led to the creation of the first bright scarlet bloom. Because of its small size, it was launched commercially with the name *P.* 'General Tom Thumb'; this was after the American dwarf, only 75 cm (31 in) tall, who had just visited Britain and whose nickname it was; his real name was Charles Stratton. By 1874, the growing and showing of Pelargoniums had aroused such interest that under the auspices of the Royal Horticultural Society a specialist Pelargonium Society was formed.

Not everyone, however, was so keen on this strident mixture of colours, and by the 1870s the gardens both at Kew and the Crystal Palace at Sydenham were thought by some to be overly vulgar and crude. Other public places fared better. John Gibson, whom we met earlier working at Chatsworth, was by now the superintendent in charge of Hyde Park in London, and was much admired for his creativity in the use of subtropical foliage in the bedding schemes.

The new and quite different use of plant material had originated in the public parks around Paris. Hyde Park was given the *jardin anglais* treatment sometime in the late 1850s, and by 1859 they were using between 30,000 and 40,000 bedding plants to create floral patterns. The exuberance of colour was considered to be especially beneficial to those workers not lucky enough to have gardens of their own. Special mention was made of the beauty of the flower beds bordering the carriageway leading towards Marble Arch. Today, a hundred and fifty years later, the shape of the beds, the planting, the design, the colour schemes and the plants themselves are all a sparkling reminder of that earlier craze.

The production of thousands of seasonal bloomers usually took place on site, with country houses, great estates and royal parks all vying with one another to see how many plants could be produced to accommodate their intricate garden designs. Nurseries, on the other hand, concentrated on the growing of individual specialities and newly introduced plants, and it was this search for the new which led to an interesting development. A number of the larger nurseries expanded their field of operations by employing their own plant hunters to seek out undiscovered and preferably commercially viable plants.

Perhaps the premier nursery involved in this enterprise was the Veitch Nursery. John Veitch (1752–1839) founded what was to become the family firm in 1808 when he set up a small nursery at Killerton in Devon. Later, in 1832, Veitch and his son, James Veitch Senior (1792–1863), moved the nursery to Exeter. In 1853 it was decided to expand further, this time with a base in London, and the old established nursery of Knight and Perry in Chelsea was purchased and developed by James Veitch Junior (1815–69) as The Royal Exotic Nurseries.

The reputation of the nursery, and of the younger James in particular, was confirmed when the Royal Horticultural Society instituted the Veitch Memorial Medal to commemorate his life in horticulture, 'for his skill as a cultivator, his genius as an organiser, and his enterprise in sending forth collectors to various parts of the globe'. The firm grew still further, with branches around London, most notably at Coombe Wood near Kingston-upon-Thames. On the death of James Veitch Senior in 1863, the two branches of the nursery separated and were run by different members of the family.

The Veitch family could have had no idea what effect their decision to send abroad plant hunters to collect exclusively for them would eventually have. Their commercial policy of garnering new exotics from all corners of the globe to sell

from their nurseries proved to be beneficial both for the family and for British gardens generally. In all, twenty-two plant collectors were employed over a period of sixty-five years to support the enterprise.

The list of botanists, plant collectors and other people who became interested in sending home new material to the nursery demonstrated the wide variety of people who were intrigued by the natural world – or who were simply interested in earning a few extra pounds.

Amongst the first to send back horticultural material were two brothers who were already well-seasoned travellers and plant collectors. They were James (*c.* 1784–1863) and Thomas Drummond (*c.* 1790–1835). Thomas, whom we met previously collecting seed from Texas, had also been, in 1825, assistant naturalist with Sir John Franklin's Second Arctic Expedition. On the first exploration, 1819–22, the surgeon-naturalist Sir John Richardson had discovered a small evergreen Mountain Avens which was subsequently named *Dryas drummondii* (see p.157) in honour of Thomas. James became Curator of the Botanic Gardens in Cork and in 1829 travelled to Western Australia, where he was appointed government botanist for that area.

The new Veitch policy began in earnest in the 1840s with, coincidentally, two brothers again taking the stage. They were the Cornishmen William (1809–64) and Thomas Lobb (1820–94). William worked all over South America between 1840 and 1848 and in California and Oregon from 1849 to 1857; his brother specialised in collecting orchids, working in India, Burma, Java, Malaya, Borneo and the Philippines from the beginning of the 1840s onwards. They are both remembered in a number of plants with the suffix *lobbiai*.

Richard Pearce (d.1868), also from the West Country, collected plants *c.* 1859–66 throughout Central and South America; he found *Eucryphia glutinosa* in 1859, one of only two Eucryphia species discovered in South America, the others all being from Australia. His name was remembered in *Pearcea hypocyrtiflora*, which was subsequently changed to *Isoloma hypocyrtiflorum*; this Greek description of the name was discarded also, as it had earlier been designated to describe a fern, so the entire Isoloma genus of about 50 species was renamed Kohleria to commemorate Michael Kohler, a nineteenth-century natural history teacher from Zurich. Very few of the Kohlerias which were collected at this time remain in cultivation in Britain. Coming from the tropical regions, their rather Gloxinia-like features require considerable nurturing in England's fickle climate.

Frederick Burbidge (1847–1905), a gardener at Kew, collected plants on the island of Borneo for a year (1877–8) for Veitch's Nursery. He introduced the rare but spectacular *Nepenthes rajah*, the Tropical Pitcher Plant, which has some of the largest leaves in the genus – up to 1.8 m (6 ft) long. The pitcher itself is anything up to 35 cm (14 in) long and has been known because of its size to entice rats on a one-way journey into the interior of its flower. Because it is so rare, commercial trading in the plant is severely restricted. Burbidge later became curator of the garden at Trinity College, Dublin, and later Keeper of the College Park. He wrote ten books, ranging from general horticulture to botanical drawing and orchids. Like the preceding Veitch collectors, he too was commemorated, by the Burbidgea genus in the *Zingbereacea* family, which he discovered when travelling in Borneo. There are only two plants in the genus, and the one he found in the north-west of the island was *Burbidgea nitida*. It sounds spectacular, with its bright shining orange-scarlet flowers and long leathery glistening leaves, but, having been brought to England in 1879, the plant seems to have retreated back to the tropics and is no longer grown here.

Charles Curtis (1853–1928), from Barnstaple in Devon, collected plants all his working life from around the Indian Ocean, and for twenty years was the superintendent of the Botanic Gardens at Penang in Malaya. First, however, in 1878, he began a six-year period working for Veitch in Madagascar and Mauritius, and later in the East Indies. In 1897 Curtis had a pink and white epiphytic orchid named for him, *Cirrhopetalum curtisii* (now subsumed into the *Bulbophyllum* family, along with the rest of the genus).

In the 1880s, Curtis was joined in the East Indies by yet another 'Veitchite', David Burke (1854–97), who collected throughout the area until his death in the Molucca Islands off Indonesia. The special plant named for him was *Didymocarpus burkei*, a genus of about 80 species of perennials, nearly all with violet-blue flowers, which inhabit south and east Asia and through to Australia and Madagascar; like the orchid mentioned above, it does not appear to have settled here and has drifted back home.

A Señor Endres collected orchids from Costa Rica for Veitch's Nursery between *c.* 1868 and 1873. His name, together with that of Viscount Milton, a keen orchid buff, was remembered in *Miltonia endresii*, but again, that plant has disappeared. At about the same time, 1870–3, George Downton was active in Central America and later in Chile as a Veitch orchid devotee. So was James Chesterton (d.

1883). He is credited with having found and sent home *Miltonia vexillaria* in 1872, yet another vanished species. Apparently, he had to combine his orchidmania with being a valet, so presumably he must have always checked at interviews that his prospective 'gentleman' was a traveller, and to the desired locations.

Herr Gustav Wallis (1830–78), who worked in Brazil and tropical South America, collected both for Veitch and later for the Chelsea nursery of William Bull – who, in their turn, also sent out two trained orchid collectors to South America: John Carder (d.1908) and Edward Shuttleworth (1829–1909).

Two Veitch brothers, John and Harry, the sons of James Veitch Junior, and a cousin, Peter, joined the Far Eastern foray for plants. John Gould Veitch (1839–70) made a whistle-stop tour of Japan, China and the Philippines in 1860, then four years later collected in Australia and the Pacific islands. His commemorative plant (named for his more famous father as well) was Veitchia, a select genus of very tall palms native to Fiji (or 'Feejee', as its name was written on some old maps) and the New Hebrides. He was only thirty-one when he died, so neither his younger brother nor his cousin had the opportunity to compare notes with him when their turn came to go abroad.

John's brother, Harry James Veitch (1840–1924) – later Sir Harry – was a great traveller, collecting plants as he went – what Veitch could not? – but hunting out garden-worthy plants in a systematic fashion was not what Harry Veitch was most interested in. However, while travelling in Japan in the 1880s, he did make a collection of 'Sato Zakura' flowering cherries. On his return, he ruthlessly culled all but one of them, a double rose-pink pendulous late flowerer which he deemed to be worthy of commercial cultivation. It arrived on the scene in 1882 and is still available; for a long time it was known as 'James H. Veitch' after his nephew, but has now been renamed *Prunus* 'Fugenzo'.

In 1875, Peter Christian Veitch (1850–1929), John and Harry's cousin, had begun his two-year journey collecting for the firm in Fiji, Australia, New Zealand, New Guinea and Borneo. His special plant was *Spathoglottis petri*, an orchid from the South Seas which had lilac and purple markings and looked as if it should be highly perfumed, but alas was not. It was introduced by him in 1877, and could still be found about twenty years later, but is no longer available.

By this time, travel in Japan and China was much less restricted, and Charles Maries (*c.* 1851–1902), a foreman at the Veitch Nursery, was sent out there in 1877. One of his first tasks was to concentrate on collecting conifer seed from Japan; the craze for conifers had been triggered in Britain by the introductions of the David Douglas imports from the west coast of America a few years previously. Maries visited Yokohama and travelled to some of the Japanese islands for nearly a year, collecting as he went. He is responsible for bringing to our gardens the autumnal-tinted shrub *Enkianthus campanulatus*, and several Abies. *Abies veitchii* was one he sent back to England; it was thus named since it was first seen and noted by John Veitch when he was climbing Mount Fuji in 1860. However, it was Maries who collected the seed, as he did of another three Abies. *A. homolepis*, with its long purple cones, is known as the Nikko Fir; while from the far northern islands of Sakhalin and Kurile (now Russian territory) came *A. sachalinensis* and *A. mariesii*, the Maries Fir.

Whilst still in Japan, Maries also discovered the first two Lacecap Hydrangeas, *Hydrangea* 'Mariesii' and *H.* 'Veitchii'. By 1879 he was plant hunting in China where he found the Chinese Witch Hazel, *Hamamelis mollis*, and the beautiful species lily *Lilium speciosum* var. *gloriosoides*, the flower of which is white and flushed-rose, spotted with crimson and scarlet. He found and collected a great deal of seed of the pretty but hairy *Primula obconica* near Ichang, a town on the great Yangtze River which, in the twentieth century, was to feature so prominently in the story of the most famous of all the Veitch collectors, Dr Ernest Wilson. Maries spent the rest of his working life in northern India, becoming in 1882 the Superintendent of Gardens to the Maharajah of Durbhungah, and subsequently holding the same post for the Maharajah of Gwalior.

The last but one of the Veitch-employed plant hunters was the greatest. E. H. Wilson (1876–1930) had won the Queen's Prize for Botany from Birmingham Technical College, and the Botanic Gardens at Kew wanted him to head an expedition to China. However, funds were short, so the Veitch Nursery agreed to employ him. It was probably the best commercial investment they ever made, and his introductions (some 1,000 species in all) changed the look of the domestic garden in Britain (see p.226).

The final collector to be employed directly by the firm was a Westmorland man, William Purdom (1880–1921), who like so many of his Veitch colleagues, had earlier received his training at Kew. Despite having a reputation for being quiet and unassuming, he found himself at the centre of a political row. Purdom and the other gardeners at Kew considered they

were grossly underpaid, receiving only eighteen shillings a week for six twelve-hour days. There was much discontent over this, since the amount was not considered to be a living wage, and was less than employees in the royal parks received.

In 1905 they formed themselves into 'The Kew Gardeners Employees' Union', with Purdom taking an active role. Things came to a head in October of that year when the Director at Kew, Sir William Turner Thiselton-Dyer, summarily dismissed Purdom. Following intervention, by the President of the Board of Agriculture, which was then responsible for the running of the Gardens, Purdom was reinstated.

Purdom must have proved a good plantsman over the following years, since he was later chosen by the Veitch Nursery to be sent to collect in China, a country with which he fell in love. Between 1909 and 1912 he collected for both Veitch and the Arnold Arboretum in America (see p.227).

As a mark of the Veitch influence and reputation, a representative of the nursery was sent to the horticultural section of the International Exhibition held in Paris in 1867; this was the young William Robinson (1838–1935) who was later, through his writings, to change the whole philosophy of gardening in Britain.

During this period when the Veitch Nursery was at its height, there were perhaps only two or three of their collectors whose expeditions were deemed horticulture failures. Carl Kramer went to Japan, then Costa Rica and Guatemala, and failed to find any plant of note for the nursery. In 1877 Christopher Mudd was sent to South Africa but was recalled with negligible results. Even Gustav Wallis, who had collected successfully in Brazil for Veitch, could not find plants of any great garden value for the nursery when he travelled to the Philippines in 1870. In 1914, as the lease on the Coombe Wood nursery in Kingston-upon-Thames ran out, and with no younger family members coming forward, Sir Harry Veitch decided to auction off the contents of the famous London nursery and shut the firm down. The Exeter branch carried on until 1969, when it was sold.

South America occupies about an eighth of the total land mass of the world, and some of it, particularly central Chile and the foothills of the Andes, has a climate very similar to that of the Mediterranean, Stretching right to the very southern tip of South America, Tierra del Fuego, there are a large number of indigenous plants that are quite at home in our climate. Horticulturally, South America had been feeding the gardens of Europe since the sixteenth century, but it

Hydrangea veitchii was discovered by Charles Maries in Japan in 1877–8. He later worked in India where he was the Superintendent of Maharajah's of Gwalior's Gardens.

was in this, the nineteenth century, that the true extent of its flora was realised and became a veritable botanical fiesta.

Already the three Verbenas (their ancient Latin name), which were being used to make such an impact in the new style of flower beds, had established themselves in Britain. *Verbena x hybrida*, which became the bedding plant par excellence, was the result of mixing the parentage of *V. incisa*, introduced in 1826, and *V. peruviana*, which was collected the following year by the naval surgeon John Gillies (1792–1834). The third plant, *V. phlogiflora* (syn. *V. tweedii*), arrived in 1834 and was named for John Tweedie (1775–1862), who travelled from the Royal Botanic Gardens in Edinburgh to Buenos Aires in 1825. In 1837 Tweedie collected a fourth species, *V. platensis*; this plant releases its perfume at night.

A number of other Verbenas had been discovered – in total there are about 250 species, nearly all found in the Americas – but it was the development of these four which quickly became such an asset. By the 1840s, Jane Loudon could write that 'It is now rare to see a garden or a balcony without them.'

The other great bedding standby from South America was the Calceolaria, the Pouch or Slipper Flower, as it is sometimes called, which also comes from a large family. With its extraordinary puffball features and zany colours, it exudes exotic foreignness even though it has inhabited Britain for well over two hundred years. The pale yellow *Calceolaria pinnata* was the first to arrive, in 1773 – a rather quiet beginning bearing in mind what was to come: its territory stretches from Chile to Jamaica. Then a second species, with the rather more restricted range of Patagonia and the Falkland Islands, was collected by Benjamin Hussey in 1777 and named for Dr Fothergill, a London doctor who had a celebrated botanic garden in West Ham. This Calceolaria, *C. fothergillii*, would have certainly startled those who first viewed it in flower as it is a bright sulphurous yellow with red spots. One with purple stripes and spots, *C. corymbosa*, came from Chile in 1822, as did *C. integrifolia* (yellow-flowered) which was found growing on Chiloe island by Alexander Cruckshanks; in 1826, he also found the purple-violet *C. purpurea* in Chile. The arrival of these last two species meant that the development of the bedding Calceolaria was all set to complement the Verbenas in the carpeting of the High Victorian garden.

In 1840, as we have seen, the thirty-one-year-old William Lobb sailed from Falmouth in Cornwall as the first plant hunter specifically employed by the Veitch Nursery. His destination was Rio de Janeiro, then the capital of Brazil. The country was by now well established as a coffee producer. *Coffea arabica*, a native of Ethiopia and the mountains of Yemen, had first been introduced to the East and West Indies, and had reached Brazil by 1727. Thousands of acres of forest were felled to make way for the plantations, and by the time Lobb arrived, great swaths of land had become derelict.

As the Cornishman journeyed across the Atlantic, he must have wondered what plants were waiting for him in Brazil. As soon as he got into his collecting stride, both the quality and the quantity became clear, and he started to send home a remarkable range of plants to the Veitch Nursery for whom he so assiduously collected.

Two of the first he found were species of the Peruvian Lily, *Alstroemeria inodora* and *A. chilensis*; the second of these has pink and tawny flowers and is quite hardy. They had been named by Linnaeus for a friend of his, Baron Claus Alstroemer, who had sent him seed of the 'Lily of the Incas' (*A. pelegrina*) in 1753. The Organ Mountains proved a good hunting ground, and Lobb's eye was caught by *Barbacenia squamata*, now called *Vellozia squamata*, and named in honour of a Capuchin monk called Vellos rather than Señor Barbacena, Governor of Minas Gerais in south-east Brazil. Lobb found as well the lovely pendulous Angelwing Begonia, *Begonia coccinea*, with its coral-red flowers, which makes an excellent house plant if given enough light. It has also proved itself a good parent to many of the later hybrids.

Lobb collected two climbers from Brazil: a passion flower, *Passiflora actinia*, with its large banded white blue and red markings, and the pink Mandevilla, *Mandevilla splendens*. Henry Mandeville (1773–1861), after whom this was named, was a keen horticulturist as well as being His Britannic Majesty's Minister at Buenos Aires. He had dispatched to the Horticultural Society in 1837 the first and most fragrant of all the species, *M. laxa*. It was known then – and still is – as Chilean Jasmine; whilst it does look rather similar to jasmine, having beautiful creamy-white trumpet blooms, it appears to inhabit Argentina, Bolivia and Peru but not Chile. A swan-neck epiphytic orchid was another of Lobb's finds, this time from Brazil in 1841; according to the records, this was the first of the Lobb orchids to flower in Britain. *Cycnoches pentadactylon* is a curiosity rather than a great beauty. Its petals are pale yellow tinged with green, complemented with broad chocolate blotches; the lip, with the same mix of colour, is divided into five. It looks rather like a hand, hence its second name.

The surgeon-naturalist attached to HMS *Sulphur* on its six-year global voyage was Richard Brinsley Hinds (c. 1812–47), and it is his name which is commemorated in the shrub *Hindsia longiflora* which William Lobb collected in 1841 in Brazil. Three years later, he discovered its relative *H. violacea*. As there were only three members in the family, they have been subsumed in the much larger Rondeletia genus. Another Lobb find, *Hypocyrta strigillosa*, has changed its name too, becoming *Nematanthus strigillus*; it is a trailing shrub with orange-yellow trumpet flowers.

By 1842 Lobb was in Peru, where he collected a shrub belonging to the very large South American *Lythraceae* family. Some members of this had already been sent to England, where sheltering in the greenhouse was their only means of survival. William's find was *Cuphea cordata*, with its small scarlet flowers; it is rather a discreet bloomer but is related to the much more dramatic Cigar Flower, *C. ignea*, whose blooms replicate a long glowing cigar with a rim of white ash. This was collected in 1845 from Mexico. The Brazilian Firecracker was another plant that had a change of Latin name: when Lobb collected it, it was originally called *Manettia bicolor*, but is now *Manettia luteorubra*, named in honour of Saverio Manetti (1723–85), a medical man in charge of the Botanic Garden in Florence. Its vernacular name, although descriptive of the flower, is again a misnomer, as the plant only appears to grow wild in Uruguay and Paraguay.

The orchid *Oncidium curtum* discovered at the same time does not seem to have changed its name; instead, it has disappeared from cultivation. Many orchids collected during the nineteenth century are today no longer commercially available, reflecting changes in fashion and worldwide concern that the natural environment should be protected. The changes show up dramatically in, for instance, RHS publications of various dates. The 1956 four-volume *Dictionary of Gardening* lists over 280 Oncidiums (there are over 450 species in the genus), but the *RHS A–Z Encyclopedia of Garden Plants* published forty years later, names fewer than 10 species.

A shrub originally noted by someone else – in this instance Charles Darwin on his voyage on the *Beagle* in 1835 – was introduced by Lobb in 1849. It is the glossy-leaved, orange-flowered *Berberis darwinii* (*berberis* is the Arabic name for

William Lobb brought the seed of this magnificent tree, *Sequoiadendron giganteum*, to England in 1853. Its shape and height make it instantly recognisable in the landscape growing as it does to a height of between 25–80 m (80–260 ft).

the barberry), which William gathered when he visited the island of Chiloe. He also collected two evergreen shrubs from Chile which still make an impact on our gardens today. One has leaves closely resembling a holly and scarlet tubular flowers tipped with yellow, and is the only species in its genus, *Desfontainea spinosa*. It was named to honour René Desfontains (1750–1833), Professor of Botany at the Jardin des Plantes in Paris, who wrote the massive *Flora Atlantica*. The other is a semi-evergreen known as the Chilean Fire Bush, *Embothrium coccineum*, a glorious shrub or small tree which thrives in sheltered West Country gardens. Its scarlet flame-shaped petals makes a startling show in early spring and create the illusion, as do so many of these South American beauties, that Britain basks in a warm and balmy climate.

Another native of Chile, the *Lapageria rosea*, an evergreen climber, was discovered by Lobb and brought home in 1847. The flowers are stunning, being about 7 cm (3 in) long in a rich crimson rose, and hanging down pendulously when they appear in late summer/early autumn; they need to be trailed around a cool conservatory or on a shady wall outside (fierce sunlight does not become them). There is too a much rarer white-flowered variety, *L. alba*, which was discovered by Richard Pearce, collecting for Veitch's nursery, in 1860. As becomes such a stunner, the genus was named in honour of the Empress Josephine, whose maiden name, de la Pagerie, was also the name of the estate where she lived and grew up on the island of Martinique.

William Lobb moved on to Oregon and California in 1849 and continued to collect for the Veitch Nursery, particularly in the hills of the Sierra Nevada in California, until 1857. From the western slopes, he was responsible for the introduction of the lugubrious gentle giant, *Sequoiadendron giganteum*. In the United States it was known as Big Tree, but in Britain, where it arrived in 1853, it was called Wellingtonia in honour of the Iron Duke, the 1st Duke of Wellington, who had died the previous year. A Big Tree found growing in California was so enormous, even by American standards, that it eventually received the nickname 'General Sherman', after the American Civil War general who was considered one of the greatest commanders of his day. This particular tree is believed to be the world's largest living organism. It is also one of the oldest: an authenticated ring count of a similar tree felled in 1934 revealed that it had reached the amazing age of 3,200 years old, while others of its kind appear to regularly reach ages of over 2,000 years. (The title 'oldest' was recently challenged in the White Mountains of California by the discovery of a Bristle Cone Pine, *Pinus aristata*, which is thought to be of an even greater age, close to 5,000 years old; this species was introduced to Britain in 1863.) William Lobb paid his last visit home in 1853, carrying the seeds of the Wellingtonia to his Veitch employers. He never returned to his native Cornwall, but settled on the west coast of America.

Others continued the search for plants in the Sierra Nevada and the Rocky Mountains, usually suffering as much aggravation and privation as David Douglas had earlier been warned to expect. Exploration of the vast country was gradually being assisted by the extension of the railway system, an Irishman William Bell (1841–1921), who was Vice-President of the Denver and Rio Grande Western Rail Road Company, found time to botanise and write books on the subject as the tracks were being laid from Colorado southwards in the late 1860s. The first transcontinental railroad was completed in May 1869.

One of the first people to benefit from the easier travel arrangements was a remarkable Yorkshirewoman, Isabella Lucy Bishop, *née* Bird (*c.* 1832–1904), who travelled extensively in North America, normally on her own, something that was unusual in itself; she was also very petite, being only four foot eleven inches tall. She must have had an overwhelming desire to travel and been very sure of herself as she visited the most outlandish places collecting plants, keeping a journal as she went. Back in England in 1879 Mrs Bishop published her extraordinary story in her book *A Lady's Life in Rocky Mountains*. Later, in the 1880s, her travels took her to Persia (Iran), and in 1892 her exploits were recognised by the Royal Geographical Society, who made her their first lady Fellow.

Whilst the rush of plants continued from the New World, Asia and South Africa, Australasia was gradually beginning to fulfil the wonderful botanical promise hinted at in the 1770s. Neither New Zealand nor Australia has disappointed British gardeners with their horticultural introductions.

In the year 1858, a remarkable-looking plant arrived in Britain, from a very small genus, Clianthus, which belongs in the *Leguminosae* family. *Clianthus formosus* is known as the Glory Pea or Sturt's Desert Pea. The latter was named for Charles Sturt (1795–1869), who explored southern Australia between 1828 and 1845. He discovered the River Darling and the Lower Murray, but later lost his sight due to hardship and exposure. The plant had been originally discovered long before 1858, probably by the first white man ever to set foot on Australian soil, William Dampier (1625–1715). Originally a marauding pirate with a bent for botany, Dampier became an Admiralty commander when he was appointed to the specially built explorer HMS *Roebuck* in 1699. He was instructed to look at the prospects for colonising Western Australia and came back, after being shipwrecked, greatly impressed with the country. One of his botanical finds was the Glory Flower, which was named in his honour *C. dampieri*. However, when the flower, with its crimson and black blooms, eventually arrived in Britain in 1858, it was renamed *C. formosus* and joined its only relative, *C. puniceus*, from New Zealand, which had been getting used to our climate since 1831. This species has slightly larger flowers and is known as Lobster Claw or Parrot's Bill.

It was in 1824 that the first British settlers sailed to New Zealand, taking with them cereal seeds and acorns from our native oak to plant on their arrival; the original oak tree survives at Paihia, on North Island. At the same time the French were also making a settlement, at Akaroa on the Banks Peninsula on South Island; being more pragmatic, they chose to take with them the walnut, which not only reminded them of home but produced something edible as well.

Although botanically the flora of New Zealand was quite different from anything the settlers had seen before, it was not until 1863 that there was any attempt to develop botanic gardens. The first of these were at Christchurch and Dunedin, both on South Island; neither was considered of great importance. Wellington Botanic Garden on North Island followed in 1868 and was the only botanic garden in New Zealand to succeed during the nineteenth century.

New Zealand houses most of the 100 species or so of Hebe, the evergreen shrubs in the *Scrophulariaceae* family. Like the confused naming surrounding the South African Geraniums and Pelargoniums in the previous century, so the Hebe and its alter ego the Veronica still seems to be in the pending tray of the taxonomists. In 1789 the original proposal had been made to call the shrubby members of the Veronica family Hebe (the name of the Goddess of Youth, cup-bearer to the gods). However, they continued to be called Veronicas in England until late in the twentieth century, though New Zealanders had embraced the change of name much earlier. Most of the New Zealand Hebes are confined to South Island: there is one which is specific to Chatham Islands, and one or two from south-east Australia and New Guinea. The Veronica genus, the speedwell, is in the same family, but most of the 250 species are European natives.

One of the first Hebes to arrive here, in 1860, was *Hebe hulkeana*, a rather dainty pale-lilac-flowered shrub which belies its hulk-like name, since it needs a bit of nurturing against a warm wall if it is to survive. It was collected from South Island by the Director of the Botanic Gardens at Melbourne, Sir Ferdinand Jakob Heinrich von Mueller (1825–96), who went on to become the President of the Australian Association for Advancement of Science. He was a most assiduous plant collector of Australian plants and an horticultural ambassador for the Antipodes. During his sixteen years as Director of the Melbourne Botanic Gardens, he created what was considered to be the greatest herbarium in the southern hemisphere. *Hebe cupressoides* also came from South Island in 1880; it is what is now called a 'whipcord' Hebe (having tiny scale-like leaves which lie flat against the stem), and is a small plant suitable for the rock garden. Six years later, also from South Island, came *H. odora*: it was first called *Veronica anomala* (meaning 'abnormal') then *Hebe anomala*. It has very glossy leaves and grows into a 1 m (3 ft)

high shrub with white flowers held in dense racemes.

The Daisy Bush or Tree Daisy is another group of shrubs of Antipodean origin, most of whose members lie on the cusp of toleration in Britain. Some of the 130 species have leaves which look similar to those of the olive tree, hence its Latin name *Olearia*. Most of them are white-flowered; not Persil-white, but tinged with grey, so that, as the writer Alice Coats so aptly remarks, it looks as if they have been 'inadequately laundered'. Conversely, the first one to arrive, from south-east Australia in 1793, had large mauve aster-looking blooms and was decidedly tender: this was *Olearia tomentosa*, which only managed a foothold in Britain in the far south-west on the island of Tresco in the Scillies. A few more robust species arrived later, including in 1816 *O. paniculata*, which has been used in the making of hedges and windbreaks, as has the New Zealand Holly, *O. macrodonta*. The latter was an 1886 introduction. The plant's Maori name is Arorangi, and it can withstand wind but not much frost. In favourable conditions it can grow to about 6 m (20 ft), and the wood has been used as a veneer. It was brought to Britain by the Revd William Colenso (1811–99), who worked for the Church Missionary Society and helped Sir Joseph Hooker with his *Flora Novae-Zelandiae* (1844).

A number of smaller isolated islands lie scattered around Australia and New Zealand, some having their own particular plants: Lord Howe Island, with its own unique palm, Norfolk Island and its Norfolk Island Pine; and the Chatham Islands. These outposts have all developed, like other isolated islands throughout the world, their own botanical nuances in their plants. *Astelia chathamica* is an excellent example. It was collected from the Chatham Islands in 1853, either by David Lyall (1817–95), the surgeon-naturalist on board the survey vessel HMS *Acheron*, and who later collected the Mount Cook Lily, *Ranunculus lyallii*, or by Ernest Dieffenbach (fl. 1820–40), who was the appointed naturalist to the New Zealand Company. Both men visited the Chatham Islands in the first half of the century. The Chatham Islands Astelia differs from its New Zealand cousin in both its leaves and its flowers. Its leaves are very much more silvery than the duller ones of *A. nervosa*, and its flowers are pale yellowish-green compared to *A. nervosa*'s blooms of greenish-yellow to brownish-purple.

The Chatham Island Forget-me-not is a single species found only on the islands, and because of its distant relationship with our own forget-me-not has the Latin name of *Myosotidium hortensia*. It reportedly grows on the sandy shore surrounded by seaweed and rotting fish, but its looks belie its seemingly odd habits. It has delicious eye-bright blue flowers and luscious shiny elephant-ear leaves, and although it appears to be difficult to please, it can be grown in this country. It seems to have been originally collected by John Enys (1837–1912), a magistrate in New Zealand who was so taken with the beauty of the plant that in 1858 he arranged to have it sent to his old home in Cornwall where the plant flourished. The scene of him discovering this unique plant was recorded for posterity in a painting, and it too returned to England where it was hung in the drawing room of Enys' family home in Cornwall.

Chatham Island was again explored twenty-eight years later, in 1886, when the first of three unique species of Olearia was collected: this was the robust and, in favourable conditions, fast-growing *O. traversii*. It was followed in 1910 by the other two, which were collected by another Cornish visitor, Captain Arthur Dorrien-Smith, for his burgeoning garden on Tresco, in the Isles of Scilly. These were *O. semidentata* with a most lovely and tender lilac and purple flower, and *O. chathamica*, which is similar but with broader leaves and only single flower-heads of pale violet. *O. avicennifolia* was found in the same year on Stewart Island, just off the southern tip of New Zealand's South Island; it makes a good dense hedge and has small whitish fragrant flowers.

Unabated discovery and collecting of all things natural continued in Great Britain in such a crescendo of frenzy from all around the world that it was as if some awful cataclysmic event were about to happen (as indeed it was in the coming century) which would halt the floral flow into our gardens. As the nineteenth century drew to a close, a profound garden revolution also began.

Like all good revolutions, it started with the barest ripple of wind on the surface of the water, and it is only now, a hundred years later, that we are beginning to understand how much of a fundamental change has taken place. The 'natural revolution', as it could be called, arrived, like a great flood tide that sweeps into a bay and then quietly seeps into every nook and cranny, swirling around rocky outposts and eventually covering our childhood sandcastles. We are now able to see that nearly all the great Victorian gardening ideas – and, indeed, the Edwardian influences as well – have been submerged before the refreshingly simple idea of the development of a natural garden.

The revolution began with the Irishman William Robinson, who, in 1870, expounded his ideas on horticulture in a book entitled *The Wild Garden*, subtitled *or the Naturalization and Natural Grouping of Hardy Exotic Plants with a Chapter on the Garden of British Wild Flowers*. He had already talked to the Revd Samuel Hole (first president of the National Rose Society) about the 'superiority of the natural to the artificial system', and what he was keen to change was the ethos, the style and the habits of the entire gardening scene. Plants, foreign or otherwise, could be grown in what he called 'mixed borders' of half-hardy and tender material, although Christopher Lloyd, in the book he wrote in 1957 entitled *Mixed Borders* called such designs 'makeshift shilly-shallying, a mongrel form of art' which he believed was a compromise between a shrubbery and an herbaceous border – a sort of floristic synergy. An herbaceous border was just that – a hardy herbaceous display – while of course a shrubbery was filled only with shrubs.

What Robinson wanted to encourage was a different way of growing plants, a more free and easy style rather than the regimented whaleboning of corset-control that so dominated the High Victorian garden. So contrived was its style and so sanitised its plants that the whole system had found its way into a cul-de-sac from which in design terms there was no way forward. Robinson's ideas had been crystallised originally by John Ruskin (1819–1900), the author, art critic and befriender of the Pre-Raphaelite brotherhood, whose writings drew attention 'to what in nature was most wonderful and lovely'.

A second Robinson book which greatly influenced Britain's gardening life – and still does – was *The English Flower Garden*, with a sub-title of *Home Grounds of Hardy Trees and Flowers*. This was first published in 1883 and proved immensely popular. There have been sixteen editions (the last in 1956) and innumerable reprints. In it, Robinson pulls no punches about his disregard for carpet bedding and mosaic culture, thinking it crude and absurd 'and in its extreme expressions is ridiculous'. In all his writings he was for ever keen to show his readers the ceaseless store of delight and happiness that there was in creating 'artistic' gardens, and the sense of spiritual renewal that could be discovered in making a garden glow with beauty. He probably expected the nation to react enthusiastically, since it was perpetually hungry for horticultural inspiration, but he surely would have been astounded to see the influence his thrilling creed of creativity has had on the development of the modern garden.

Plant Introductions in the period 1800–1899

1800 *Allium cernum* Nodding or Wild Onion. E. USA.

Vanilla planifolia (syn. *V. fragrans*) Vanilla. West Indies, Mexico; see p.160.

c. 1800 *Fallopia japonica* Japanese Knotweed. Japan; see p.13.

Matthiola longipetala **subs.** *bicornis* (syn. *M. bicornis)* Night-scented Stock. Greece to S. W. Asia; see p.69.

1801 *Dicksonia arborescens* Tree Fern. St Helena.

1802 *Geranium ibericum* Caucasus, N. E. Turkey, N. Iran.

Scabiosa caucasica Scabious, Pincushion Flower. N.E. Turkey, Caucasus, N. Iran.

Tanacetum coccineum (syn. *Chrysanthemum coccineum, Pyrethrum occineum, P. roseum*) Painted Daisy, Pyrethrum. Caucasus, S.W. Asia.

1803 *Achillea filipendulina* (syn. *A. eupatorium*) Yarrow. Caucasus.

Astelia nervosa New Zealand.

Campanula sarmatica Caucasus.

1804 *Begonia grandis* subsp. *evansiana* China to Malaysia, Japan.

Cunninghamia lanceolata (syn. *C. lanceolata* var. *sinensis*) Chinese Fir. China.

Dahlia coccinea Mexico.

Dahlia pinnata Mexico.

Dahlia rosea Mexico.

Euonymus japonicus Japanese Spindle Tree. Japan, China, Korea.

Kerria japonica Jew's Mantle. China, Japan.

Lilium lancifolium (syn. *L. tigrinum*) Tiger Lily. E. China, Korea, Japan; see p.185.

Lilium monadelphum (syn. *L. szovitsianum*) Turk's-cap Lily. N.E. Turkey, Caucasus.

Nepeta racemosa (syn. *N. mussinii*) Catmint. Caucasus, Turkey, N. and N. W. Iran.

Pittosporum tenuifolium Kohuhu. New Zealand.

Pittosporum tobira Japanese Mock Orange. China, Korea, Japan.

Pittosporum viridiflorum South Africa; see p.177.

1805 *Puschkinia scilloides* Lebanese Squill. Caucasus, Turkey, Lebanon, N. Iran, N. Iraq.

Tetrapanax papyrifer (syn. *Aralia papyrifer, Fatsia papyrifer, Tetrapanax papyriferus*) Rice-paper Plant. S. China, Taiwan.

1806 *Calycanthus fertilis* Carolina Allspice, Spice Bush. Collected by John Lyon from S.E. USA.

1807 *Rosa banksiae* 'Alba Plena' (syn. *R. banksiae* 'Alba', *R. banksiae* 'Alba Plenia') Double White Banksian, Banksia Rose. China.

1808 *Paeonia lactiflora* (syn. *P. albiflora, P. japonica* of gardens) E. Siberia, Tibet China. Reintroduction (see p.181) from Canton by Reginald Whitley, a nurseryman from Fulham in London.

Rosa acicularis Arctic Rose. Canada, N. USA, N.E. Asia. Reintroduced (see p.131).

1809 *Trifolium stellatum* Starry Clover. W. Mediterranean. Reintroduced (see p.131).

1810 *Galega orientalis* Goat's Rue. Caucasus.

c. 1810 *Rhododendron arboreum* Himalayas, Kashmir, S.W. China, S. India, Sri Lanka (see illustration p.236).

1811 *Oenothera macrocarpa* (syn. *O. missouriensis*) Ozark Sundrops. C. USA. Found by Thomas Nuttall, Curator of the Botanic Garden at Harvard (1822–34), and John Bradbury, plant collector for Liverpool Botanic Gardens, on Captain Hunt's expedition up the Missouri.

1812 *Gaillardia aristata* Blanket Flower. W. Canada, W. USA to New Mexico.

Pieris floribunda S.E. USA; see p.229.

1813 *Yucca glauca* Central USA. First offered for sale in a catalogue of 'curious American plants' published by John Fraser (fl. 1790s–1860s) who had a nursery in Sloane Square, London; see 1824.

1814 *Agave americana* (syn. *A. altissima*) Century Plant. Mexico. Revd James Yates (1789–1871) grew the first plant to flower, in Salcombe, Devon. He made a collection of cycads.

1815 *Ficus elastica* India Rubber Plant, Rubber Tree, India Rubber Fig, Rubber Plant. E. Himalayas, Assam, Burma, Malaya, Java; (see p.25).

Sinningia speciosa (syn. *Gloxinia speciosa*) Florists' Gloxinia. Brazil

1816 *Camellia welbankiana* China.

Dicentra spectabilis Bleeding Heart, Dutchman's Breeches, Lyre Flower. Siberia, China, N. Korea; see p.178.

Olearia paniculata Daisy Bush. Australia.

Wisteria sinensis (syn. *W. chinensis*) Chinese Wistaria. China.

1817   *Helleborus purpurascens* S.E. Poland, E. and N. Hungary, W. Ukraine, Slovakia.

Ornithogalum congibracteatum (syn. *O. caudatum*) False sea onion. N. and E. Cape, South Africa (see illustration p. 94).

1818   *Beaufortia sparsa* Swamp Bottlebrush. Western Australia; see p.155.

Ceanothus azureus (syn. *C. coeruleus*) Azure Ceanothus. Mexico. Noted by Robert Brown in the garden of Malmaison home of the Empress Josephine, and brought over to grow on at Loddiges Nursery.

c. 1818   *Pelargonium x ardens* (*P. fulgidium x P. lobatum*) Glowing Geranium. Garden origin; see p.156.

1819   *Hedychium gardnerianum* Kahili Ginger. N. India, Himalayas.

1820   *Dombeya wallichii* E. Africa, Madagascar; see p.140.

Eryngium giganteum Miss Willmott's Ghost. Caucasus, Iran; see p.244.

Linum grandiflorum Flowering Flax. N. Africa; see p.163.

Malva alcea var. *fastigata* Hollyhock Mallow. S. Europe.

1821   *Brassica pekinensis* Chinese Cabbage. China.

1822   *Calceolaria corymbosa* Chile.

Calceolaria integrifolia (syn. *C. rugosa*) Chiloe Island, Chile, Mexico.

Dryas drummondii North America.

Grevillea rosemarinifolia New South Wales, Victoria, Australia.

1823   *Boronia alata* Western Australia.

Fuchsia magellanica (syn. *F. macrostemma*) Peru, Chile, Argentina to Tierra del Fuego; see p.153.

Lonicera hirsuta E. North America.

Mahonia aquifolium Oregon Grape. W. USA.

1824   *Boronia anemonifolia* Australia, Tasmania.

Boronia fraseri New South Wales. Named for John Fraser, who kept a nursery in Sloane Square in London.

Boronia polygalifolia Australia, Tasmania.

Cotoneaster frigidus Himalayas. The parent of some vigorous hybrids.

Erysimum asperum Western Wallflower. British Columbia to Washington, Minnesota, Kansas; see p.40.

1825   *Eschscholtzia californica* California Poppy. Oregon to coastal California.

Laburnocytisus adamii (*Chamaecytisus purpurens x Laburnum anagyroides*) France; see p.101.

Ribes sanguineum Flowering Currant. W. North America. Found by David Douglas at Fort Vancouver on the banks of the Columbia River, Oregon. The fort was then under construction by the Hudson's Bay Company.

1826   *Anemone vitifolia* Afghanistan to W. China, Burma.

Calceolaria purpurea Chile.

Clematis montana Great Indian Virgin's Bower. W. and C. China, Himalayas.

Gaultheria shallon Sacal, Shallon. W. North America.

Lupinus polyphyllus W. North America.

Paeonia brownii American Wild Peony, Western Peony. W. North America.

Pinus ponderosa Ponderosa Pine, Western Yellow Pine. W. North America.

Pseudotsuga menziesii (syn *P taxifolia, P. douglasii*) Douglas Fir. British Columbia to California.

Verbena incisa S. South America.

1827   *Camassia quamash* (syn. *C. esculenta*) Quamash. Canada, USA.

Drimys winteri (syn *Wintera aromatica*) Winter's Bark. Mexico, Chile, Argentina; see p.114.

Herbertia pulchella Chile, S. Brazil, Texas.

Penstemon cardwellii Washington, Oregon. Collected by David Douglas.

Pinus lambertiana Sugar Pine. California, Oregon.

Ribes munroi Flowering Currant. W. North America.

Verbena peruviana (syn. *V. chamaedrifolia, V. chamaedrioides*) S. Brazil to Argentina.

1828   *Garrya elliptica* Silk-tassle Bush. California, Oregon. Found by David Douglas.

Gaultheria mucronata (syn. *Pernettya mucronata*) Tierra del Fuego, Falkland Islands, Chile, Argentina. Reintroduction (see p.169).

1829   *Bougainvillea spectabilis* Brazil (see p.169)

Musa acuminata (syn *M. cavendishii*). Banana, Plantain. S. E. Asia to N. Australia.

Umbellularia californica California Laurel, Headache Tree. S. Oregon, N. California.

1830   *Abies grandis* Giant Fir, Grand Fir. British Columbia, Oregon to Idaho.

Erythronium grandiflorum Dog's-tooth Violet. W. USA.

Hosta plantaginea China.

Hosta sieboldii (syn. *H. albomarginata*) Japan.

Wisteria floribunda Japanese Wistaria Japan.

1831   *Abies procera* (syn. *A. nobilis*) Noble Fir. Oregon, Washington.

Alstroemeria aurea (syn. *A. aurantiaca*) Chile.

Clianthus puniceus Glory Pea, Lobster Claw, Parrot's Bill. North Island of New Zealand.

Picea sitchensis Sitka Spruce. W. North America; see p.80.

Pinus contorta Beach Pine, Lodgepole Pine, Shore Pine. N.W. North America.

1832   *Pinus coulteri* Big-Cone Pine, Coulter Pine. California, Mexico.

Pinus sabiniana Digger Pine. California.

1833   *Pinus radiata* Monterey Pine, Radiata Pine. California.

1834   *Verbena phlogiflora* (syn. *V. tweedii*) S. South America.

c. 1834   *Tradescantia fluminensis* Wandering Jew. Brazil to N. Argentina; see p.133.

1835   *Cosmos atrosanguineus* (syn. *Bidens atrosanguinea*) Mexico; see p.17 and 178.

Cosmos diversifolius (syn. *Bidens dahlioides)* Mexico; see p.177.

Phlox drummondii Annual Phlox. E. Texas; see p.157.

1836   *Oxalis tetraphylla* (syn. *O. deppei*) Good Luck Plant, Lucky Clover. Mexico.

1837   *Amherstia nobilis* Orchid Tree, Pride of Burma. Burma.

Ceanothsus thyrsiflorus Blue Blossom. California, Oregon; see p.156.

Mandevilla laxa (syn. *M. suveolens, M. tweediana*) Chilean Jasmine. Peru, Bolivia, Argentina.

Verbena platensis S. South America.

1838   *Cupressus macrocarpa* Monterey Cypress. California; see p.72.

Fatsia japonica (syn. *Aralia japonica, A. sieboldii*) Japanese Aralia Japanese Fatsia. South Korea, Japan. Introduced by Philipp von Siebold.

1839   *Herbertia drummondiana* Texas.

1840   *Lathyrus pubescens* Chile, Argentina; see p.237.

Mandevilla splendens (syn. *Dipladenia splendens*) S. E. Brazil.

Passiflora actinia Passion Flower. Brazil.

1841   *Alstroemeria chilensis* Peruvian Lily. Chile.

Alstroemeria inodora Peruvian Lily. Chile.

Begonia coccinea Angelwing Begonia. Brazil.

Cycnoches pentadactylon Swan-necked Orchid. Brazil.

*Nematanthus strigillos (*syn. *Hypocyrta strigillosa)* Brazil.

Rondeletia longiflora (syn. *Hindsia longiflora*) Brazil.

Vellozia squamata (syn. *Barbacenia squamata)* Brazil.

1842   *Cuphea cordata* Peru.

1843   *Desfontainia spinosa* (syn. *D. hookeri*) Colombia to Straits of Magellan.

Drimys lanceolata (syn. *D. aromatica, Tasmania aromatica*) Mountain Pepper. S.E. Australia. Tasmania; see p.115.

Manettia luteorubra (syn. *M bicolor, M. inflata*) Brazilian Firecracker. Uruguay, Paraguay.

1844   *Abelia chinensis* (syn. *A. rupestris*) China.

Anemone hupehensis var. *japonica* Garden origin. China.

Forsythia viridissima China.

Kohleria bogotensis (syn. *Isoloma bogotense*) Colombia.

Lardizabala biternata Chile. Reintroduced 1927 (see p.237).

Pelargonium 'General Tom Thumb' Garden origin.

1845   *Cuphea ignea* (syn. *C. platycentra*) Cigar Flower. Mexico to Jamaica.

Edgeworthia chrysantha (syn. *E. papyrifera*) Paper Bush. China.

Victoria amazonica Amazon Water Lily, Royal Water Lily. Tropical South America.

1846   *Embothrium coccineum* Chilean Fire Bush, Flame Flower. S. Chile.

Fortunella margarita Kumquat. China.

1847   *Erysimum x allionii* Siberian Wallflower. Garden origin; see p.41.

Lapageria rosea Chilean Bellflower. Chile.

1848   *Escallonia macrantha*. Chiloe Island, Chile Introduced by William Lobb.

1849   *Berberis darwinii* Chile, Argentina.

Rhododendron campylocarpum East Nepal, Sikkim. Introduced by Sir Joseph Hooker.

Rhododendron falconeri Sikkim. Introduced by Sir Joseph Hooker.

Rhododendron thomsonii Himalayas. Introduced by Sir Joseph Hooker.

1850   *Gaultheria nummarioides* China, Himalayas.

c. 1850   *Brachychiton acerifolius* (syn. *Sterculia acerifolia*) Flame Kurrajong, Flame Tree. Queensland, New South Wales.

Erysimum x kewense 'Harpur Crewe' (syn. *Cheiranthus*

cheri 'Harpur Crewe') Garden origin; see p.41.

Lavandula x intermedia (*L. angustifolia x L. latifolia*) English Lavender. Garden origin; see p.105.

1851 *Eucryphia cordifolia* Ulmo. Chile. Discovered on the island of Chiloe.

Fremontodendron californicum Flannel Bush. California, Arizona.

Torreya californica California Nutmeg. California. Introduced by William Lobb. See also p.248.

1852 *Ligustrum sinense* Chinese Privet. China. Collected by Robert Fortune.

c.1852 ***Chaenomeles cathayensis*** Flowering Quince. China; see p.61.

1853 ***Astelia chathamica*** (syn. *A. nervosa* var. *chathamica*) Chatham Islands.

Chamaecyparis nootkatensis (syn. *Cupressus nootkatensis*) Nootka Cypress. Alaska to Oregon; see p.72.

Sequoiadendron giganteum Wellingtonia, Big Tree, Giant Redwood, Sierra Redwood. California.

Sequoia sempervirens Coastal Redwood. California, Oregon.

1854 ***Chamaecyparis lawsoniana*** (syn. *Cupressus lawsoniana*) Lawson Cypress. W. North America.

Dendromecon rigida Tree Poppy. California, Mexico.

1855 *Cortaderia selloana* (syn. *C. argentea*) Pampas Grass. Temperate South America.

Kohleria amabilis (syn. *Isoloma amabile*) Colombia.

1856 *Osmanthus fragrans* Brazil. Reintroduction (see p.170).

Syringa oblata var. *dilatata* Lilac. Korea. Introduced by Robert Fortune after he had seen it growing in a garden in Shanghai.

1857 *Penstemon jaffrayanus* California. Introduction by William Lobb.

1858 *Brachychiton discolor* Kurrajong, Queensland Lacebark. Queensland, New South Wales.

Clianthus formosus (syn. *C. dampieri*) Glory Pea, Sturt's Desert Pea, Handsome Pea. N. Australia.

Myosotidium hortensia (syn. *M. nobile*) Chatham Islands Forget-me-not. Chatham Islands.

Pieris formosa China, Himalayas; see p.230.

1859 *Buxus microphylla* Small-leaved Box. Japan; see p.29.

Cinchona succirubra Red Peru Bark. Peru, Ecuador.

Eucryphia glutinosa Chile.

c. 1859 ***Berberis linearifolia*** Chile, Argentina; see p.237.

Fabiana violacea Chile; see p.237.

Mutisia decurrens Chile.

Mutisia oligodon Chile, Argentina; see p.237.

1860 ***Cinchona calisaya*** Yellow Bark. Peru and Bolivia.

Hebe hulkeana New Zealand.

Lapageria alba Chile.

1861 ***Abies homolepis*** (syn. *A. brachyphylla*) Nikko Fir. Japan.

1862 ***Clematis x jackmanni*** (*C. lanuginosa, C. x hendersonnii x C. viticella*) Garden origin. Received the First-Class Certificate from the RHS in 1863. The first of the many cross-bred Clematis to be produced by George Jackman & Son of Surrey.

Lilium auratum Golden-rayed Lily. Japan; see p.49.

Parthenocissus tricuspidata Boston Ivy. Japan, China, Korea.

1863 ***Pinus aristata*** Bristle Cone Pine. Arizona, New Mexico, Colorado.

1864 *Aquilegia caerulea* W. North America. Seed first collected from the Rocky Mountains.

1866 ***Dianella tasmanica*** S.E. Australia, Tasmania.

1867 *Epimedium perralderianum* Barrenwort, Bishop's Mitre. Algeria.

Gunnera manicata (syn. *G. brasiliensis*) Colombia to Brazil.

1868 *Osmanthus decorus* (syn. *Phillyrea decora*) Georgia, N.E. Turkey.

1869 ***Chaenomeles japonica*** (syn. *C maulei*) Japonica, Japanese Quince, Maule's Quince. Japan; see p.61.

Hedera colchica Bullock's Heart Ivy, Persian Ivy. N. Iran, Caucasus.

1870 ***Boronia megastigma*** Brown Boronia, Scented Boronia. Western Australia.

Boronia serrulata Sydney Rock Rose. New South Wales.

c. 1870 *Dianthus* 'Old Blush' (syn. 'Souvenir de la Malmaison') Garden origin.

1872 *Griselinia littoralis* Broadleaf. New Zealand. The plants at Kew were all killed during the severe winter of 1908–9.

Miltonia vexillaria Ecuador, Colombia.

1873 *Allium unifolium* (syn. *A. murrayanum*) Oregon, California.

Bromelia balansae Heart of Flame. Colombia, Brazil, Bolivia, Paraguay, N. Argentina.

1874 *Alchemella mollis* Lady's Mantle. Caucusus, Turkey.

1875   *Cornus kousa* Dogwood. Korea, Japan. It was first discovered on the Japanese island of Kyushu. Received a First-Class Certificate from the RHS in 1892. See also p.234.

Lilium pardalinum Panther Lily. California; see p.50.

Miltonia endresii Central America.

1876   *Galanthus reginae-olgae* (syn. *G. corcyrensis*) Snowdrop. Sicily, Greece.

Hosta sieboldiana Japan.

1877   *Actinidia kolomikta* E. Siberia, E. Asia; see p.227.

Alstroemeria pelegrina 'Alba' Lily of the Incas. Peru, Chile.

Spathoglottis petri Pacific Islands.

1878   *Abies sachalinensis* Sakhalin Fir. Sakhalin Islands, Kurile Islands, N. Japan.

Tulipa kaufmanniana Water-lily Tulip. Kazakhstan, Uzbekistan, Tajikistan, Kyrgyzstan. Found by Albert Regel, physician in Turkestan and son of Edward Regel, Director of Imperial Botanic Gardens in St Petersburg.

1879   *Abies mariesii* Maries Fir. Japan.

Abies veitchii Veitch Fir. Japan.

Burbidgea nitida N.W. Borneo.

Hamamelis mollis Chinese Witch Hazel. W. & W. C. China.

Hydrangea 'Mariesii' Lacecap Hydrangea. Japan.

Lilium speciosum var *gloriosoides* Taiwan, E. China, Japan.

Primula rosea Rosy Himalayan Primula. N.W. Himalayas.

Ranunculus lyallii Mount Cook Lily, Giant Buttercup. New Zealand.

1880   *Enkianthus campanulatus* Japan.

Hebe cupressoides South Island of New Zealand.

Hydrangea 'Veitchii' Lacecap Hydrangea. Japan.

Papaver rhoeas Shirley Series. Garden origin; see p.33.

Primula obconica China.

1881   *Nepenthes rajah* Tropical Pitcher Plant, Monkey Cup. Borneo.

1882   *Heuchera sanguinea* Coral Bells. N. Mexico, Arizona.

Prunus 'Fugenzo' (syn *P.* 'James H. Veitch') Japan.

1883   *Fallopia baldschuanica* (syn. *Bilderdykia baldschuanica, Polygonum baldschuanicum*) Mile-a-Minute Plant, Russian Vine. Tajikistan, Afghanistan, W. Pakistan; see p.13.

Philadelphus microphyllus Mock Orange. S.W. USA.

1884   *Arisaema amurense* N. Asia.

1885   *Viburnum carlesii* Korea, Tsushima Island, Japan. Named to honour William Carles of the British Consular Service in China. He also collected plants in Korea 1883–5.

1886   *Hebe odora* (syn. *H. anomala, H. buxifolia* of gardens) New Zealand.

Olearia macrodonta Arorangi, New Zealand Holly. New Zealand.

Olearia traversii Chatham Islands

1887   *Eucomis pallidiflora* Giant Pineapple Flower. South Africa; see p.152.

1888   x *Cupressocyparis leylandii* (syn. *Cupressus leylandii*) Leyland Cypress. Garden origin; see p.72.

Ostrowskia magnifica Giant Bellflower. Uzbekistan, Tajikistan. Collected for the Veitch Nursery. Named for Michael von Ostrowsky, Russian Minister of Imperial Domains, c. 1884, a patron of botany.

1889   *Grewia biloba* (syn. *G. parviflora*) Korea; see p.122.

Picea omorika Serbian Spruce. Bosnia, Serbia.

1890   *Allium cyaneum* (*A. purdomii*) China; see p.231.

Dicentra macrantha E. China It was originally found by Dr Augustine Henry and is rare both in the wild and in cultivation.

c. 1890   *Didymocarpus burkei* Malaya.

1892   *Nemesia strumosa* South Africa. It was introduced by Messrs Sutton of Reading.

1893   *Vitis coignetiae* Korea, Japan. Named for a Frenchwoman, Mme Coignet. First introduced by Anthony Waterer's nursery through the East India merchants Jardine and Matheson.

1894   *Gerardia tenuifolia* Mexico; see p.102.

Rosa moyesii W. China. Introduced by Antwerp E. Pratt who first found it at 2,769 m (9,000 ft) while climbing in the Tibetan border region. It was re-introduced in 1903 by Dr. E. H. Wilson and first flowered in Britain at the Veitch Nursery in 1908. Named after Revd J. Moyes, a missionary in W. China.

1895   *Corydalis scouleri* N.W. North America. Named for John Scoular, Professor of Geology at Glasgow.

1896   *Muscari botryoides* Grape Hyacinth. C. and S. E. Europe.

1897   *Bulbophyllum curtisii* (syn. *Cirrhopetalum curtisii*) Molucca Islands of Indonesia.

1898   *Acca sellowiana* (syn. *Feijoa sellowiana*) Fruit Salad Bush, Pineapple Guava. Brazil, Uruguay.

1899   *Pyracantha angustifolia* Firethorn. W. China. Introduced to Kew by a Lieutenant Jones.

A *Plethora of Plants*

1900–2000

Significant dates

1901	Edward VII (1901–10)	1931	National Trust for Scotland founded
1904	Miss Ellen Willmott of Warley Place in Essex (1858–1934) becomes first woman to be elected to the Linnaean Society	1936	RHS Rock Gardens & Rock Plants Conference
			Edward VIII abdicates; George VI (1936–52)
1906	Hidcote Garden in Gloucestershire begun by Lawrence Johnston	1938	RHS Ornamental Flowering Trees and Shrubs Conference
1908	*Colour in the Flower Garden* by Gertrude Jekyll (1843–1932) published	1939–45	Second World War
	First Allotment Act	1951	*RHS Dictionary of Gardening* (4 vols.) published; editor Fred J. Chittenden
1909	John Innes Horticultural Institution founded	1952	Elizabeth II (1952–)
1910	George V (1910–36)	1956	*Flowers and their Histories* by Alice Coats (1905–78) published
	The Genus Rosa by Ellen Willmott published	1965	The Garden History Society founded
1912	First RHS Flower Show held in Chelsea	1970–80	*Trees and Shrubs Hardy in the British Isles* by W.J. Bean, revised in 4 vols by D. L. Clarke
1913	Suffragettes invade Kew Gardens, attack the Orchid House and burn down the Tea Pavilion	1972	*Hilliers' Manual of Trees and Shrubs* published
	RHS begins National Diploma of Horticulture	1974	Town and Country Amenities Act (includes first direct reference to historic gardens in British legislation)
1914	Panama Canal opened		
1914–18	First World War	1979	Founding of National Council for the Conservation of Plants and Gardens
	Trees and Shrubs Hardy in the British Isles (2 vols.) by W. J. Bean (1863–1947) published	1988	International Climate Conference held at Rio de Janeiro, Brazil
1927	Institute of Landscape Architects founded; first President Thomas Mawson (incorporated into Landscape Institute in 1978)	1989	First Hampton Court International Flower Show
	National Gardens Guild formed; first President Lord Noel Buxton	1996	*RHS A–Z Encyclopedia of Garden Plants* published: editor-in-chief Christopher Brickell
1928	RHS Primula Conference	2000	Opening of National Botanical Garden of Wales in Carmarthenshire
1929	Alpine Garden Society founded		Building of the Eden Project, Cornwall

Left: In the two World Wars allotments provided a source of much-needed food. In the Great War as a result of the 'Every-man-a-Gardener' campaign, there were one and a half million plots. The number of allotments in the Second World War dropped slightly but the 'Dig for Victory' campaign was so successful that it was estimated to have produced well over a million tons of food.

As WE GAZE INTO THE INFINITY of the third millennium and the beginnings of the twenty-first century, there surely cannot have been in all our garden history such a hundred years as that of the twentieth century, the last part of which began to feel like the final day of the Chelsea Flower Show. By its end, the plant gathering of previous centuries had become anachronistic and our whole attitude to the natural environment both in Britain and around the world had undergone a fundamental change. This change in emphasis first began after the watershed of the First World War and slowly developed as the century progressed. Today, stewardship and conservation of the environment govern all aspects of the plant world, an attitude that is far removed from that of the beginning of the century, with its restless, pretentious labour-intensive gardens in which almost the most important consideration was how many thousand bedding plants would be needed to create the increasingly complicated patterns.

Yet alongside the dinosaur of the plant equivalent of 'ribbon development' came the naturalistic plant philosophy propounded by William Robinson (1838–1935). His theories of growing in a more relaxed way native plants mixed with introduced material, existed side by side with the rigidity of the established order for at least two or three decades early in the century. Converts to these exciting new ideas were already making their mark, and a pioneer in this informal but artistic planting style was Gertrude Jekyll (1843–1932), whose own gardening ideas greatly expanded the Robinsonian theories. Her writings on style, plants and design held the gardening world in thrall for most of this period – her most important book was *Colour in the Flower Garden*, published in 1908 – and just as the *beau monde* had embraced the earlier stylistic decrees of the Victorian age, so this revolution, begun by Robinson late in the previous century, gradually swept through our gardens and landscapes.

At about the same time as Robinson's gardening ideas were burgeoning, a similar revolutionary theory was occurring in the allied field of town planning, where ideas first proposed four centuries earlier by Thomas More in his book *Utopia* were to find expression in a new social order. In 1888, Edward Bellamy (1850–98), an American writer, had theorised about fundamental changes which were occurring in society. In turn, his writings were read in Britain by Sir Ebenezer Howard (1850–1928) who, profoundly influenced by the American's ideas, began to develop his own far-reaching theory on improving mass housing. At the beginning of the

nineteenth century, the population of mainland Britain had been just over ten million; a hundred years later it was forty million. The soaring increase in population, combined with the rapid industrialisation of towns and cities, made whole areas scrabbling stew-pots of stress. Concern for improving the daily life of the worker was one of the foremost ideals at this time, and Sir Ebenezer's spectacular suggestion was to create what he called 'garden cities'.

The original idea was that these were to be built not as suburbs attached to already existing cities or large towns but as settlements in their own right, surrounded completely by countryside; that particular scheme, however, was soon abandoned. Sir Ebenezer published his ideas in 1898 in his book *Tomorrow*, and the following year founded The Garden City Association. His original theories and ideas were elaborated, and by 1903 Letchworth Garden City in Hertfordshire had begun to be developed. Hampstead Garden Suburb was started in 1907, and is the epitome of the garden city, with its principles of domestic and public planning. This ethos of public benefit to improve the quality of the 'labouring poor' was also expressed in the passing of the first Allotment Act in 1908, which provided for local authorities to make land available in urban areas. The idea of space in which flowers and vegetables could be grown was instantly appealing, and a further six Acts were to follow, the last one in 1950.

One of the architects involved in the planning of Hampstead Garden Suburb was Sir Edwin Lutyens (1869–1944), who designed two churches and the Institute in the centre of the development. He and Gertrude Jekyll had already begun their long and happy collaborative friendship in garden and landscape design. They had first worked together in 1893 on designing a garden for Adeline, widow of the 10th Duke of Bedford, at her home, Woodside, on the Chenies estate in Buckinghamshire. In 1904 they were asked by the Honourable Edward Portman to create a formal garden at his home, Hestercombe in Somerset. The result of their collaboration, the wonderfully contrived Great Plat, with its views to the Blackdown Hills in the distance, was a most successful combination of Lutyens' crisply defined stonework of rills, fountains, steps and walls, and Jekyll's beguiling use of plants. What few people realised at the time was that to the north of Hestercombe House there lay the most incredible landscape filled with follies, temples, a cascade and lakes. Created by Coplestone Warre Bampfylde (1720–91), who inherited the estate from his father in 1750, it had surprisingly been allowed to drift into obscurity. It lay

Sir Edwin Lutyens and Gertrude Jekyll were commissioned in 1904 to design the garden at Hestercombe near Taunton, Somerset.

totally overgrown and almost forgotten until Philip White, the director of the Hestercombe Gardens Project, uncovered it in the 1990s. This garden has metamorphosed into the most beautiful and serene landscape, totally in harmony with the Lutyens and Jekyll creation.

Two years later, in 1906, Major Lawrence Johnston (1871–1958) began creating his garden at Hidcote Manor in Gloucestershire. An American by birth, he was a superb and knowledgeable plantsman. His own designs for this garden and for his home at Menton in the south of France made bold architectural statements using plant material. Crisp hedges of yew, holly and beech were planted to divide the spaces instead of using stone or brickwork. This horticultural idea of small 'rooms' within the garden devoted to a particular theme was not new, but now incorporated modern developments: for instance, a single-coloured garden (the

most well known being the White Garden created in the 1930s at Sissinghurst Castle in Kent by Vita Sackville-West), or a water garden, or planting from only one continent, or to a design which reflected a country – French and Dutch gardens being the most popular. These contained gardens could be sumptuously planted in great detail but, as at Hidcote Manor, the strong overall design for the whole has been the linchpin of its great success.

The expansion of 'gardening for all' really did mean just that, because for the first time, women and children were being allowed or encouraged – depending on your point of view – to participate in garden work. Although women for centuries had been considered good weeders (there are records of payments for this work as early as the fourteenth century), they rarely seem to have been allowed to do much else.

The first gardening book written especially for women was the *Countrie Housewife's Garden*, published in 1617 by William Lawson (fl. 1570s-1620s). In it he wrote of planting and designs, and gave advice about herbs and vegetables. There was then almost a hundred-year gap before Charles Evelyn wrote in 1707 *The Ladies' Recreation*. This book concentrated more on flowers and their beauty rather than on gardening practicalities and must have been popular, since it was reprinted twice in 1717 and again in 1719. The first book written by a woman especially for female readers was by Jane Loudon (1807–58) in 1840, and was entitled *Gardening for Ladies*; she followed this in 1841 with the immensely popular *The Ladies' Companion to the Flower-Garden*.

The idea of 'women' getting their hands dirty had never bothered anyone before but the thought of 'ladies' doing manual and apparently menial work was totally abhorrent in a male-dominated society. This was vividly illustrated at the Empire's home of gardening, when Kew very reluctantly agreed to admit women as journeymen-gardeners in 1896. Just three braved that male bastion. The Director at the time was Sir William Turner Thiselton-Dyer (1843–1928), an autocrat who insisted that everyone employed at Kew should wear a uniform. He designed his own fetching outfit, but could never persuade the Kew Fire Brigade, all volunteers, to even consider the notion.

The three brave females naturally had to have a uniform too, and were kitted out in clothing 'unlikely to provoke their male colleagues', which meant they laboured in the most unbecoming – and, one would have thought, the most uncomfortable – attire: brown bloomers, woollen stockings, waistcoat, jacket and peaked cap; certainly not an outfit to excite any of the opposite sex. Sir Joseph Hooker, then in his eighties, wrote in 1902 of his disgust at the very idea of employing female gardeners: 'not ladies in any sense of the word', he called them. During the First World War, however, at Kew as in so many other places of employment, women replaced conscripted male gardeners, and by 1917 the number of women employed had risen to more than thirty.

In 1902 the first gardening book written specifically for children was published, aimed at encouraging them into the garden. It was called *Children's Gardens* and was written by Lady Alicia Rockley (1865–1941). This first tentative step was followed in 1908 by Gertrude Jekyll's *Children and Gardens*. Although children and gardening are not necessarily always compatible, Jekyll wrote most engagingly and simply of the adventures they might have in the garden, especially where

water and mud are concerned; she also drew light-hearted attention to the actual make-up of plants and flowers. This theme of encouraging young people into the great outdoors was later carried on by the delightful poetry and illustrations of *The Flower Fairy* books written by Cecily Mary Barker during the 1920s – and still in print today.

All these new ideas were able to take advantage of the floral feast which was cascading on to our shores, and for a short time, in the early 1900s, England was ablaze with a pyrotechnic 'flower-work' display unequalled before or since. It had begun during the previous century with the relaxation of the attitude of the Chinese and Japanese authorities to Europeans travelling and trading in their countries. As we have seen, reports and then plants began to filter back, whetting the gardening appetite of the horticultural world, and, just as in the seventeenth century, when the four Johns (Parkinson, the two Tradescants and Evelyn) influenced their garden world, so now, as the Edwardian 'glitterati' socialised around their carpet-bedded gardens, four plant hunters, each of whom was to have a revolutionary influence on the garden and landscapes of the new Elizabethan Age, ranged over south-east Asia.

Two of them, Ernest Wilson and George Forrest, were born within three years of each other in the 1870s and died within two years of each other in the early 1930s. Both men collected plants from China and Tibet, each having myriad 'Boys' Own' adventures. They were acquainted with one another but never collected together. The third plant hunter was a Yorkshireman, Reginald Farrer, an inspirational botanical writer whose main interest was alpine and rock plants. Before the First World War he made a journey to China (accompanied by William Purdom, the last collector for the Veitch Nursery; see p.208), and later, in 1919, a second trip, this time to Burma where he subsequently died. The fourth plant hunter, Frank Kingdon Ward, began his exploration of south-east Asia just before Reginald Farrer but continued into the 1950s, making some twenty-five major plant-hunting sorties in all, mostly around the borderlands of south-west China, Burma, India and Tibet. Fortunately, the four men published accounts of their expeditions and their plant discoveries, so we are in a supremely privileged position to acknowledge the debt we owe them when we plant their botanical finds in our gardens.

All four journeyed into what was, to the British, virtually unknown botanical country. Only the French missionaries of

TWO MISSIONARY PLANT HUNTERS

The Jesuit missionary **Père Jean Pierre Armand David (1826–1900)** was sent to China in 1862 and, amazingly, made three very early explorations into the interior of China, Tibet and Inner Mongolia. He was responsible for the introduction into France of about 250 new plants. He was the first westerner to see the Giant Panda, and also wild silkworms (shades of James I and the planting of mulberry trees in the seventeenth century; see p.135). A Chinese deer was also named in his honour, *Elaphurus davidianus*, or Père David's Deer, which he discovered in the Emperor's hunting park in Peking (Beijing) in 1865. Apart from plants, his consuming interest was birds, and he published an illustrated account of Chinese ornithology. He retired to France in 1874.

Père Jean Marie Delavay (1834–95) was a man of the mountains, having been born in the Haute-Savoie in France. He went to China in 1867, but unlike Père David, he never returned to France. He lived first in the coastal province of Guangdong in south-east China, and then for many years, until the time of his death, in north-west Yunnan. He was a tireless explorer, constantly collecting plants to make up his herbarium, in which there were eventually over 200,000 specimens, all of which he sent to the Museum d'Histoire Naturelle in the Jardin de Plantes in Paris. Many were completely new genera, and 1,500 were found to be new species. He also sent an overwhelming number of packets of seeds home for growing on; some of these unfortunately apparently remained unopened. In his honour, a number of shrubs and trees have the name delavayi attached to them, including a Magnolia which Ernest Wilson introduced to Britain in 1900.

Davidia involucrata painted by Siriol Sherlock. Père David, in whose honour this tree is named, arrived in Beijing in 1862 to work as a Lazarist priest, but was exempted from his missionary duties to pursue his natural history work.

the eighteenth and nineteenth centuries had penetrated into the hinterland of China to any extent, and little of what they had discovered was known in Britain. Several of them, however, made careful study and were dogged and keen collectors of the natural world. Some were very keen botanists and two in particular, Père Armand David and Père Jean Delavay, made significant contributions to our botanical knowledge of the area.

Père David's most famous botanical discovery was a tree which was eventually named *Davidia involucrata*, the Pocket-handkerchief Tree, Ghost Tree or Dove Tree, so called because of its large white bracts, which look for all the world like either a white handkerchief, ghostly figures or a fluttering white dove. He sent back the botanical description and an herbarium specimen, and it was promptly named in his honour.

Some years after Père David left China, Dr Augustine Henry (1857–1930) arrived in the same area. Born in County Antrim, Henry qualified in Edinburgh as a doctor, and in 1881 joined the Imperial Chinese Maritime Customs Service. In addition to being a physician, he was a dendrologist and keen plant collector. He accrued immense knowledge of the native horticulture and recognised the important floral potential of the Chinese landscape. He continually urged the British authorities to send collectors to China to take advantage of the botanical harvest, but no one arrived during almost the whole of his twenty-year residency, and it was not until he was about to retire to England that the first collector was appointed. (Dr Henry was quite right about the floral potential: recent estimates show that China houses approximately one-eighth of the world's total flowering plant population – a staggering 30,000 different species.) By the time Henry returned to England in 1900, he had sent some 158,000 herbarium specimens to the Royal Botanic Gardens at both Kew and Edinburgh, enough to lay a substantial scientific foundation for further horticultural achievements.

The Irishman Augustine Henry became the greatest authority on Chinese flora.

Augustine Henry was a polymath: he came home just once during the twenty years he spent in China, in 1889, and during his leave took the opportunity of being called to the Bar, becoming a member of Middle Temple. After his final return to Britain, he was first Reader of Forestry at Cambridge University and then, in 1907, was appointed Professor of Forestry at the Dublin College of Science (later incorporated into the National University of Ireland). He died in Ireland in 1930, the same year as Ernest Wilson.

The link between the missionary Père David and Dr Augustine Henry is *Davidia involucrata*. When Henry first arrived in China in the 1880s he was stationed at Ichang (Yichang) on the upper Yangtze River (Jinsha Jiang) in central China. He already knew of Père David's description of the tree, and was fortunate to discover one growing deep in

the countryside some miles from Ichang. He collected seeds from the tree and dispatched them to Kew, where, however, for some inexplicable reason, they were never grown on – a strange omission.

In 1899 Ernest Henry Wilson (1876–1930) travelled to China on behalf of the Veitch Nursery, specifically to bring home the seeds of the *Davidia involucrata*. The Royal Horticultural Society had been keen to send someone to China to collect viable seed, but it was the Veitch Nursery who eventually employed the twenty-three-year-old man. It was assumed that the tree was not in cultivation and the nursery was keen to acquire it, so Wilson's instructions were simply to procure the seed and then return to England. The only person known to have seen the tree was Dr Henry – still working in China – and it was vital, therefore, for Wilson to meet him to get as much information as possible about its location and description. By the time Wilson arrived, Henry – now Acting Chief Commissioner for Customs, with the official status of Mandarin – was stationed in Szemao (now Simao) in Yunnan province in south-west China, on the borders of French Indo-China (Vietnam) and Burma.

It took the young Wilson six months to journey from England to China via the United States; here he took the opportunity to visit the Arnold Arboretum to learn about the latest techniques in plant collecting, packaging and transportation. He docked in Hong Kong and then travelled the 1,600 km (1,000 miles) to Szemao. This involved travelling through the borderland with Indo-China where the journey turned into a nightmare. First he was held prisoner as a suspected spy, then was nearly drowned when travelling up the 500 km (310 mile) Song Hong or Red River with a boatman high on opium. As if this was not enough, the Boxer Rebellion was just about to erupt, and the local people were aggressive and volatile.

Wilson finally reached his destination in September 1899, and the two men spent several days together, during which time Wilson learned much about the botanical features of China. All he gleaned about the tree which was the sole objective of his terrible journey, however, could be put on the back of the proverbial postage stamp. The challenge facing Ernest Wilson was truly daunting; it would surely be as difficult as finding a needle in a haystack.

Undeterred, Wilson set off to find the tree. This involved another lengthy and circuitous journey, since there was no direct route to Ichang. He travelled north-east back through China to the port of Shanghai, this part of the journey alone was about 2,400 km (1,500 miles). In Shanghai he boarded a boat to take him the 1,600 km (1,000 miles) up the Yangtze River (the lower Yangtze River is now called the Chang Jiang and the upper river the Jinsha Jiang) to Ichang, which he reached some six dangerous months later in March 1900. A further three weeks elapsed before he finally reached the region where Henry had sighted the tree. Only then could he start to explore the countryside and try to discover it.

Despite the lack of any real information, the young Ernest Wilson amazingly reached the location which had been described to him by Dr Henry where he certainly found the *site* of the Davidia, but, almost like a scene from classical tragedy, he was just too late: the tree had been felled – to provide timber for a nearby house. Stoically, all that Wilson wrote in his diary about the events of that day, 25 April 1900, was: 'I did not sleep during the night . . .' Quite so.

Wilson retreated eastwards to Ichang, where he decided to remain for a few months collecting and botanising, even though the Veitch Nursery had specifically instructed him to focus on discovering this one tree. His failure to carry out these orders was most fortuitous, since, quite unexpectedly and like a bolt from the blue, Wilson found a Davidia in full bloom on 19 May 1900, and then later, in a grove, several more. He thought the tree was 'most beautiful' and, not surprisingly, it became his lifetime favourite plant. From the quantity of seed he brought home to England to be distributed by the Veitch Nursery, some 13,000 seedlings were grown on. What a needle to find in a haystack.

However, the drama of returning to Britain with the seed was rather like the race to the South Pole – someone else had got there first. A parcel of thirty-seven seeds had already been sent to the famous Paris nurseryman Maurice de Vilmorin (1849–1918) by the missionary Père Paul Guillaume Farges (1844–1912) even before Wilson received his commission from the Veitch Nursery in 1899. Unbeknown to them, Vilmorin successfully raised the plant in his Arboretum des Barres near Paris. This young tree bloomed there for the first time in 1906, while it was not until five years later that Wilson's Davidia first flowered in the Veitch Nursery at Coombe Wood, by which time Wilson had made his last expedition to China to collect the gorgeous Regal Lily, and was living in the United States.

Earlier in May 1900, before he made his exciting discovery and almost as an aside, Wilson found *Actinidia deliciosa*, or Wilson's Chinese Gooseberry; today we know and enjoy it as the commercially grown kiwi fruit.

There are about 40 species of the hardy Actinidia, all of them natives of Asia, and vigorous climbers. The kiwi fruit was first commercially developed in New Zealand, hence its Maori name, but it is now also grown in the United States. The fruit contains about ten times more vitamin C than a lemon does. The earliest Actinidia – from the Greek *aktis*, 'a ray', referring to the styles of the flower – was collected by Robert Fortune in China in about 1846, but only as an herbarium specimen. The earliest species to be grown here was *Actinidia kolomikta*; it had been collected on behalf of the Veitch Nursery from the Japanese island of Hokkaido by Charles Maries in 1877. Its range extends to the forests of Amur in eastern Siberia, which gives the fruit the name by which it is known in Russia, the Amur Gooseberry. Several of the shrubs, including *A. kolomikta*, are particularly favoured by cats, as are Valerian and catmint.

Ernest Wilson's first journey was such a success and he returned with so many excellent garden-worthy plants that in 1903 he was engaged a second time by the Veitch Nursery for a further two years. In 1907 and again in 1910 he returned to China, collecting this time on behalf of the Arnold Arboretum. He became so identified with the area and was such a successful plant collector that he was often referred to as 'Chinese' Wilson, but his last journey to China nearly cost him his life and left him with a permanent limp. For reasons which will shortly become clear, he used to call it his 'lily limp'. From 1914 onwards he made botanical journeys to Japan, India, New Zealand, Australia and Africa, and in 1919 he settled in the United States, where he was appointed first Assistant Director and then in 1927 Keeper of the Arnold Arboretum in Massachusetts.

It has been estimated that Wilson introduced into Britain between 2,000 and 3,000 different species of seed, and many

---

THE ARNOLD ARBORETUM

The Arnold Arboretum, part of the campus of Harvard University in Massachusetts, was founded in 1872 by Asa Gray, the eminent botanist, with a bequest from Hames Arnold. His generous gift was to be 'applied for the promotion of Agricultural or Horticultural Improvements' and for the Trustees to collect, grow and display 'as far as practicable all the trees, shrubs and herbaceous plants, either indigenous or exotic, which can be raised in the open air'.

more herbarium specimens; from the seeds, at least 1,000 new plants have been introduced into cultivation. Possibly the most gorgeous is the Regal Lily, *Lilium regale*, the plant which was the indirect cause of Wilson's 'lily limp'. This amiable and accommodating lily, as the writer Alice Coats called his introduction, was first grown in 1905 under the name *L. myriophyllum* (meaning 'many leaves'), but even though it was easy to cultivate and sweetly scented, it did not become as popular as Wilson thought it should.

He was so keen for people to share his enthusiasm for this splendid lily that on his fourth expedition to China, in 1910, he travelled yet again from Shanghai to the borders of Tibet, where he had first found the flower, a trek of over 3,200 km (2,000 miles). The site was a remote mountain valley, and the journey to it was through some of the most difficult and desolate country. As Wilson himself said of the route undertaken, it was 'absolute *terra-incognita*'. It is a mark of his enthusiasm that he braved this arduous journey again just so that the western world could share in the delights of the Regal Lily. Its gentle beauty and graceful habit absolutely defy its natural home; Wilson recorded in his diary that 'no more barren and repelling country could be imagined', but when the lovely lily burst into flower, the landscape was transformed, as he then noted, from 'a lonely semi-desert region into a veritable fairyland'. It was on the return journey that, in trying to escape one of the frequent landslides, Wilson broke his leg. The remaining rigours of the journey, the delay in treatment and the subsequent infection setting in resulted in his almost having to have his leg amputated; in fact, he nearly died. In due course, he returned to America where the infection was finally cured and the leg saved, but Wilson was left, for the rest of his life, with his 'lily limp'.

Such a plethora of wonderful plants for our gardens was collected by Ernest Wilson that one can imagine that almost every garden in the United Kingdom probably unknowingly boasts at least one of them.

Despite Sir Joseph Hooker's earlier introductions (see p.194) there is one plant above all others which, for sheer quantity, altered the landscape of our gardens in the first half of this century – and to which our four plant collectors in Asia were devoted – and that was the Rhododendron. The name is derived from a Greek word meaning 'rose tree', and the plant was thus named by Linnaeus. Its homeland is the western Himalayas, where 1,000 or so species are locked into the remote snowbound regions. Like a glacier beginning to melt, the trickle of rhododendrons into this country began

during the nineteenth century; in the early part of the twentieth century the trickle turned into a stream, and the stream soon became a torrent.

George Forrest (1873–1932), our second explorer and plant hunter in Asia, was swept along on this flowing tide of rhododendrons. During his time in China he gathered a staggering collection of 5,375 Rhododendron seeds. Possibly one of the best and most distinctive he found was *Rhododendron sinogrande*, from the steep forested hillsides of Yunnan province in south-west China, in 1913. Grand it certainly was, with its amazingly large shining dark green leaves with their silver or sometimes fawn undersides. Because of the enormous size of the leaf, it is one that can be more easily recognised than some others. The first known flowering in Britain was at Arisaig on the west coast of Scotland in the 1930s – in gardening, patience is often a virtue.

Forrest, a Scotsman, was thirty when he first set foot in China in 1904, having already travelled to Australia and back when still in his teens. His first expedition to China, which ended in 1907, was followed by another five, the last being cut short by his sudden death in 1932. Like Wilson and his adventurous first journey, Forrest only just escaped marauding tribesmen and man-hunting dogs on his initial expedition. His sponsor, Arthur Kilpin Bulley (1861–1942), the cotton-broker and founder of Bees Nursery Ltd, wished him to collect mainly alpines, since he was keenly developing his extensive garden at West Kirby on the Wirral (now the Liverpool University Botanic Garden).

Forrest was a plant collector in quite a different mould from Wilson. Whereas the latter collected all his seed and plant material personally, Forrest trained local collectors, who were nearly all recruited from one village high in the Lichiang range in Yunnan province, close to the borders of Burma and Laos and well to the south of Wilson's main collecting area. These local men ranged far from the main encampments, and over the seven expeditions and nearly thirty years of Forrest's involvement with China, they became expert plantsmen and collectors. For his second expedition in 1910, keen garden owners in this country were invited to become involved by contributing towards the cost of the expedition, the reward being a share of the plant material brought back. These arrangements worked smoothly as it was a small but knowledgeable and enthusiastic gardening syndicate that supported Forrest in his collecting.

Forrest is held in particular esteem in Cornwall, where the great plantsman John Charles Williams (1861–1939) of

This painting by Annie Ovenden of *Rhododendron sino-grande*, shows clearly the wonderfully smooth indumentum of the underside of the huge leaves which are such a feature of this tree.

Caerhays Castle was an early syndicate member; his garden on the south Cornish coast is positively littered with the progeny of Forrest's collecting (as well as with plants collected by Wilson and later by Frank Kingdon Ward). Sir John Ramsden of Muncaster Castle in Cumbria was another member of the syndicate. All over the country there are plants, usually trees and shrubs, growing from the original seed brought home by these early twentieth-century plant hunters, and I always find it a unique and humbling experience to view the results of such intrepid journeys.

One of George Forrest's introductions, a shrub with which most people are familiar, was *Pieris formosa* var. *forrestii*, with its spectacular brilliant red spring growth and slightly fragrant lily-of-the-valley hanging flowers. It was discovered on the 1910 expedition high in the mountains between the mighty Salween (Nu Jiang) and Shweli (Longchuan Jiang) rivers in Yunnan province. The name 'Pieris' is also the family name of the Cabbage White butterfly – perhaps the flowers resemble their flutterings?

The first Pieris, *Pieris floribunda*, came to England from

The mountainous Yunnan Province of south-west China proved one of the richest sources of imported plants.

the south-east United States in 1812 and was found by the Scotsman John Lyon (1765–1814), who explored and collected extensively in that area. Lyon returned to England in 1806 and again in 1812, when he is reported to have brought with him 'cargoes of plants'. In 1858, *Pieris formosa* was found in the eastern Himalayas and brought back to this country. Forrest's discovery had much larger flowers and was first thought to be a distinct species; it was only later, when the botanists had looked at it under the microscope, that it was realised it was a varietal example of the earlier species. Some years later, in 1919, Ernest Wilson discovered a further species, *Pieris taiwanensis*, this time in Formosa (now Taiwan) while he was working there for the Arnold Arboretum.

An alpine with impact which was found by Forrest in 1911 was *Gentiana sino-ornata*, with its deep blue trumpet flower; it was another plant from the botanically rich area of Yunnan province, and was discovered at about 4,600 m (15,000 ft). It joined some 400 other Gentian species which are found throughout the world, except Africa; nearly all of them are known for their 'gentian blue' colour. However, one of the earliest Gentians we learn about growing in England, *G. lutea*,

a tall yellow-flowered species from Europe; see p.117. The 'Father of British Botany', herbalist William Turner (*c.* 1508–68), reported that the chewing of the root helped overcome the effects of 'bitings of mad dogs and venomous beasts'. Linnaeus named the family for Gentius, King of Illyria (*c.* 500 BC), who is supposed to have discovered the antiseptic qualities of the root. He obviously knew a thing or two, because it is still used in brewing to make a tonic bitters.

In 1911, while George Forrest was still plant hunting in Yunnan province, fighting and general mayhem broke out. There had been unrest in the area since 1902 when Tibet was opened up to the western world as the result of a British expedition led by Colonel Sir Francis Younghusband (1863–1942). The expedition had been mounted, in part, to ensure that Britain, as opposed to Russia, was the prime influence in Tibet, and thus in a better position to protect the 'back door' of India. The following year, Younghusband was a member of the Sikkim-Tibet Boundary Commission and was later appointed British Resident in Kashmir. The anti-Christian rage which the lamas felt at the violation of their holy city, Lhasa, spilled out into the surrounding country, and Forrest found himself in the middle of it. He was hunted through the upper reaches of the Mekong (now Lancang Jiang) Valley for over eight days, suffering appalling hardship and nearly starving to death until he was rescued and taken in by some sympathetic villagers. He lost everything, including a whole season's worth of photographs, seeds and over 2,000 plants. In comparison to his native companions on the expedition, however, he was lucky – nearly all of them lost their lives.

Forrest recuperated for a couple of months and then set off again, this time with his friend George Litton, British Consul at Tengyueh (Tengchong) in Yunnan province. They found a couple of beautiful Primulas: one a bluish-violet colour which was named in honour of his companion *Primula littoniana* (now *P. vialii*), and *P. forrestii*, yellow-flowered with an orange eye. Much later in 1939, a pale purple Primula with a yellow eye was discovered almost in the same area. At first it was named for the ill-fated Colonel Younghusband, *P. younghusbandiana*, but was subsequently renamed *P. caveana*.

It is surprising and oddly reassuring that despite the cataclysmic wars which took place during the first half of the twentieth century, the comparatively inconsequential drama of exploration and plant hunting was able to continue in some parts of the world. For instance, on the very eve of the First World War, Sir Ernest Shackleton (1874–1922) set off to explore the Antarctic and to try to reach the South Pole. Reginald Farrer and William Purdom were departing to hunt for new plants in China at the same time. Ernest Wilson, who, from 1906, had been working for the Arnold Arboretum, continued to botanise in Japan from 1914 on their behalf. For his part George Forrest remained in China virtually without a break.

Reginald Farrer (1880–1920) was a rock and alpine man *par excellence*, so it is strange that, unlike the other early twentieth-century Himalayan explorers, no Primula was named for him, although he collected a number. He was an amateur in the true sense, and was inspired to collect for the beauty of a plant and not for botanical reasons alone. After studying at Oxford he visited Japan (where he became a Buddhist) and wrote his first book, *The Garden of Asia: Impressions from Japan*. He was a prolific and lyrical writer about his travels and the plants he discovered. He began his plant exploring much nearer home, in the European Alps, during the first years of the century, and in 1912, at the age of thirty-two, began to write his encyclopaedic two-volume *The English Rock Garden*, which was finally published in 1919, the delay

caused by his being in China until 1916, and then, on his return to England, working for the Ministry of Information until the end of the First World War.

The journey that Farrer and William Purdom (1880–1921) undertook into north-west China, to Kansu (now Gansu) province near the Mongolian and Tibetan border, lasted two years from 1914 to 1916. The two men worked well together and collected avidly, sending plants and specimens back to Britain. It was extremely unfortunate that due to haphazard wartime conditions in Britain, with a shortage of gardeners and staff at Kew and other botanical establishments, most of their introductions failed to survive or indeed even germinate. However, a few struggled against the odds; these included an Allium, *A. cyathophorum* var. *farreri*, which has a deep violet-blue dainty flower and grows readily. A second Allium discovered by the two men was originally named for Purdom, *A. purdomii*, but this turned out to be a species which had already been discovered, in 1890, and named *A. cyaneum*, so, following the nomenclature rules of always using (or reverting to) a plant's earliest name, William Purdom's Allium was subsumed into *A. cyaneum*. It was already established in England well before the First World War.

One of the most gorgeous of their finds was *Buddleja alternifolia*. This shrub is now so familiar that it seems amazing the impact that it has made in such a short time. Perhaps Farrer writing so seductively about it had something to do with its immediate popularity: 'It sweeps in long streaming cascades . . . like a gracious small-leaved weeping willow when it is not in flower, and a sheer waterfall of soft purple when it is.' What a picture those words conjure up.

Gentiana farreri, a Gentian collected from further west in Kansu province, also bears Farrer's name, as does *Geranium farreri*, a dwarf pink geranium. This is a true alpine and needs to be grown in a pot or an alpine house. Quite different, and found in the south of the province, was a species rose which blooms in June and has pale pink or white flowers which develop into luscious coral hips; its pretty fern-like leaves turn purple and crimson in the autumn. The particular variety usually offered was selected from a batch of Farrer's seedlings sent to his friend and fellow plantsman E. A. Bowles (1865–1954), and was *Rosa persotosa* var. *farreri* (now *R. elegantula*). It is sometimes remembered as the Threepenny-bit Rose because of the size and shape of the flower – although the eight-sided bronze threepenny bit is hardly remembered by anyone nowadays.

Farrer and Purdom ended their joint explorations in 1916,

Buddleja alternifolia **is one of the 100 species of Buddleja which are at home growing on the riversides and the rocky and scrub areas of Asia, Africa and North and South America.**

As soon as the war was over and he could be released from his desk, Reginald Farrer returned to the area he loved, and in 1919 he journeyed to Upper Burma on the border with China. His companion on this second journey was Euan Hillhouse Methven Cox (1893–1977), of later Rhododendron fame, and for over a year the two of them were absorbed in collecting, pressing the herbarium specimens and writing up their field notes. In addition, Farrer wrote many letters and articles about their adventures. During the summer of 1919 Frank Kingdon Ward came to stay with them for a few days. It must have been one of the very rare occasions that any of the four Sino-botanists came into contact with one another, even though Forrest, Farrer and Kingdon Ward must have crossed and recrossed each other's exploratory paths. Kingdon Ward had already been exploring in roughly the same area for the previous eight years and was to continue for another thirty years.

After they had gone their separate ways again, Farrer and Cox discovered a beautiful juniper, known locally as the Chinese Coffin Tree. For obvious reasons, its timber was very valuable, and with its blue-green drooping branches, *Juniperis coxii* is considered to be one of the finest conifers introduced into Britain in the twentieth century.

Farrer also collected new species of Rhododendron; as becomes the discoveries of an alpinest, two of these were quite small, while a third was tiny and only suitable for the rock garden. In fact, its name is probably larger than the plant itself: *Rhododendron campylogynum* (meaning with a curved or bent ovary). It has rose-purple nodding waxy flowers. The second Rhododendron, *R. sperabile*, has a lovely thick spread of soft downiness on the underside of its leaves, and deep crimson bell-shaped flowers. The citron-yellow flower of the third species is reflected in its name, *R. caloxanthum*, which means 'beautiful yellow'.

Reginald Farrer never spared himself during his expeditions, despite the fact that he always found the climate trying, sapping his strength and energy. He had suffered poor health since childhood (when he had had to be educated at home), and during this second trip he was overcome by exhaustion and died. His death at the age of forty deprived the gardening world of a writer who could inspire and enthuse with his words. There is no doubt that his two books about his travels, *The Eaves of the World*, written in 1917, and *The Rainbow Bridge*, published posthumously in 1921, cannot fail to have encouraged the growing of plants from the Asian continent in our gardens, and his lyrical writing about the

Farrer returning to a Civil Service desk in London for the rest of the war and Purdom remaining in China to become a forestry adviser to the Chinese government, a post which he held until his death in 1921.

landscape and the plants he saw is still inspiration today. He was buried in Upper Burma with a simple brass memorial marking the site. It reads: 'In loving memory of Reginald John Farrer of Ingleborough, Yorkshire. He died for love and duty in search of rare plants.'

The 'Blue Poppy man', Frank Kingdon Ward (1885–1958), is the fourth of the great Asiatic influences in our twentieth-century gardens, and the last in the line of great plant hunters and explorers which stretches back like a trail of convolvulus to the Tradescants in the seventeenth century. His longevity in the field and his assiduous collecting techniques, together with his writings, have meant that his plant introductions have captivated gardeners and have quickly been found space in the herbaceous border.

The Blue Poppy, *Meconopsis betonicifolia*, is a plant which fulfils the romance and drama of growing and collecting plants, and has the added advantage of being easily recognised too. The translation of the first Greek name just means 'poppy-like', and *betonicifolia* derives from its resemblance to the leaves of betony. Not very inspiring, so it is just as well it has such a fragile beauty, with its 'Virgin Mary' blue petals.

Part of its original attraction must have been that it proved such a challenge to bring into cultivation. Kingdon Ward first saw it in flower at about 4,900 m (16,000 ft) against the edge of a glacier (how lovely the blue must have looked) on the Yunnan–Tibetan border in 1924. In fact, it had been described about forty years earlier by the missionary priest Père Jean Marie Delavay on his travels in north-western Yunnan; see p.225. Later, in 1922, the Political Officer for Sikkim, Lieutenant Colonel F. M. Bailey, found and picked a similar poppy in south-eastern Tibet which aroused great interest, but no one was able to persuade the seed to germinate. The seed packets, numbered KW 5784, and KW 6862, sent home by Kingdon Ward in 1925 proved to be of much sturdier stock, and it is from this single importation of seed that all the subsequent seed now on the market has been derived. Convinced that the Blue Poppy would be a great garden plant, Kingdon Ward wrote a detailed analysis of its good points, ending with the phrase: 'It will be perfect.' It is.

The following year, 1926, late in the afternoon of 1 June, he found a Primula near the northern frontier of Burma and Assam, and wrote in such an ecstatic and moving way that one cannot help but be transported to the very site. 'I stood there transfixed . . . in a honeymoon of bliss, feasting my eyes on a masterpiece. The vulgar thought – is it new? – did not at this moment occur to me . . . It was enough for me that I

Frank Kingdon Ward first saw the Blue Poppy, *Meconopsis betonicifolia*, growing wild in its Tibetan homeland in 1925.

had never set eyes on its like!' The stunning find was *P. agleniana*, named after one of the three classical Graces but nicknamed by Kingdon Ward the Tea Rose Primula because of the large 'rosy' globe of the bloom. He found it at about

Primula rosea, the Rosy Himalayan Primula, was introduced to Britain from the north western Himalayas in 1879.

Cornus kousa var. *chinensis* is such a beauty that it is surprising that it remained undiscovered until 1950 when the seed of this variety was collected by Frank Kingdon Ward.

3,000 m (10,000 ft) on a very steep scree slope. Although it does not seem to be commercially available now, some of the original seed scattered beneath beech trees at Caerhays Castle in Cornwall germinated at least once, though nothing came of them the second year. (Such is the fate of so many introduced plants.)

During the same year, Kingdon Ward also found at least another two Primulas, both of which are still being grown in English gardens. The first was *P. alpicola* which, because of its many-coloured flower, Kingdon Ward referred to as 'Joseph's Sikkimensis Primula', after the biblical coat of many colours. The second was perhaps the most important of all in the packet numbered KW 5781: the fragrant yellow Giant Cowslip, *P. florindae*, named in honour of his first wife. The couple divorced and many years later Kingdon Ward remarried. His second wife, Jean Macklin, has been immortalised by a lily with beautiful rose-pink trumpet flowers, *Lilium mackliniae*, introduced in 1946.

After the Second World War (in which he volunteered,

spending part of it secretly exploring viable routes from India through Burma to China for the 14th Army to use), Kingdon Ward, though advancing in years, was still in the field continuing his explorations, and in 1950 he was back again, collecting this time in northern Assam. Although he did not realise it, this was to be his most dangerous assignment. It was a domestic group consisting of Kingdon Ward, his wife Jean and two Sherpa porters who embarked on the exploration. On the evening of 15 August 1950 they were camped near Rima, then in Assam (now Ch'a-yu China), about half a mile from the Lohit River, planning the next day's route, which would take them up the Lati Gorge. For some weeks previously there had been earth tremors, and now, suddenly, an earthquake of devastating ferocity struck, lasting a full five minutes. It was of such intensity that at a seismological station in Dorking in Surrey, the needle on the recording machine vibrated so violently that the trace was unreadable. Miraculously amidst all the destruction – in Assam over a thousand people lost their lives – the group survived without

Frank Kingdon Ward is here shown preparing for a day's plant hunting in Tibet in 1933.

injury; they later discovered their camp had been only fifty miles from the epicentre. After being rescued by the Assam Rifles, they carried on plant collecting with typical British grit before returning to England early in 1951.

The reward for their intrepid behaviour – and our gain – was the introduction of the beautiful cream-flowered shrub *Cornus kousa* var. *chinensis*, the seed of which they gathered and returned with in packet no. KW 19300. It differs from *Cornus kousa*, which was collected in Japan in 1875, in having larger and sturdier bracts.

Despite their horrific experiences, Kingdon Ward and his wife journeyed from England again in 1952, this time for northern Burma, where they spent a productive time finding a number of Rhododendrons. Kingdon Ward's sponsors were the Royal Horticultural Society, the Natural History Museum and also the New Zealand Rhododendron Society. The Burmese government, who gave permission for the couple to travel, also insisted that two Burmese botanists were included in the party and that a duplicate set of the plants collected was presented to the Burmese government.

Frank Kingdon Ward, the last of our Sino-botanists, died in England after a short illness in 1958. Like all the other great plant collectors, particularly prior to the outbreak of the Second World War, Kingdon Ward collected quantity as well as quality – as yet, little thought was being given to the conservation of the natural habitat – and his introductions cover the widest range of plants: alpines, trees and shrubs, including honeysuckles, gentians, Berberis, acers, all came under his penetrating garden gaze. But as so often with the Sino plant collectors, there is one genus above all others which stands out for its gardening presence, and that is the Rhododendron.

The glut of new Rhododendron species which was released into numerous parks and woodlands in the first half of the twentieth century enabled enthusiastic gardeners to set about breeding, crossing and hybridising with an abandonment which echoes the tulipomania of the mid-seventeenth century. It is as if the landscaped park of the eigh-

Rhododendron arboreum was first brought to England as early as 1810 by Dr Francis Hamilton of the Bengal Medical Service. The species was also collected by Sir Joseph Hooker when he visited Sikkim during 1849.

teenth century had been especially created and was just waiting for the introduction of Rhododendrons two centuries later, so well did they fit into the Brown and Repton concept. John Charles Williams – known to everyone in the horticultural world as J.C. Williams – made a huge number of crosses at his two residences in Cornwall, Caerhays Castle on the south coast and Werrington Park in the north-east of the county, on the edge of Bodmin Moor. As we have seen, Williams was part of the syndicate which sponsored George Forrest's expeditions, and he paid Forrest extra each time he sent him back a new species of Rhododendron. The gardens at Exbury in Hampshire, the home of Lionel Nathan de Rothschild (1882–1942), were almost wholly devoted to the breeding of over 1,000 new varieties, of which some 500 were named and registered. In 1928 Mr de Rothschild founded and was the first president of the Rhododendron Association. Two years prior to that, the first specialist Rhododendron Show had been held at the RHS's Floral Hall in London.

The Sussex homes of the Loder family at Leonardslee, Wakehurst Place and High Beeches, the Aberconways of Bodnant in North Wales, the newly designed gardens at Cragside in Northumberland and the old established Cornish gardens of Heligan, Trewithin, Penjerrick and Pencarrow, among many others, all took part in the frenzy of woodland planting. This frenzy was not confined to England but was pursued with great success north of the border, where planting on Scottish estates, especially on the west coast, went ahead with great verve. Of particular note were the Duchess of Montrose's estate at Brodick Castle on the Isle of Arran, Inverewe in Wester Ross where Osgood McKenzie and his daughter Marie Sawyer created a sumptuous Rhododendron garden, and Glendoick in Perthshire.

Glendoick is the home of the horticultural Cox dynasty, who are particularly famous for their Rhododendrons and Azaleas. Euan Cox, as we have noted, travelled to China with Farrer in 1919, after which he returned home to establish the Glendoick gardens, and began writing extensively about his passion for Chinese plants and their cultivation. His son Peter (1934–), who, for his horticultural achievements, has received the RHS Veitch Memorial Medal, has with his own son Kenneth continued the family tradition of searching for and collecting seed in the wild, and breeding new and interesting Rhododendrons and Azaleas.

Rhododendrons were really the last of the great plant imports and the overwhelmingly long list of their delights could go on and on. It was another enthusiast, John Barr Stevenson (c. 1881–1950), of Tower Court in Ascot, who, with his wife, managed to keep the largest collection of these new species growing during the difficult days of the First World War and after. His collection of Rhododendrons was considered so important that, following his death in 1950, a number of them were successfully moved to the Valley Gardens in Windsor Great Park and are enjoyed annually by thousands of enthusiasts.

With the amount of hybridising and breeding which took place in the latter part of the twentieth century, there must now be a Rhododendron to suit every garden situation, but considering the Himalayan size which some of the early plantings have already grown to in quite a short space of time, they need watching, since, as Mae West once said – about something quite different – 'You ain't seen nothin' yet.'

China, because of the revolution of 1911 and subsequent continuing unrest, gradually receded once again into its armoured shell of isolation, and botanists and plant hunters had to look elsewhere. In fact, there were still plants waiting to be discovered much closer to home, on the European mainland, and we have to go back to the early years of the century to note those that would grow happily in our British gardens.

Clarence Elliott (1881–1969), a great traveller and plant hunter, spent much of 1908 and 1909 exploring the Mediterranean island of Corsica. He came home with *Helleborus corsicus* (now *Helleborus argutifolius*, meaning 'with sharp-toothed leaves') and *Thymus herba-barona*, the Caraway Thyme. There was also *Morisia monanthos*, the single species of a genus belonging to the *Brassicaceae* family. This was a small, ground-hugging bright sunshine-yellow perennial, named in honour of Professor Moris (1796–1869), the author of a *Flora of Sardinia*. All three plants were indigenous to the two islands of Corsica and Sardinia. Later in 1909, following explorations in the Mediterranean, Elliott embarked on a journey to the Falkland Islands. He expressed surprise when he found they 'were a little further away than I imagined . . . but it was a pleasant month's voyage'; not quite the sentiments of the British servicemen when they visited the islands in 1982. On his return the following year,

he botanised in the Alps with his friend Reginald Farrer, who was shortly to leave for China.

Nearly ten years after the end of the First World War, during 1927 and 1928, Clarence Elliott travelled to South America with William Balfour-Gourlay (c. 1879–1966), who was collecting plants in Chile for the Royal Botanic Gardens of both Kew and Edinburgh. This resulted in the bulb *Leucocoryne ixiodes*, the fragrant pale blue Glory of the Sun, becoming available, as well as *Calceolaria darwinii*. From this expedition too came the Chilean-named *Puya alpestris* with its exotic metallic blue-green flower spike. It is a remarkable sight to see it when it is blooming in the gardens of Tresco Abbey on the Isles of Scilly, with the local birds trying to collect the nectar from the flowers in the manner of their South American counterpart, the humming bird.

Between 1925 and 1927 Harold Comber (1897–1969), the son of James Comber (c. 1866–1953), the distinguished head gardener at Nymans in Sussex, the home of the Messel family, collected in South America. Three of the plants which he sent home from Chile had, in fact, previously been introduced by Richard Pearce (d.1868) whilst he was collecting plants for the Veitch Nursery; see p.206. They had obviously found life in nineteenth-century Britain not quite to their liking, so it is good to be able to record their reintroduction and the fact that they are still enjoying the British weather. The three delightful reintroductions were *Mutisia oligodon*, a climber with silky pink daisy-shaped flowers; *Berberis linearifolia*, an especially lovely species with a profusion of apricot-coloured flowers; and *Fabiana violacea*, a heath-like shrub and a member of the potato family. Comber brought home two further plants which had originally been discovered in Chile in the previous century. One was *Lathyrus pubescens*, which had originally been found by John Tweedie (1775–1862) in 1840. Although the plant had already gained an RHS Award of Merit in 1903, it had never really made the grade until Comber, on his return, brought it to the fore of the nursery and gardening world of Britain. The other plant was again one first collected for the Veitch Nursery, this time by George Thomas Davy (fl. 1840s) near Concepción in 1844: *Lardizabala biternata*, a shiny-leaved climber named for a Spanish naturalist, and today still clinging on by its tendrils, being available from only one British nursery.

Although the number of new discoveries was declining, possibilities still abounded and throughout the 1920s and 1930s, and then again during the years following the Second World War, the place to be seen in the plant-collecting world was without doubt the Himalayas. It was a Mecca for explorers, expeditions and plant collectors. The Grand Trigonometrical Survey of India (the mapping of the great Himalayan range) was continuing, there were expeditions to attempt to conquer Everest, and people like Francis Sydney Smythe (1900–49), who was both a mountaineer (he had taken part in three of the Everest climbs) and a plant collector, walked and botanised in the central Himalayas during the summer of 1938.

Frank Ludlow (1885–1972) also knew the Himalayan scene, having served there during the First World War and then entered the Indian Education Service. He was very much enamoured with the mountains and was appointed during the Second World War to be British Resident in Lhasa, Tibet. In 1932 he met George Sherriff (1898–1967), the British Consul in Kasgar (Kashi, in north-west China), who was also a military man, having served with the Mountain Artillery on the North-West Frontier during the First World War before joining the Consular Service. The two men made a series of seven expeditions through Tibet and Bhutan up until 1947 and succeeded in finding at least 27 new Primulas (including one named *Primula sherriffae*, which they discovered in 1935 in south-east Bhutan), 38 new Saxifrages and 23 new Gentians.

As we saw earlier with one of the last of Kingdon Ward's expeditions, there were a number of sponsors supporting the plant hunters, as well as active participation by the host countries; plant hunting was changing direction, and it was during the 1930s that this began to happen in earnest. Gradually the emphasis shifted from the search for garden-worthy plants and shrubs to those offering economic, medicinal and scientific benefits. Explorations and expeditions were beginning to be sponsored and funded by the Royal Botanic Gardens of Kew and Edinburgh, or by British or foreign universities, or by government departments of either the host or the visiting country. The other major change brought about by improved communications, particularly air travel, was the reduction in the length of time expeditions took.

It was in 1930 that the very first plant was delivered by air to England. This was *Primula sonchifolia*, a delicious lavender blue beauty which had been discovered in 1911 by Frank Kingdon Ward when he was travelling in western China and south-eastern Tibet on behalf of the Bees Nursery. It proved to be a difficult plant to domesticate and was not introduced into this country until 1926. Its historic journey by air from Burma in 1930 was so that it could be exhibited at a show in London.

Perhaps the most fundamental change of all was the relationship with the host country. Instead of an individual or a group of plant hunters arriving in the country unannounced and taking off into the wide blue yonder, collecting what they could and as much of it as could be transported ('pillage' is a word that perhaps comes to mind), now increasingly expeditions were joint enterprises. If there was not an invitation from the host government, then permission to mount the expedition was always sought in advance. As has been seen with Kingdon Ward's travels in Burma in 1952, local officials and representatives of botanic gardens or universities would often join the expedition. The end result was not only a collection of plants but also the production of a highly academic and noteworthy study. Seeking remedies from the plant kingdom for the appalling ills of humanity has always been an inspirational and worthy cause. Often the results, and the investment in people and equipment, can be of lasting and real benefit to the area the expedition has visited. A number of countries in the tropics and the southern hemisphere became the focus of much scientific botanical work, which, from small beginnings between the two world wars, expanded greatly from the 1950s onwards.

This new approach to plant collecting is exemplified by a number of people. Noel Sandwith (1901–65), the botanist in charge of the American section of the Kew Herbarium, collected plants from British Guiana (Guyana) in 1929 with the Oxford University Expedition. He went there again in 1937, this time with the Imperial College of Science. John Hutchinson (1884–1972), a botanist with responsibility for the African section of the Kew Herbarium, explored South Africa during 1929 thanks to a grant from the Empire Marketing Board. The following year, when collecting with General Jan Smuts (1870–1950), the South African statesman, he found an interesting new member of the Daisy family which he named, in honour of his distinguished companion, *Pteronia smutsii*, the first part of which means 'with winged nerves'. Sarawak was explored in 1932 by another Oxford University Expedition, this time to collect orchids, of which at least a dozen new species were found

Just prior to 1939, the British Museum mounted an expedition to the mountains of East Africa, the plant collector and writer Patrick Synge (1910–82) being a member. An exciting and exotic area called the Mountains of the Moon was explored; Mount Kilimanjaro is the highest of the peaks at

Primula sonchifolia is found high up near the snow line in the alpine meadows of Tibet, S.E. China and Burma, the area which was the favourite hunting ground of Frank Kingdon Ward.

5,890 m (19,340 ft). The vegetation is unlike anywhere else, but unfortunately the gigantic Lobelias and Senecios which grow there with such peculiar abandonment do not take kindly to our equable climate, and few have settled here. Only one, *Lobelia gibberoa* (meaning 'hunchbacked') is named in the current *RHS Plant Finder*. Synge tells of seeing this particular plant with its flower-spire reaching a cathedral-like 8.8 m (29 ft) high, and with leaves 'two or even three feet long'. However, for it to survive in England it has to have special treatment. The specimen growing at Kew is contained in the Temperate House and is almost a dwarf in comparison with the free-growing African plants.

Lurking still somewhere in Britain might be the fine scarlet lily, *Choananthus cyrtanthifloris*, seen by Synge during the same expedition. It had been collected first in 1924 from its one known area, the 5,115 m (16,794 ft) Mount Ruwenzori. With its funnel-shaped flower, the lily received an Award of Merit when it was presented to the RHS, but it seems to have returned to the warmth of Africa. On Mount Elgon, 4,334 m (14,177 ft) high, on the borders of Uganda and Kenya, Patrick Synge came across a large white balsam, *Impatiens elegantissima*, with 'flowers several inches across, hovering on a long pedicel like a giant white butterfly, marked with deep crimson and with spurs several inches long': how lovely it sounds. The expedition returned to England with seeds from the balsam which were grown on, some of the plants survived for a few years, but that species too seems to have retreated back to the Mountains of the Moon.

South Africa, meanwhile, was a continuing source of new delights. From 1911, when he first went there to live, the great collector and botanist Thomas Pearson Stokoe (1868–1959) was avidly seeking new species. His favourite area for collecting was the Hottentots-Holland Mountains on the Cape (he was a keen mountaineer), where he identified numerous new species. At least 30 were named in his honour, among them in 1918 *Leucadendron stokoei*, a member of the *Proteaceae* family, and *Brunia stokoei*, a small heath-like shrub in the *Bruniaceae* family. Brunias are a very small genus (named in honour of one Corneille de Bruin, a Dutch traveller in the Levant in the eighteenth century), and the only one to come to Britain apart from *B. stokoei* was *B. nodiflora*, which had been collected in 1786. Both appear to have taken the steamer back to South Africa, as neither of them is in the latest catalogues.

Collecting plants in the wild for later introduction into our gardens was, as we have seen, a somewhat haphazard affair up until the nineteenth century. It was not until the Royal Horticultural Society, great estate owners, the Royal Botanic Gardens of Kew and Edinburgh, universities and commercial nurseries began commissioning plant collectors to travel to a particular continent or country to gather specific plants that the resulting collection really stood any chance of coming before a wider public. The great Veitch Nursery was a repository for much of the new material arriving in England, while in Germany there was the famous Ludwig Spath Nursery and in France Maurice Vilmorin's Arboretum des Barres, but by the 1950s all three of them had ceased to exist in their original form.

So it was fortunate that in England a nursery and arboretum which had originally been founded at Winchester in 1864 by a

In 1913 this National Botanic Garden was established on the Kirstenbosch Estate on the outskirts of Cape Town, South Africa, to replace the Dutch East India Company's earlier botanical garden.

rare and unusual and most of which have been cultivated from the vast collections of trees and woody shrubs brought home by the plant hunters earlier in the century.

As well as the search for new species, a number of countries began the process of having their flora formally recorded. This included a complete herbarium being made, each specimen being botanically drawn and the results of this enormous effort being eventually published. As we have seen in the nineteenth century, Kew Gardens had been instrumental in beginning the painstaking recording of the flora of various countries, in particular those of the Empire (though not exclusively so), and this work continued and grew in importance. From 1927 to 1936, John Hutchinson wrote, with John Dalziel, the *Flora of West Tropical Africa*, the compilation of which was continued after the Second World War by R. Keay of the Nigerian Forest Service and F. Nigel Hepper, an expert in tropical West African plants from the Kew Herbarium. The *Flora of Tropical East Africa* was begun in 1952 and *Flora Zambesiaca* in 1961; work is continuing on both these last two. The *Flore des Mascareignes* involved the Paris Herbarium, the Royal Botanic Garden at Kew and the governments of Réunion and Mauritius, and was started in the 1980s. In South America, too, there are several botanical surveys taking place. As is evident, these enterprises take years and even decades to complete but are the botanical benchmark of the country concerned.

Because of the thoroughness of the surveys, new plants often come to light for the first time. For instance, the holiday islands of the Seychelles in the Indian Ocean were first surveyed botanically in 1962 by Kew; when a further study was made in 1973, a tree thought to be extinct, *Medusagyne oppositifolia*, a genus entirely on its own, was rediscovered. Its clustered flowers are small and white and grow at the base of the elongated oval leathery leaves, while the bark is fibrous and corky. The seeds, when ripe, open up like an umbrella; it is known locally as the Jellyfish Tree.

That tree will probably never be commercially available to gardeners, but one plant that is is *Rosularia sempervivum* (meaning 'having a rosette'), which was discovered in the Elburz mountains of Iran in 1977. It belongs to the same family as the house leek (Sempervivum) and received an Award of Merit from the RHS in 1985. From the other side of the world, and now available to us all, is *Abutilon striatum*, discovered in northern Argentina by John Lonsdale of Kew in 1978. It has already had its name altered to *A. pictum* – 'being brightly coloured' – and, writes John Lonsdale, the

Veitch-trained man – Edwin Hillier (1840–1929) – was able to take up the horticultural challenge. It was Edwin Lawrence Hillier (1865–1944), Edwin Senior's son, who laid the foundations for its superiority by cultivating the heritage of woody plants, of which there had been such a vast quantity arriving from the plant hunters. This work was carried on and expanded by Sir Harold Hillier (1905–85), a grandson of the firm's founder, who was a world authority on trees and shrubs which would grow in the temperate zone. It was he who, in the 1950s, established Jermyn's, the 73 hectare (180 acre) gardens and arboretum at Ampfield, near Romsey. The gardens and arboretum are now owned by Hampshire County Council and house well over 14,000 specimens, many of them

flower is 'a vivid red with blood-red veining' (hence its original name, *striatum*). Lonsdale found it while searching for some small orchids; on extricating himself from a particularly thorny thicket, he discovered a single unusual fragile seedling whose survival was not considered hopeful. However, rather like the extra strength and will to live that premature babies seem to have, this youngster not only lived but flower-powered its way to maturity.

These specific and sometimes complicated official journeys obviously do not mean that plant hunting has stopped for good, but although there is still room for the highly trained and interested traveller and explorer enthusiastically searching for a particular genus in some remote region, there are now very strict controls both on what can be taken out of other countries, and even more on what can be imported into Britain. Conservation of the native natural environment is at the heart of the regulations, but the sad truth also is that, because of unscrupulous behaviour by rogue 'collectors', very often the location of rare, new or unusual species has to remain a secret, with field notes, photographs and map references all locked away.

Indeed, in the 1960s and 70s so much of the natural world (not only plants) appeared to be under threat that in 1973, the Washington Convention on International Trade in Endangered Species (CITES) was set up. Plant hunters and collectors, now more than ever, have to be politically aware of a country's 'open' or 'closed' policy, or of local development, natural disasters or wars, all of which can change the indigenous plant kingdom very quickly. In a way, all plants occurring naturally in the wild may be thought of as 'living fossils'. They all have a pedigree going back into the earliest times and each of them in some way makes a contribution to the living planet. That is why it is so distressing to learn about the wholesale destruction of many botanically rich areas, in Asia and South America in particular, leading us to wonder what remarkable and startling finds may perhaps only be discovered by our descendants as 'fossils' because of the devastation carried out in the twentieth century.

Even though it seems as if every corner of the globe has had the horticultural spotlight turned upon it at some time or other during the past thousand years, there are still remarkable discoveries to be made. For example, just before the Bamboo Curtain descended in the early 1950s, China had one more botanical surprise in store for the rest of the world. This was the discovery of a 'living fossil', a conifer known as *Metasequoia glyptostroboides*, the Dawn Redwood. It was first noticed in north-east Sichuan in 1941 by a forester, T. Kan, searching for firewood. Because he could not identify it, he sent some of the tree's leaves to Beijing, where they caused quite a stir; it was known there as the Water Fir. In 1947, with the help of the Arnold Arboretum, seeds were collected, some of which came to England in 1948.

What is a most remarkable and extraordinary coincidence is that also in 1941, in Japan, just prior to the discovery of the tree in China, a Japanese palaeobotanist, Shigeru Miki, was working on fossilised material sent to him by the Arnold Arboretum. He found what he considered to be a new fossil species, which he estimated would have last been seen growing three million years previously. This was exciting enough in itself, and because it looked similar to the American Coastal Redwood he gave it the name Metasequoia – 'looks like the sequoia'. A few years later, in 1946, the Chinese 'living fossil' came to be examined by Professor Hu, head of the Botanical Institute in Beijing. He thought the tree seemed similar to the Chinese Swamp Cypress, *Glyptostrobus lineatus*, a monotypic genus related to the Taxodium; for this reason it was given the second name of *glyptostroboides*. (The Chinese Swamp Cypress arrived here in 1919 but is a very rare tree in Britain, only surviving in the warmest and most sheltered of areas.)

It was ultimately agreed that the two parts of the puzzle fitted perfectly: the fossil remains and the ancient survivor found in the forest were one and the same tree, not only a remarkable horticultural and fossil find but an extraordinary coincidence of timing. Remarkably, too, considering that *Metasequoia glyptostroboides* was only just clinging on to life in the remote valley when it was first found, it has turned out to be the most accommodating of trees, easy to propagate and tolerant of a wide range of sites. A fast and ornamental grower, it is now widely available. Its story is not unlike that of the Ginkgo, another 'living fossil', which was found in Zhejiang province in eastern China in the eighteenth century and which is, again, widely available; see p.166.

Another rarity of a 'living fossil' was discovered in Australia in 1994, but this is one which may not be easily grown in Britain: the so-called Dinosaur Pine, *Wollemia nobilis*, found by David Noble in Wollemi National Park in New South Wales. The word 'wollemi' means 'look around' and was well chosen since Noble, walking in the National Park, discovered the tree by looking back and seeing a tree he did not recognise and had not previously noticed. The tree has been placed in the same family as the Araucarias, which also contains the Monkey Puzzle Tree and the Norfolk Island Pine. There are apparently about forty of these Dinosaur Pines growing, in two small groups, and intensive studies and propagation are taking place to see just what the tree's survival requirements are in various climates. A seedling raised in cultivation from the tree was put on display in 1999 at the Royal Botanic Gardens, Sydney (behind a burglar- and vandal-proof fence), and further intensive propagation is underway to ensure the survival of yet another 'original'.

Along with the new approach to plant collecting, there had been, right from the beginning of the century, stirrings and nudgings about our own environment. For example, at a meeting in 1928 of the Smoke Abatement League, the Superintendent of Manchester Parks spoke of 'the influence of air pollution on vegetation'. From little acorns do tall oaks grow, and in 1956 and then 1968, two Clean Air Acts came into effect. The introduction of the Green Belt by Sir Patrick Abercrombie in 1947 was a brilliant scheme, and although considered very modern and a direct riposte to that most fast-developing phenomenon, the rampageous city or town, it only reflected the philosophy expounded by Sir Thomas More in his 1516 book *Utopia*.

A vast amount of twentieth-century legislation can be viewed in the light of trying to keep the idyll of rural serenity close at hand, so that we do not become totally divorced from greenness and the growing of plants. The development of the public park for recreation, the more recent country park and, on a smaller scale, the community garden all reflect a widespread need to 'commune with nature', a refreshment of the spirit which Sir Thomas More would have recognised. Contact with the soil and the growing of plants is now recognised as enhancing the quality of life: green space around schools, hospitals and residential homes helps to soak up our twenty-first-century angst and raise our spirits in adversity. Prison gardens play their part too, encouraged perhaps by the recent involvement of the inmates of Ley Hill Prison in the Chelsea and the Hampton Court flower shows (with Gold Medals to show for it, too).

There has developed a wide divergence between plant collecting 'out in the field' and the plants that are available to us to grow in our gardens. The hybridisation of plants, which began in the 1830s, came of age in 1899 when an international conference on the subject was held under the auspices of the Royal Horticultural Society and the Royal Society. Such was the interest in the subject that two further conferences

on it took place in the early years of the twentieth century.

A hybrid plant is produced when two (or more) species have been cross-fertilised, and although this may occur naturally, it is now usually the result of plant breeders scientifically creating better plants to satisfy the demands of the gardening public.

Perhaps it is just as well that these hybrid plants and cultivars were being developed from the 1920s onwards, since in gardening terms, as in so much else, the thirty-one-year period from the beginning of the First World War to the end of the Second was a watershed. Suddenly thousands of gardens were left short of the one thing they could not do without – people. Gardening never recovered from the shock of that necessary abandonment and gradually took the path of low maintenance, do-it-yourself, ground cover and self-sufficiency.

A plethora of knowledgeable gardening writers exhorted garden owners to get on themselves with the planting, the mowing, the pruning and the digging in their gardens. The humorous writers W. C. Sellar and R. J. Yeatman (of *1066 and All That* fame), in their book *Garden Rubbish*, published in 1936, claimed that: 'The point is – if you don't get shivers of delight up and down the spinal cord whenever you're gardening . . . in fact, if you're not plumb-potty about the whole sobgoblinatious affair, what the Weevil do you mean by attempting to be a Gardener at all? Take it from us, it is utterly forbidden to be half-hearted about Gardening. *You have got to LOVE your garden whether you like it or not.*'

In truth, the British are plantaholics, more obsessed with their plants than with the design of their garden, whereas continentals tend to garden by design, with plants taking second place. Significantly there are apparently more greenhouses per capita in Britain (there are over two million in total) than anywhere else, although that could, of course, be due to the unpredictability of our weather rather than our love of plants.

Throughout the twentieth century there was an increasing inclination to learn more and more about gardening. Perhaps it is something to do with the remoteness of our modern lives from the growing world, but it seems that, with the gradual disappearance of the gardener and his intimacy with the floral scene, there is an urge to replace him with information gleaned from books, television and the internet. It is almost as if the quantity of knowledge can somehow make up for its quality.

An amazing salmagundi of specialist societies has come

Wollemia nobilis, the Dinosaur Pine, is the latest of the 'living fossil' trees to be discovered. It has been placed in the same family as the Monkey Puzzle Tree, and perhaps one day it might be found hardy enough to be grown in Britain.

into being during the past one hundred years. These range from the National Sweet Pea Society, founded in 1900 (the same year as the Women's Farm and Garden Association), to the British Iris Society, formed in 1924 by William Dykes (Secretary of the RHS 1920–5). The justly famous Alpine Garden Society was begun by a group of enthusiasts, including Clarence Elliott, in 1929. Elliott had been a friend and colleague of Reginald Farrer, who, with his love and specialist knowledge of alpines, would, had he not died so prematurely, surely have revelled in the formation of such a society. There is the International Camellia Society, the Hosta Society, societies for the Rose, Daffodil, Clematis, Day-lily, Fuchsia, Orchid – and many more; all have their champion groups.

Writers too have played their part in disseminating gardening philosophy, usually with erudition and entertainment, their words and visions cascading through our minds like seeds spewing from a ripe poppy head.

We have seen the impact that three of the Sino-botanists made with their writings, and lest one be thought sexist about this, there was also a tremendous trio of female writ-

ers, all of whom have had a lasting influence on the development of modern English gardens: Gertrude Jekyll (1843–1932), Ellen Willmott (1858–1934), who was the first women to be elected to the Linnaean Society, and Vita Sackville-West (1892–1962). All three women created their own style of garden and shared their thoughts and ideals with their readers in a series of articles and books which are still read today.

Munstead Wood in Surrey, the home of Miss Jekyll, showed off to perfection her skills as both a practical and an imaginative gardener. She used the space and the plants as an artist would, to create pictures – she had, after all, begun life as just that, training at the South Kensington School of Art. Her colour schemes in planting were innovative and yet never looked contrived. Both she and Ellen Willmott were among the first recipients of the 1897 Victoria Medal of Honour, which is given in recognition of special service to horticulture. Miss Willmott herself owned Warley Place in Essex – once the home of John Evelyn: she was a skilled gardener and was a patron in particular of E. H. Wilson. Consequently, there are a number of plants with the epithet *willmottae*, as, for

inspiration *par excellence* for the visiting gardening public.

Alice Coats (1905–78), gardener, writer and artist, is also important in the story of plants and their history. She wrote three books in the 1950s and 1960s which give fascinating background to many of the flowers and shrubs which are now familiar in our gardens, her wit and wisdom making them an entrancing read.

An important development during this period was the study, in its own right, of the history of the garden. This new discipline brought together a number of specialists from different fields who were all passionate about the subject and determined to ensure that gardens and landscapes were fully recognised as a major part of our heritage. This in turn led in 1965 to the formation of The Garden History Society under the presidency of Miles Hadfield (1903–82), author of the seminal work on British garden history, *Gardening in Britain* (1960). The recognition that was being sought is now acknowledged, with the Society's involvement in the planning process where it affects historic parks and gardens.

At the same time, there was a greater interest and enthusiasm being shown in the study of our older garden plants, both indigenous species and those introductions which had been successfully cultivated over the centuries. The sudden realisation that, in the drive to produce modern counterparts of old favourites, countless numbers of those earlier horticultural delights were fast disappearing spurred the beginnings of the plant heritage movement, and 1979 saw the formation of the National Council for the Conservation of Plants and Gardens, the inaugural meeting of which was held, appropriately, at the Chelsea Physic Garden. Collecting and thus saving rare, old or unusual species from individual genera has been codified under the National Plant Collections scheme, and by the end of the twentieth century there were over 600 different genera being 'collected'. These two organisations, complementing each other in their different spheres, and coupled with the County Garden Trust movement, have led to a much greater understanding and public awareness of the heritage which exists in the country's gardens.

Because of this interest, retro-gardening caught the public's imagination during the latter decades of the twentieth century, with disused and overgrown areas of land being suddenly woken like Sleeping Beauty and transformed back into the gorgeous gardens they once were. The 'lost gardens' of Heligan in Cornwall, and Bampfylde's lost landscape at Hestercombe in Somerset (see p.222) epitomise the romance of these magical transformations.

The outstanding horticulturalist Sir Harold Hillier who founded the Arboretum at Jermyn's, Romsey, Hampshire.

instance, *Rosa willmottae*, which Wilson had collected from western China in 1904. Her special love was roses, and she wrote the two-volume *The Genus Rosa* between 1910 and 1914. A plant which had arrived from the Caucasus in 1820, *Eryngium giganteum*, somehow later acquired the nickname Miss Willmott's Ghost. It is a sea holly and very prickly; since she was known to have a rather imperious and difficult nature, perhaps that is how it got its sobriquet.

The third of the trio, Victoria (Vita) Sackville-West, wrote in 1946 a most evocative epic poem entitled *The Garden*, and later a weekly column, 'In Your Garden', for the *Observer* newspaper. The garden surrounding her home, Sissinghurst Castle in Kent, which she and her husband Harold Nicolson created from 1930 onwards is truly remarkable; perhaps the

The changing face of gardening in Britain and the role of conservation and stewardship of the landscape has meant that species new to our gardens are now more often than not designer plants created to satisfy a specific purpose or fill a perceived need. The use made of gardens has altered too: Edwardian and Victorian gardens were perhaps somewhat too intimidating to relax in; now the current fashion of making the garden into an 'extra room' and using it for outdoor living rather than just for gardening is becoming well established. It is an attractive idea but is, in fact, only an extension of a notion first mooted in the eighteenth century, when there was a sudden leap over the garden wall and, hey presto, all of nature was embraced in the garden. Thus, links can be made with an earlier age and reassurance gained for the future.

THE CLASSIFICATION OF PLANTS

The classification of any new plant begins with it first being placed in what was originally called the natural order and is now called the family. The family consists of perhaps only one group but more likely a number of groups all of which show a wide range of similar features. A single group is called a *genus* (the plural form is *genera*). The individual members of a genus are known as the *species*, and it is this group of plants which are able to breed together to produce an almost identical offspring. A species name is usually in two parts: the first name is the plant's genus (or generic name) and the second part is its special epithet. Botanical generic names are usually Latin, although, as has been seen, names have been taken from other languages as well, particularly Greek. *Subspecies* are distinct and are a naturally occurring development which differs from the norm of the species – these often occur when plants grow in consistent isolated groups, say, on an island.

Still within the species, some plants can develop as *hybrids*. These are the result of cross-fertilisation between two or more species and can happen either naturally or as the result of scientific selection. There are also *intergeneric hybrids*; these have parents from two different genera. The word 'variety' is normally used quite freely to indicate any plant that is different from its species, but botanically, *varieties* are plants which differ very slightly from their normal species and form a minor subdivision of that species. A *cultivar* is the name given to an artificially raised plant that has distinctive characteristics which are stable and can be reproduced. It comes from the combination of two words, 'cultivated' and 'variety', which accurately explains precisely what it is.

Looking at the past thousand years and realising that our gardens have been the recipients of the world's floral beauty is both a humbling and an exhilarating experience. The rooted wisdom which is encapsulated in each plant we grow surely encourages us to strive further in our own lives. The recognition of the visual beauty of a flower seems almost to run parallel with the pleasure gained from listening to music, or reading, both experiences giving an external framework to our innermost thoughts and aspirations. Our quest for perfection in our plants, which has led man to explore the planet over the ten centuries of the millennium, will continue and hopefully lead to the discovery of more untold floral wonders during the next thousand years. It is worth remembering that the most modish and sophisticated plant grown for today's market had its beginning five hundred million years ago in a pool of green slime. What, in just the next five hundred years, will be the plant world's answer to that?

Perhaps the nirvana for gardeners of the twenty-first century to which they can look forward will be an endless supply of beautiful and richly perfumed plants, set in a landscape of divine inspiration, with just the right weather and climate to keep it green and pleasant. Weeds and weevils, slugs and snails and all things noxious will have learned that their role in life is not to be a pest. And in a flight of fancy, the annual fret of worrying about who will water the garden for us while we are away – perhaps visiting a distant planet – will disappear. Maybe 'virtual reality' is going to take over. In horticultural terms, the mechanics of the garden will know no bounds, but in the quiet inner serenity of our souls, the refreshment of the spirit and the glorious anticipation of floral beauty will remain the driving force for the creation of the perfect garden flower.

The whole world lives by its very nature on the cusp of the unknown, so it is inevitable that gardeners tend to look both back to the serenity of acquired garden knowledge, and forward to the garden brinkmanship of the challenge of the new. However, it could just be that we feel as John Ruskin did in the nineteenth century when he wrote that 'Flowers seem intended for the solace of ordinary humans.' He was reflecting the words which the Anglo-Saxon Abbot Aelfric used at the beginning of our thousand-year journey of plant introductions, when he wrote of the 'luffendlic stede', 'the beautiful place'. In the floral cascade which has arrived on these shores during the past millennium, have we yet found our own 'luffendlic stede' of a gardening paradise, or will it take another thousand years to perfect the beautiful place?

Plant Introductions in the period 1900–2000

1900   *Actinidia deliciosa* (syn. *A. chinensis*) Chinese Gooseberry, Kiwi Fruit. Hubei Province, China.

Camellia cuspidata S. China. Collected by E. H. Wilson for the Veitch Nursery.

Ehretia dicksonii China, Taiwan, Japan.

Magnolia delavayi China.

1901   *Rosa* 'Dorothy Perkins' USA. Raised by Jackson & Perkins and named for George Perkins' daughter. So popular did the rose prove that later a chain store was also given the same name.

1902   *Rhus potaninii* Sumach. China. Named in honour of Grigori Potanin(1835–1920), a Russian explorer of C. and E. Asia who had a vast botanical collection.

1904   *Davidia involucrata* Pocket-handkerchief Tree, Dove Tree, Ghost Tree. C. and W. China.

Rosa willmottae W. China.

1905   *Lilium regale* Regal Lily. W. China.

1906   *Daphne aurantiaca* S. W. China. Introduced by George Forrest.

1907   *Linum narbonense* 'Six Hills Variety' Blue Flax. W. and C. Mediterranean. Named for the Clarence Elliott Nursery at Stevenage in Hertfordshire.

1908   *Ceratostigma willmottiae* Blue Plumbago, Chinese Plumbago. W. China. The seed was sent by E. H. Wilson to Miss Willmott of Warley Place in Essex, who raised two plants, and it is from these that present stocks derive.

Paeonia mlokosewitschii Caucasian Peony. Caucasus. Received RHS Award of Merit in 1929.

1909   *Erysimum* 'Belvoir Castle' Wallflower. The 8th Duke of Rutland's head gardener described this wallflower as 'the best of the yellow varieties'.

Helleborus argutifolius (syn. *H. corsicus, H. lividus* subsp *corsicus*) Corsican Hellebore. Corsica, Sardinia.

Morisia monanthos (syn. *M. hypogaea*) Corsica, Sardinia.

Thymus herba-barona Caraway Thyme. Sardinia, Corsica.

1910   *Anemone hupeheusis* Hubei Province, China.

Pieris formosa var. *forrestii* S.E. China.

Olearia avicenniifolia Stewart Island off New Zealand; see p.214.

Olearia chathamica Chatham Islands; see p.214.

Olearia 'Henry Travers' (syn. *O. semidentata* of gardens) Chatham Islands; see p.214.

1911   *Gentiana sino-ornata* S.W. China, Tibet.

Primula forrestii China.

Primula vialii Sichuan, Yunnan, China.

1912   *Phlox divaricata* subsp. *laphamii* Blue Phlox, Wild Sweet William. Canada, E. USA. Named for Allen Lapham (1811–75), American naturalist.

Rhododendron campylogynum E. India, Tibet, W. China, N.E. Burma.

1913   *Meconopsis delavayi* China. Collected by George Forrest. Reintroduced (see 1990).

Rhododendron sinogrande Tibet, China, Burma.

1914   *Gentiana farreri* N.W. China.

Rhododendron caloxanthum Yunnan, China, Burma.

Rosa elegantula 'Persetosa' (syn. *R elegantula* f. *persetosa, R farreri* f. *persetosa*) N.W. China.

1915   *Allium cyathophorum* var. *farreri* (syn. *A. farreri*) China.

Balsamorhiza hookeri Hairy Balsam Root. W. USA.

Buddleja alternifolia China.

1916   *Philadelphus argyrocalyx* New Mexico. A silver-leaved Mock Orange.

1917   *Astilbe koreana* Korea Collected by E. H. Wilson. Reintroduced see 1932.

Forsythia ovata Korean Forsythia. Korea. Collected by E. H. Wilson.

Geranium farreri W. China.

1918   *Deutzia pulchra* Taiwan, Philippines. Collected by E. H. Wilson. First flowered at Kells, Co. Meath, in 1922.

Leucadendron stokoei Cape Peninsula, South Africa.

Brunia stokoei Cape Peninsula South Africa.

1919   *Glyptostrobus lineatus* Chinese Swamp Cypress. Canton (Guangzhou), S. China.

Pieris taiwanensis Taiwan.

Rhododendron sperabile N.E Burma.

Syringa x prestoniae (*S. reflexa x S. villosa*) Lilac. Garden origin.

1920   *Juniperus recurva* var. *coxii* (syn. *J. coxii*) Coffin Juniper. N. Burma.

1922   *Malus x purpurea* 'Eleyi' (*M. atrosanguinea x M. niedzwetskyana*) Crab Apple with purple flower and fruit. Garden origin.

1923   *Narcissus* 'Mrs R. O. Backhouse' Garden origin. First daffodil to be bred with a pink trumpet.

1924   *Camellia saluenensis* China; see p.165.

Choanathus cyrtanthifloris E. Africa.

Rosa 'Penelope' Garden origin. Bred by Revd Joseph Pemberton, past President National Rose Society.

1925   *Gypsophila paniculata* 'Bristol Fairy' Baby's Breath. Garden origin. RHS Award of Merit 1927; see p.169.

Meconopsis betonicifolia Himalayan Blue Poppy, Tibetan Blue Poppy. Tibet, S.W. China, Burma.

c. 1925   *Lathyrus pubescens* Chile, Argentina.

1926   *Genista lydia* (syn. *G. spathulata*) Broom. E. and S. E. Europe, Syria.

Primula agleniana E. Tibet, China, Burma.

Primula alpicola S.E. Tibet, Bhutan.

Primula florindae Giant Cowslip. S.E. Tibet.

Primula sonchifolia Burma, S. E. Tibet, China.

1927 *Calceolaria darwinii* (syn. *C. uniflora* var. *darwinii*) Argentina, Tierra del Fuego.

1927 *Berberis linearifolia* Chile, Argentina. Reintroduction (see p.218).

Lardizabala biternata Chile. Reintroduction (see p.217).

Leucocoryne ixioides Glory of the Sun. Chile.

Mutisia oligodon Chile, Argentina. Reintroduction (see p.218).

1928 *Fabiana imbricata* f. *violacea* (syn. *F. violacea*) Chile. Reintroduction (see p.218).

Puya alpestris Bromeliad. C. Chile.

1930 *Leptospermum grandiflorum* (syn. *L. rodwayanum*) Tasmania. Named originally for Leonard Rodway (1853–1936), a dental surgeon who went to Tasmania in 1890 and was the Honorary Government Botanist from 1896 until 1932.

Pteronia smutsii South Africa.

1931 *Jasminum polyanthum* W. and S.W. China. A climber collected from Yunnan Province by George Forrest and Major Lawrence Johnston, who grew it in his own garden at Hidcote Manor in Gloucestershire.

Mahonia lomariifolia S.Sichuan, Yunnan, China. Collected, introduced and grown by Major Lawrence Johnston as above.

1932 *Astilbe koreana* Korea. Reintroduction (see 1917).

1933 *Cotoneaster* 'Cornubia' Raised in Lionel de Rothschild's garden at Exbury, Hampshire, and received an RHS Award of Merit.

1934 *Primula rockii* China. Collected by Joseph Rock, and received RHS Award of Merit in 1936.

1935 *Geranium renardii* Caucasus. Named for Charles Renard (1809–86), Second Secretary of the Imperial Society of Naturalists of Moscow.

Primula sherriffae S. E. Bhutan, Manipur.

1937 *Anchusa cespitosa* (syn. *A. caespitosa*) Crete. Collected by Dr Peter Davis (1918–92), author of *Flora of Turkey and the Eastern Aegean Islands (1965–1988)*.

c. 1938 *Impatiens elegantissima* Mountains of the Moon, E. Africa.

Lobelia gibberoa Mountains of the Moon, E. Africa.

1939 *Primula caveana* (syn. *P. younghusbandiana*, *P. cana*) India, Tibet, Bhutan.

1940 *Yushania anceps* 'Pitt White' (syn. *Arundinaria anceps*, *A. jausarensis*, *Sinarundinaria jausarensis*) Anceps Bamboo. N. W. and C. Himalayas. Can grow over 15 cm (6 in) in twenty-four hours. First grown in the garden of Dr Mutch at Pitt White, Lyme Regis in Dorset.

1943 *Sorbus reducta* N. Burma, W. China. A dwarf species introduced by Frank Kingdon Ward.

1945 *Chamaecyparis lawsoniana* 'Winston Churchill' (syn. *Cupresus lawsoniana*) Lawson Cypress. Garden origin.

1946 *Lilium mackliniae* Assam, N.E. India.

1947 *Metasequoia glyptostroboides* Dawn Redwood. N. W. Hubei, China.

Paeonia lutea var. *ludlowii* (syn. *P. delavayi* var. *ludlowii*) Tree Peony. S.E. Tibet. Collected by Frank Ludlow and George Sherriff.

Rosa 'Peace' (syn. *R.* 'Gioia', *R.* 'Gloria Dei', *R.* 'Mme A. Meilland') Hybrid tea cultivar. Bred in France by Mr Meilland and named for his wife.

Saxifraga lowndesii Himalayas Named for Colonel Donald Lowndes (1899–1956) of the Royal Garhwal Rifles, botanist with H. W. Tillman's expedition to Nepal, 1950. Reintroduced (see 1983).

1948 *Camellia williamsii* 'Mildred Veitch' Raised and named by Robert Veitch & Son in honour of the last member of the family to run the nursery.

1949 *Euphorbia griffithii* Bhutan, Tibet, S.W. China. Named for William Griffith (1810–45), a botanist and surgeon who worked in India and Afghanistan.

1950 *Cornus kousa* var. *chinensis* India, W. China.

Geranium 'Johnson's Blue' Named for Arthur Johnson (1873–1956), a schoolmaster and market gardener in Wales, and an expert on geraniums.

1952 *Clematis phlebantha* W. Nepal. Collected by Oleg Polunin (d. 1985), renowned botanist and plant photographer. Raised from seed at RHS's garden at Wisley, Surrey, and distributed to gardens.

1953 *Lilium arboricola* N. Burma. Collected by Frank Kingdon Ward.

1954 *Rosa* 'Queen Elizabeth' (syn. *R.* The Queen Elizabeth') Raised by Dr W. Lammerts in the USA.

1956 *Geranium yunnanense* Burma. One of the last plants to be discovered by Frank Kingdon Ward; he died two years later.

1957 *Abutilon ochsenii* Chile. Won an RHS Award of Merit in 1962.

1958 *Rosa* 'Peter May' Hybrid tea rose. Named for the cricketer and captain of England.

1960 *Pelargonium* 'Mabel Grey' Nairobi, Kenya. Discovered in the grounds of Government House.

Rosa 'Super Star' (syn. *R.* 'Tanostar', *R.* 'Tropicana') Raised by M. Tantau of Germany. A seedling of whom one parent was Peace.

1961 *Rosa* 'Blaby Jubilee' (syn. 'Dries Verschuren') Large-flowered bush rose. Raised by Blaby Rose Garden in Leicestershire.

1962 *Daphne bhoua* var. *glacialis* 'Gurkha' Himalayas. Collected by Major T. Le M. Spring-Smyth.

Rosa 'Woburn Abbey' Floribunda Rose raised by Alfred Cobley and George Sidey of Earl Shilton in Leicestershire.

1963 *Allium regelii* Afghanistan. Collected from the Hindu Kush by the plant collector and retired Rear Admiral J. Paul Furse (1904–78).

1964 *Iris doabensis* Afghanistan. Another plant collected by Paul Furse from the Hindu Kush.

1965 *Narcissus* 'Barnsdale Wood' Bred by W. A. Noton of Oakham, Rutland.

1966 *Fritillaria alburyana* N. E. Turkey. Named for S. D. Albury who first discovered it.

1967 *Dahlia* 'Silver City' A large white decorative type raised by Brother Simplicius – a Dutchman, Gerard Peters, who settled into a religious community in Leicester following the Second World War.

1968 *Dahlia* 'Lifesize' A large decorative type with yellow flowers, also raised by Brother Simplicius.

1969 *Geranium clarkei* India. Thought to be named for Charles Baron Clarke (1832–1906), who collected many Indian plants.

1970 *Torreya californica* 'Spreadeagle' California Nutmeg. Garden origin. The species was found originally by William Lobb in 1851. This form originated at Hilliers' Nursery in Winchester, Hampshire.

c. 1970 *Rhododendron* **'Ben Morrison'** Garden origin, USA; see p.17.

Rhododendron **'Frosted Orange'** Garden origin, USA; see p.17.

Rhododendron **'Peggy Ann'** Garden origin, USA; see p.17.

1971 *Acer pycnathum* Japan. Introduced by James Harris of Mallet Court Nursery, Taunton in Somerset.

1972 *Narcissus* 'Hot Sun' Bred by W. A. Noton of Oakham, Rutland.

1973 *Daboecia cantabrica* 'Covadonga' Found by Terry Underhill of Devon while in the Picos d'Europa Mountains of N. Spain.

1974 *Daboecia x scotica* 'Goscote' (*D. azorica x D. cantabrica*) Discovered in the garden of Derek Cox in Syston, Leicestershire.

1975 *Helleborus orientalis* 'Helen Ballard'. Garden origin. Raised in Hertfordshire.

1976 *Chamaecyparis lawsoniana* 'Berrydown Gold' Raised in Devon by Berrydown Nurseries, Gidleigh in Devon.

1977 *Rosularia sempervivum* Iran.

Euphorbia schillingii Nepal. Collected and introduced by A. D. Schilling (b. 1935), Deputy Curator at the Royal Botanic Gardens, Kew, and in charge of their gardens at Wakehurst Place in Sussex 1963–91.

1978 *Abutilon pictum* (syn. *A. stritatum*) N. Argentina, Brazil.

Dianthus 'Dartmoor Forest'. Raised in Bovey Tracey, Devon by Cecil Wyatt.

1979 *Fargesia dracocephala* Bamboo. China. The genus is named for Paul Guillaume Farges (1844–1912), a French missionary.

1982 *Dahlia* 'Majestic Kerkrade' Raised by Harry W. Hooper of Exeter who won many awards at Dahlia shows and trials.

1983 *Saxifraga lowndesii* Himalayas. Reintroduction by Ron MacBeath of the Royal Botanic Gardens, Edinburgh (see 1947).

1984 *Fremontodendron* 'Pacific Sunset' Flannel Bush. Garden origin. Orginated in California and N. Mexico. The first one was found by Major-General John Fremont, after whom it was named in 1842.

Rosa 'Cardinal Hume' Bred by R. Harkness & Co.

1985 *Geranium x cantabrigiense* (*G. dalmaticum x G. macrorrhizum*) Garden origin.

1986 *Fargesia denudata* Bamboo. China. The panda's favourite food.

1987 *Qiongzhuea tumidinoda* Bamboo. Yunnan to Sichuan, China. Mature stems of this plant are used to make walking sticks.

1988 *Aster* 'Brixham Belle' Raised by Sutton Seeds of Torquay, Devon.

1989 *Sinocalycanthus chinensis* Chinese Allspice. C. China. Collected by Roy Lancaster.

1990 *Meconopsis delavayi* China. Re-introduced by Ron McBeath of the Royal Botanic Gardens Edinburgh (see 1913).

1992 *Syneilesis intermedia* Taiwan. Collected by Bleddyn Wynn-Jones, seed and plant collector.

1993 *Lychnis coronaria* 'Hutchinson's Cream' Found in a garden in Quorn, Leicester.

1995 *Bergenia emeiensis* China. Introduced by the Japanese plant explorer, Mikinori Ogisu.

1996 *Hosta* 'Lady Isobel Barnett' Raised by Diana Grenfell of Apple Court Nursery in Hampshire who became interested in Hostas after she had heard Lady Isobel Barnett speaking on the radio about them.

Lathyrus odoratus 'Geoff Hamilton' Raised by Diana Sewell and named for the BBC Television presenter for seventeen years of *Gardeners' World*, who died in 1996.

1997 *Gazania rigens* (*G. splendens*) South Africa.

1998 *Aubrieta* 'Purity' Garden origin. First pure white Aubrieta to come on the market.

1999 *Daphne vermionica* Greece. Discovered on Mount Vermion by the Alpine Garden Society Expedition.

Helleborus argutifolius 'Silver Lace' Bred by R. D. Plants Ltd of Axminster, Devon.

2000 *Lavandula stoechas* 'Roxlea Park' Bred in New Zealand; see p.17.

Lavandula viridis 'Silver Ghost' Bred in New Zealand.

Streptocarpus 'Crystal Ice' Raised by Dibleys Ltd of Ruthin, N. Wales who declared it the 'Plant of the Century

Select Bibliography

Bean, W.J.: *Trees & Shrubs Hardy in the British Isles*. 3 vols. 4th ed. London 1925

Blunt, Wilfrid and Raphael, Sandra: *The Illustrated Herbal*. London 1979

Brickell, Christopher and Sharman, Fay: *The Vanishing Garden*. London 1986

Coats, Alice M.: *Flowers and Their Histories*. London 1968

Coats, Alice M.: *Garden Shrubs and Their Histories*. London 1963

Cox, E. H. M.: *Plant Hunting in China*. London 1945

Culpeper, Nicholas: *Culpeper's Complete Herbal*. 1653

Desmond, Ray: *Dictionary of British and Irish Botanists and Horticulturists*. London 1977

Desmond, Ray: *Kew. The History of the Royal Botanic Gardens*. London 1995

Elliot, Brent: *Victorian Gardens*. London 1986

Evelyn, John: *Acetaria: A Discourse of Sallets*. 1699, reprinted Devon 1996

Evelyn, John: *Sylva or a Discourse of Forest Trees and Pomona* 1669

Farrer, Reginald: *Rainbow Bridge*. London 1921

Fisher, John: *The Origins of Garden Plants*. London 1982

Gorer, Richard: *The Flower Garden in England*. London 1975

Gorer, Richard: *The Growth of Gardens*. London 1978

Grigson, Jane: *Jane Grigson's Fruit Book*. London 1982

Grigson, Jane: *Jane Grigson's Vegetable Book*. London 1978

Hadfield, Miles, Harling, Robert, Highton, Leonie: *British Gardeners*. London 1980

Handel-Mazzetti, Heinrich: *Naturbilder aus Sudwest China*. Vienna 1927. Translated by David Winstanely and published as *A Botanical Pioneer in South West China*. Essex 1996

Hanmer, Sir Thomas: *The Garden Book of Sir Thomas Hanmer Bart*. 1659

Harvey, John: *The Availability of Hardy Plants of the Late Eighteenth Century*. Garden History Society, 1988

Harvey, John: *Early Nurserymen*. Chichester 1974

Harvey, John: *Mediaeval Gardens*. London 1981

Jellicoe, Geoffrey and Susan, Goode, Patrick, Lancaster, Michael: *Oxford Companion to Gardens*. Oxford 1986

Lancaster, Roy: *Travels in China*. Suffolk 1989

Leapman, Michael: *The Ingenious Mr Fairchild*. London 2000

Leith-Ross, Prudence: *The John Tradescants*. London 1984

Le Rougetel, Hazel: *The Chelsea Gardener: Philip Miller 1691–1771*. London 1990

McCracken, Donald P.: *Gardens of Empire*. London 1997

McLean, Teresa: *Medieval English Gardens*. London 1981

Mitchell, Alan F., Schilling, Victoria E., White, John E. J.: *Champion Trees in the British Isles*. 4th ed. Edinburgh 1994

Morewood, William: *Traveller in a Vanished Landscape*. London 1973

National Council for the Conservation of Plants and Gardens: Devon Group: *The Magic Tree: Devon Garden Plants, History and Conservation*. Devon 1989

National Council for the Conservation of Plants and Gardens: Leicestershire Group: *Garden Plants of Leicestershire and Rutland* (Editor: Graham Jackson), Leicestershire 1996

Pakenham, Thomas: *Meeting with Remarkable Trees*. London 1996

Platt, Karen: *Plant Names A-Z*. Sheffield 1999

Platt, Karen: *The Seed Search*. 2nd ed. Sheffield 1997

Pratt, Anne: *Flowering Plants of Great Britain*. 4 vols. London 1855

Robinson, William: *The English Flower Garden*. London 1883

Robinson, William: *The Wild Garden*. 1870 reprinted London 1979

The Royal Horticultural Society: *Dictionary of Gardening*. 5 vols. 2nd ed. (Editors: Fred Chittenden & Patrick M. Synge. Oxford 1956

The Royal Horticultural Society: *A-Z Encyclopedia of Garden Plants*. (Editor-in-Chief Christopher Brickell). London 1996

The Royal Horticultural Society: *Plant Finder 1996–2000*. (Editor Tony Lord) London 1996

The Royal Horticultural Society: *Plant Finder 2000–2001*. (Principal Editor Tony Lord) London 2000

Smith, John: *Dictionary of Economic Plants*. London 1882

Stern, William T.: *Stern's Dictionary of Plant Names for Gardeners*. London 1997

Stobart, Tom: *The Cook's Encyclopaedia*. London 1980

Synge, Patrick M.: *In Search of Flowers*. London 1973

Vickery, Roy: *A Dictionary of Plant Lore*. Oxford 1995

Wilson, Ernest: *A Naturalist in Western China*. 1913, reprinted London 1986

Index